E-Manufacturing and E-Service Strategies in Contemporary Organizations

Norman Gwangwava
Botswana International University of Science and Technology, Botswana

Michael Mutingi
Namibia University of Science and Technology, Namibia

A volume in the Advances in E-Business Research (AEBR) Book Series

IGI Global
DISSEMINATOR OF KNOWLEDGE

Published in the United States of America by
IGI Global
Business Science Reference (an imprint of IGI Global)
701 E. Chocolate Avenue
Hershey PA, USA 17033
Tel: 717-533-8845
Fax: 717-533-8661
E-mail: cust@igi-global.com
Web site: http://www.igi-global.com

Library of Congress Cataloging-in-Publication Data

Names: Gwangwava, Norman, editor. | Mutingi, Michael, editor.
Title: E-manufacturing and E-service strategies in contemporary organizations / Norman Gwangwava and Michael Mutingi, editors.
Description: Hershey, PA : Business Science Reference, [2018]
Identifiers: LCCN 2017022425| ISBN 9781522536284 (hardcover) | ISBN 9781522536291 (ebook)
Subjects: LCSH: Manufacturing industries--Management. | Service industries. | Electronic commerce.
Classification: LCC HD9720.5 .E226 2018 | DDC 658.8/72--dc23 LC record available at https://lccn.loc.gov/2017022425

This book is published in the IGI Global book series Advances in E-Business Research (AEBR) (ISSN: 1935-2700; eISSN: 1935-2719)

British Cataloguing in Publication Data
A Cataloguing in Publication record for this book is available from the British Library.

Advances in E-Business Research (AEBR) Book Series

ISSN:1935-2700
EISSN:1935-2719

Editor-in-Chief: In Lee, Western Illinois University, USA

MISSION

Technology has played a vital role in the emergence of e-business and its applications incorporate strategies. These processes have aided in the use of electronic transactions via telecommunications networks for collaborating with business partners, buying and selling of goods and services, and customer service. Research in this field continues to develop into a wide range of topics, including marketing, psychology, information systems, accounting, economics, and computer science.

The **Advances in E-Business Research (AEBR) Book Series** provides multidisciplinary references for researchers and practitioners in this area. Instructors, researchers, and professionals interested in the most up-to-date research on the concepts, issues, applications, and trends in the e-business field will find this collection, or individual books, extremely useful. This collection contains the highest quality academic books that advance understanding of e-business and addresses the challenges faced by researchers and practitioners.

COVERAGE

- Mobile Business Models
- Interorganizational information systems
- E-business systems integration
- E-Business Management
- Electronic communications
- Trust, Security, and Privacy
- Intelligent Agent Technologies
- E-CRM
- Collaborative commerce
- Online consumer behavior

IGI Global is currently accepting manuscripts for publication within this series. To submit a proposal for a volume in this series, please contact our Acquisition Editors at Acquisitions@igi-global.com or visit: http://www.igi-global.com/publish/.

The Advances in E-Business Research (AEBR) Book Series (ISSN 1935-2700) is published by IGI Global, 701 E. Chocolate Avenue, Hershey, PA 17033-1240, USA, www.igi-global.com. This series is composed of titles available for purchase individually; each title is edited to be contextually exclusive from any other title within the series. For pricing and ordering information please visit http://www.igi-global.com/book-series/advances-business-research/37144. Postmaster: Send all address changes to above address.

Titles in this Series

For a list of additional titles in this series, please visit:
https://www.igi-global.com/book-series/advances-business-research/37144

Mobile Platforms, Design, and Apps for Social Commerce
Jean-Éric Pelet (ESCE International Business School, Paris, France)
Business Science Reference • ©2017 • 411pp • H/C (ISBN: 9781522524694) • US $215.00

The Internet of Things in the Modern Business Environment
In Lee (Western Illinois University, USA)
Business Science Reference • ©2017 • 340pp • H/C (ISBN: 9781522521044) • US $205.00

Key Challenges and Opportunities in Web Entrepreneurship
Alexandru Capatina (Dunarea de Jos University of Galati, Romania) and Elisa Rancati (University of Milan – Bicocca, Italy & Beta Consulting, Italy)
Business Science Reference • ©2017 • 291pp • H/C (ISBN: 9781522524663) • US $205.00

Apps Management and E-Commerce Transactions in Real-Time
Sajad Rezaei (Taylor's University, Malaysia)
Business Science Reference • ©2017 • 379pp • H/C (ISBN: 9781522524496) • US $210.00

Driving Innovation and Business Success in the Digital Economy
Ionica Oncioiu (Titu Maiorescu University, Romania)
Business Science Reference • ©2017 • 321pp • H/C (ISBN: 9781522517795) • US $200.00

Social Media Listening and Monitoring for Business Applications
N. Raghavendra Rao (FINAIT Consultancy Services, India)
Business Science Reference • ©2017 • 470pp • H/C (ISBN: 9781522508465) • US $205.00

Analyzing the Strategic Role of Social Networking in Firm Growth and Productivity
Vladlena Benson (Kingston University, UK) Ronald Tuninga (Kingston Business School, UK) and George Saridakis (Kingston University, UK)
Business Science Reference • ©2017 • 525pp • H/C (ISBN: 9781522505594) • US $220.00

For an entire list of titles in this series, please visit:
https://www.igi-global.com/book-series/advances-business-research/37144

IGI Global
DISSEMINATOR OF KNOWLEDGE

701 East Chocolate Avenue, Hershey, PA 17033, USA
Tel: 717-533-8845 x100 • Fax: 717-533-8661
E-Mail: cust@igi-global.com • www.igi-global.com

Table of Contents

Detailed Table of Contents

Internet of things (IoT) is a concept of providing uniquely identifiable objects connectivity to the internet. Under the roof of IoT, it is predicted that above 28 billion devices will be connected to the internet by the year 2020. When billions of things connect, it will be difficult to manage and analyze huge amount of data as each object will send and retrieve data. Smart manufacturing is an emerging concept where the manufacturing process is supported by technology and the required information is made available during the manufacturing process to get the flexibility and the product as per customer changing needs. Internet of things (IoT) may provide a good platform to enhance the manufacturing process into smart manufacturing. The advantages of smart manufacturing include the higher quality of a product, improved productivity, increased energy efficiency, enhanced scalability in manufacturing process, etc. This chapter presents in depth the IoT and smart manufacturing concepts, their requirements, relevance, and available solutions.

Consumers are spending more and more time on the web to search information and receive e-services. E-commerce, e-government, e-business, e-learning, e-science, etc. reflect the growing importance of the web in all aspects of our lives. Along

techniques to analyse multidimensional data. This chapter seeks to introduce a comprehensive and robust visualisation model and framework for adoption. The visualisation model consists of four major layers, which include acquisition and data analysis, data representation, user and computer interaction, and result storage.

The aim of this chapter is to provide insights into the causal factor of people's internet banking acceptance. It endeavors to address a research need for extending the technology acceptance model (TAM) by adding contextual variables and its implication on e-customer satisfaction and e-customer loyalty. As trust and service quality has never failed to be a significant predictor in e-commerce research, this study proposes to integrate trust and service quality into the TAM to get a better understanding of e-banking adoption. A research model reflecting the effect of e-trust and internet banking service quality dimensions on TAM constructs is proposed.

As the awareness of the importance of healthcare service quality and pressures from stakeholders continue to grow, healthcare service providers have no option except to develop appropriate service quality evaluation procedures. Patient satisfaction is imperative, and has become a critical issue especially in e-health services. Although healthcare service providers have become aware of the need for improving customer experience through provision of customer-centric services, virtually none of the existing e-health quality evaluation frameworks are grounded on customer-centric metrics. In this chapter, a critical analysis of existing evaluation initiatives is presented from the context of e-health services. Critical areas of e-health service are investigated to determine quality dimensions that influence customer experience. From this analysis, a customer-centric evaluation framework is proposed, comprising four e-health service quality scales. The framework provides a platform for continuous improvement in e-health service.

Product lifecycle management (PLM) is concerned with managing all the processes of product transition from cradle to the grave. Phases of a product life are inception, engineering design, manufacture, service, and disposal. A product passes through different systems, organisations, and users as it completes the life cycle. Global markets of today have increased the complex nature of a product path. All the life cycle phases rely upon product data for efficient management. In order to ease the strain of managing products throughout the lifecycle, a common product data schema is needed. Currently many platforms for product design use different proprietary schemas that make it difficult to have smooth lifecycle management. The chapter illustrates applications of an open source, XML-based schema for product lifecycle management. The main focus is on the inclusion of the Cloud in order to have new generation cloud product life cycle management (CPLM). The main driver of CPLM is cloud-model-based systems engineering (CMBSE). Within the framework of CMBSE are cloud-based design (CBD), cloud manufacturing (CM), and cloud-based maintenance (CBM). The three subsystems of CMBSE can be combined to form a single term, cloud-based design, manufacturing, and maintenance (CBDMM). Cloud computing, manufacturing, and maintenance are not new concepts, but many enterprises have not yet embraced them because of lack of complete seamless integration across various levels and processes in the product life. Many systems are still being run in silos of automation. CPLM is a service-oriented (SOA) model comprised of a pool of technologies such as cloud computing (CC), IoT, virtualization, and service-oriented technologies to support collaboration, sharing, and management across PLM phases.

Preface

Major innovations have taken place in information and communication technologies (ICTs). Trends in the internet era have seen technology savvy companies adopting cloud-based solutions to out compete their rivals. Manufacturing and service delivery have evolved from traditional methods to modern trends such as E-Manufacturing, E-Services, and later Cloud Manufacturing that heavily rely on the cloud. Rapid adoption of smart phones by the increasingly knowledgeable customers triggered the current trend of web 2 technologies where most services are also available through mobile applications that run on smart phones. These trends offered manufacturers and service providers opportunities to stay more in touch with their customers than before. Manufacturing is now heavily data driven, thanks to many innovations that allow instant acquisition of data for rapid integration into digital product generation and translation into physical component manufacturing. 3D scanning technologies allow digitization of components for manufacture whilst e-data acquisition systems are used during the product use phase to gather current state data about product performance for use in e-maintenance, and recently cloud based maintenance strategies. This enables manufacturers to maintain a long term relationship and maximise on revenues whilst providing quality service. Parts management systems have also evolved from static databases completely detached from the original manufacturers to e-based systems that contain digital data about the original component design. Service industries have considerably shifted from the brick and motor model to the click based online stores. This has also been backed by significant improvements in the supply chain management systems. Online payments systems have matured in the long run to enable secure handling of payments through the online stores. Security is also a major worry in the dispensation of Cloud Computing (CC). As the cyber space continue to evolve, adopters of advanced technology need to find ways of shielding themselves and their customers against cybercrime.

The book explores ways of integration of the cloud into the health, banking, manufacturing, and product life cycle management (PLM). Cyber security issues are also covered in-depth in order to guide organisations on safeguarding themselves as well as their customers against cybercrime.

PURPOSE AND TARGET AUDIENCE OF THE BOOK

Main objectives are to share and contribute new knowledge and emerging trends in the subject area of e-systems evolutions that have significantly revamped the way product and service providers do business. Manufacturing and service industries are adopting cloud computing (CC) strategies where almost everything is treated as a service (XaaS). These are relatively new areas in many enterprises, despite of wide research for the past decade. Compiling an edited volume in this rare area opens up eyes of many industrial practitioners, scholars, and researchers alike across the world. The modern way of doing business revolves around ubiquitous network of computer systems running smoothly throughout in order to ease business approach, deliver optimum solutions and bring agility into the business world. Practical cases are contributed in the book to benefit the readership audience.

This book serves the following broad audience:

- Researchers in academia and Industry R&D Practitioners
- Business Executives and Operations Personnel responsible for adopting new technology and business systems
- Students studying towards Management Information Systems, Business Information Systems, Manufacturing Information Systems at both undergraduate and post graduate levels
- Systems developers and systems integrators who are interested in bench marking against other proved installations or adopting readily available models.

Norman Gwangwava
Botswana International University of Science and Technology, Botswana

Michael Mutingi
Namibia University of Science and Technology, Namibia

Introduction

There is a paradigm shift in the modern approach to business due to the evolution of Cloud Computing (CC). Enterprises need instant feedback, traceability, uniformity and transparency in order to enhance decision making. Collaboration, integration, agility, optimization, and cost effectiveness are seen as the major pillars for today's business success. Cyber Physical Systems (CPS) and Internet of Things (IoT) technologies such as RFID, wired and wireless sensor networks, and embedded systems enable the digitization and virtualization of shared resources and capabilities in the services, manufacturing and supply chain management for access through the cloud resource pool. These technologies extend the internet into the real world, embracing everyday objects. Today's business space is characterized by smart environments/spaces and self-aware things such as smart transport, products, cities, buildings, rural areas, energy, and health. The publication gives an in-depth coverage of various strategies being adopted by the manufacturing and service industries in the modern-day era of cloud computing (CC). Globally, customers are continuously becoming highly knowledgeable, with ability to interact with modern product and service interfaces. Nowadays almost everything is offered as a services (XaaS). The service-oriented architecture (SOA) is the major model of implementation of the various service layers in cloud computing, viz Platform as a Service (PaaS), Infrastructure as a Service (IaaS), Hardware as a Service (HaaS), Software as a Service (SaaS), and Design as a Service (DaaS). In order to remain competitive, enterprises need to be agile, cost effective, and optimal in all their product and service delivery processes. As a cost saving strategy, many enterprises and their employees have hired the enterprise cloud services, and put much important information in the cloud computing applications. The pay-as-you-go model for CC makes it widely adopted by enterprises. CC makes it possible to deliver enterprise information systems (EISs) where each party can contribute to the establishment of the advanced system, while delivering superior integration capabilities with any party's legacy systems and web-based intranets applications. This calls for Interoperability, thus technology neutral, but business specific representation of functions are becoming the de facto standard. SOA is implemented using Web Service technologies and neutral languages, particularly XML, in order to facilitate open communication.

The ubiquitous computing era presents enterprises with huge quantities of data, known as Big Data. The data is gathered through smart sensors among physically networked objects. Generally the term e-services is used to describe a variety of internet based electronic interactions ranging from basic services, such as the delivery of news and stock quotes, to smart services, products, clothes etc. These electronic-services involve various types, delivery systems, advanced information technologies, methodologies and applications of online services and they are provided by e-government, e-business, ecommerce, e-market, e-finance, and e-learning systems. Big Data is transforming healthcare, banking, business, product life cycle management – PLM (design, manufacturing, maintenance, etc.), and ultimately the whole society. PLM is a strategic business approach that applies a consistent set of business solutions that support the collaborative creation, management, dissemination, and use of product definition information. Various computer aided technologies (CAx) are used during the entire product life. E-Maintenance, backed by e-data acquisition systems can now be directly offered by original equipment manufacturers or offered by third parties who are running systems directly linked to the manufacturers. E-Health refers to the use of information and communication technologies in the provision of healthcare services. It is a field that integrates the aspects of medical informatics, public health as well as business.

Adopting CC technologies help enterprises to increase their performance and competitiveness through integration and collaboration. Continuous improvements in EISs have created profitable opportunities for modern businesses to develop more advanced processes. Many business growth opportunities are embedded in the cloud. Enterprises need strategies, methods, tools, and solutions which can enable them to comply with the demands of CC. Instant feedback, traceability, uniformity and transparency are required in order to enhance decision making. Information visualisation, data analysis, technology acceptance and quality issues are crucial in the CC era. These focus on the concept of internet-based customer experience. Enhancing customer experience in the modern era where everything is treated as a service (XaaS) is crucial in order to boost confidence among the clientele.

More promising trends continue to emerge in the field of internet based systems. Voice commerce (vCommerce) where the voice replaces the screen and the key board as a user interface to the product and services. Commercial websites will incorporate voice recognition technology and the end user does not need to type in requested data. An extension to vCommerce is the inclusion of internet of things (IoT) to form IoT vCommerce. Sectors currently making promising efforts to adopt the technology include; banking, automotive industries, electronic and household goods manufacturers. Specific services on the line-up include processing payments through voice, voice assistance on the production floor, and voice as a security layer. Another trend on the rise is Artificial Intelligence and Machine Learning.

APPLICATIONS OF E-SYSTEMS IN MANUFACTURING AND SERVICE INDUSTRIES

Enterprises have become internet dependent as a strategy for success in the current dynamic market trends. Rapid developments and ready availability of the internet led to the ubiquitous growth of e-systems across various business sectors. However, the manufacturing and the health sectors took longer to effectively become part of the trend. Major reasons being the complex nature of modeling manufacturing information systems, standardization, and regulatory nature of the sectors. E-Systems is the term used to describe online information systems for various business applications. The term must not be confused with e-commerce, which mainly relates to buying and selling online. The "e-" in both case stands for "electronic", which refers to internet and electronic data interchange. E-Systems have a broader meaning, that also includes e-commerce, among internal processes such as production, inventory management, product research and development, finance, customer relationship management (CRM), human resource management (HRM). Adoption of e-systems in other sectors led to the evolution of e-banking, e-health, e-government, and e-manufacturing. This involves new business models, which continue to change at par with technological changes. Modern businesses that have embraced e-systems have become more efficient and cost effective, whilst those that resisted change have lost relevance and eventually faded away due to stiff competition. E-Systems have enabled enterprises to be more integrated; vertical integration, cross-business integration, and technology integration. These strategies were not feasible in the past. Processes and services that would take days and months to be fulfilled now just get accomplished in a matter of few clicks.

Examples of e-systems applications in services are addressed as e-services. Brief descriptions of some popular e-services are given below;

E-Government refers to the use of internet based information and communication technologies (ICTs) to improve the management and administration of activities of public sector organisations. Several public sector organisations across the world have embraced internet-enabled systems to interact with the outside groups (people and other organisations). Government activities are very broad in nature, which include central statistics, public sector utilities (power and water), land management, public health, national registry services, revenue authorities, transport, communications and many others. The demand for these services is very high since all citizens depend upon them. Whilst the general model is not for profiteering, most of the services are not offered for free, but require sustainability. Traditional manual processes are cumbersome, slow, poor service quality and pose a high risk of revenue loss and abuse. The e-government model is very efficient, cost effective, and of better quality. E-Government systems enable the public sector to improve processes (e-Administration), connect citizens (e-Citizens and e-Services), and build external interactions (e-Society). Efficient e-government systems also contribute to better economies due to efficient revenue collection, ports management, and better opportunities for citizens through international connections. Governments across different nations have become accustomed to developing e-government strategies as a way to be more relevant, make their societies happier, provide opportunities as well as to curb risks associated with the cyber space. Such strategies normally address issues of connectivity, infrastructure, integration, regulations and policies.

E-Banking is a method of banking in which the customer conducts transactions electronically via the Internet. It is also known as internet banking or online banking. Customers for banks or other financial institutions can conduct a range of financial transactions through the institution's website. Recently many banks and financial institutions joined the mobile applications band wagon, in efforts to achieve financial inclusion. The world (and money) are increasingly becoming electronic. Cash is only used for small daily purchases, but other forms of payment such as mobile money and cryptocurrency are evolving and becoming more popular. Managing personal money without going to the branch means one can conduct business anytime, and anywhere, as long as there is internet connectivity. Traditionally one had to visit the branch and wait in queues in order to perform a transaction. However, many people remain sceptic to use online banking and some cannot access reliable internet connection, especially in many parts of the developing world. Banks and other financial institutions have significantly reduced paperwork and lowered their rates by going online. Other services such as loan applications which are paperwork intensive can now be processed online. Seamless integration of financial services have seen many institutions being linked to mobile phones, which results in customers enjoying the benefit of instant messages for any transactions on their accounts. The mobile platform can also be used for real-time inquiries for account balances and mini statements. Text messaging has been widely embraced for its simplicity and fastness since it does not require internet connectivity, but just network coverage. It is also easy to learn compared to e-banking.

Security is a major concern in e-banking. Some of the security threats include invasion of individuals' privacy and theft of confidential information. Some of the methods used to ensure a high level of security are: (1) identification and authentication, (2) encryption, and (3) firewalls. Encryption blocks hackers against viewing messages transmitted between networked computers, databases, and servers;

even if the message is intercepted over the internet, it will be unreadable and nearly useless as long as it's encrypted. Most browsers are adopting the encryption standard called Secure Socket Layer (SSL). Firewall are software or hardware barriers between the corporate network and the external Internet, to protect the servers and databases from outside intruders.

E-Health refers to the use of information and communications technologies in healthcare. Due to the broad nature of e-health, some use the term health IT in place of e-health. E-Health is the convergence of medical informatics, public health and business in delivering health services (healthcare, surveillance, education, and research) through internet based technologies. This leads to better services, which are agile and cost effective.

The European Commission (2003) defines eHealth as: "The use of modern information and communication technologies to meet needs of citizens, patients, healthcare professionals, healthcare providers, as well as policy makers."

Key application areas of e-Health are (Openclinical, 2005):

- Electronic Medical Records (including patient records, clinical administration systems, digital imaging & archiving systems, e-prescribing, e-booking).
- Telemedicine and telecare services; health information networks.
- Decision support tools.
- Internet-based technologies and services.

The benefits of e-health records include: better health care by improving all aspects of patient care, including safety, effectiveness, patient-centeredness, communication, education, timeliness, efficiency, and equity. E-Health systems also reduce staff stress, retaining and attracting medical staff.

Technology acceptance is a major issue to consider when introducing e-health in the medical sector. Resistance may arise from hospital staff when they cannot adopt IT applications easily. This can be addressed by involving healthcare personnel throughout the design and implementation phases. The healthcare personnel may not be willing to learn new technology, or have fears that the system may take away their jobs, so they insist to keep working in traditional paper-base system. Cases of health sectors that adopted e-health illustrate that the system increase physician-patient relationship and also the system does not eliminate hospital staff jobs however it leads to improvements in their workflow.

E-Manufacturing is the use of internet based technologies to coordinate all the manufacturing related activities. Manufacturing covers broad activities such as order, design, planning, manufacturing, workshop floor control, assembly, delivery, maintenance, services, and marketing (Cheng and Bateman, 2008). Customers' wants or the customer's voice is very important to the manufacturers for design input, quality of manufacturing, service delivery, and maintenance during product use and final disposal. This necessitates a customer interface to hear the customers' concerns in real-time and take suitable action instantly, where necessary. Internet based solutions close the gap between demand and supply of a product (Saha and Groover, 2011). E-Manufacturing gives enterprises agility to react quickly to the changes in market, technology, and customers by operating as a virtual enterprise (Ramarao, 2016). Aspects covered under e-manufacturing address how manufacturing organizations do business, including;

- Design of manufacturing and business strategy
- Sales and marketing
- e-Procurement
- Supply chain collaboration
- Shop-floor operations and automation
- Enterprise application integration
- e-business
- Collaborative engineering

Manufacturing enterprises continue to evolve their processes and expand their business offerings as technology capabilities become more advanced. Connectivity and intelligence are brought by the internet, web-enabling, tether-free technologies and computational tools to meet the demands of e-business/e-commerce practices. E-Manufacturing promotes globalisation, mobility, collaboration, agility and cost efficiency. Internet is globally accessible, which enables manufacturers to support their engineers, partners and customers with access to information, regardless of their physical location. A variety of other tools for supporting group meetings, sharing of ideas and current trends are used to boost collaboration in design, manufacture, marketing and maintenance. Manufacturing practices such as mass customization, engineer-to-order, make-to-order, and first available slot became a reality due to wide adoption of internet based solutions. The whole engineering supply chain is becoming more efficient as internet technologies improve. Koc and Lee (2002) conceptualised the Integration of e-manufacturing with e-business and e-maintenance. The integration facilitates monitoring the plant floor assets, predict the variation of product quality and performance loss of any equipment for dynamic rescheduling of production and maintenance operations, and synchronize with related business services to achieve a seamless integration between manufacturing and higher level enterprise systems. Plant assets have become smart, intelligent, reconfigurable and adaptive. E-Maintenance strategy is adopted to integrate and synchronize the various maintenance and reliability applications to gather and deliver asset information as and when needed. Interconnectivity of the islands of maintenance and reliability information is embodied in E-Maintenance (Shivanand, et al., 2008). An effective e-maintenance strategy helps organisations to be more productive, realiable in product and service delivery, cost effective, and ultimately stay above the competition. E-Maintenance relies on e-diagnostics and e-prognostics systems for maintenance and repair of equipment and assets. This rults in reliable equipment and low cost operations. E-Diagnostics and e-prognostics systems are built on distributed sensor networks that performing signal acquisition, processing and fusion of the results on data servers. Optimal data storage and manipulation are crucial in condition monitoring taxonomy.

THE CLOUD COMPUTING (CC) ERA

The National Institute of Standards and Technology (NIST) define cloud computing as “a model for enabling ubiquitous, convenient, on-demand network access to a shared pool of configurable computing resources (e.g., networks, servers, storage, applications, and services) that can be rapidly provisioned and released with minimal management effort or service provider interaction ”(Mell and Grance, 2011). Individuals and corporate end users access the cloud through internet and access spans over different enterprises and platforms. Cloud services are based on technologies such as HTTP, HTML, XML, JavaScript, or other protocols.

Figure 1 illustrates the cloud computing metaphor. The entire provider-managed suite of hardware and software can be thought of as an amorphous cloud (Wikipedia, 2017).

There are many advantages offered by cloud computing (Michael, 2009);

- Improved performance of PCs,
- Lower maintenance issues,
- Lower hardware requirements for users,
- IT Infrastructure costs are lower,
- Lower software cost,
- Increased computing power,
- Unlimited storage capacity,
- Instant software updates,
- Improved compatibility between operating systems,
- Increased data security,
- Accessibility from a range of devices

Cloud computing enables on-demand consumer access to server time and network storage, broad access to network, pooling of resources, flexible provision of services,

Figure 1. Cloud computing metaphor
Source: Wikipedia (2017)

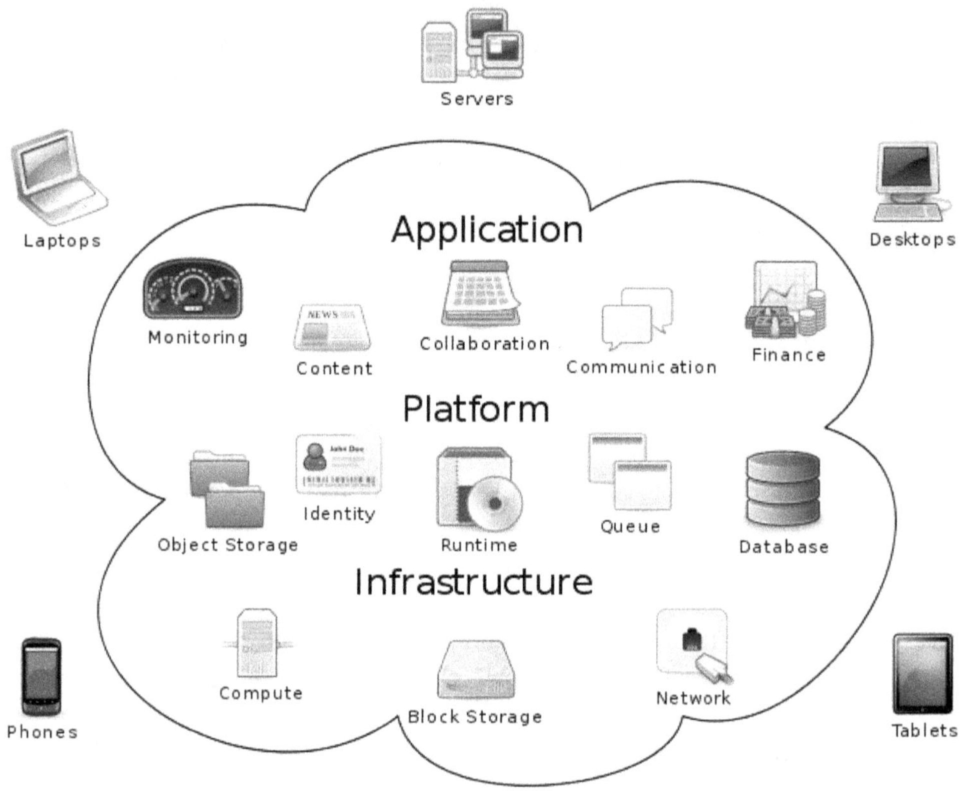

and measurement of services. Cloud systems make use of a metering capability to control and optimize the allocation of resources according to the type of service such as storage, and data processing.

Cloud computing is a game changer that offers new opportunities for outsourcing data, computation power, software, infrastructure, platform, hardware and many other requirements as they continue to be added on to the list. Individuals, SMEs, and large enterprises; whether private or public, significantly benefit from CC technologies. The entry level constraints to business have become minimal as a result of CC.

There are many players in the provision of CC services, which include; Google, Inc., for provision of the popular search engine and an array of business applications known as Google Apps; Salesforce.com, which provides a business application for managing sales and marketing; Amazon Web Services, offering access to amazon computing to web developers; Dropbox, for storage space; Autodesk, and Siemens, which offer cloud based PLM solutions to product designers and manufacturers. The list continues to balloon as new enterprises join the lucrative business opportunity.

Companies such as Bluehost have completely transformed the traditional approach to web development and hosting by their flexible hosting solutions and the Wordpress development platform. Google Apps are also a great news to many start-ups that no longer need to struggle to put up a professional business appearance to their public customers. The main investment that enterprises need to be worry about today is reliable internet connectivity, since many softwares are now offered as SaaS. The SaaS model support the delivery of complex applications over the internet, and provides reliable storage. Web browsers, such as Google Chrome, Internet Explorer, Firefox, etc., can be used to access software provided as SaaS. Cloud service providers either charge users for their services, through subscriptions and usage fees, or provide free access to the services and charge companies for placing advertisements in the services. End users do not need to be concerned about purchasing the underlying infrastructure to run the software. Many software companies have shifted from the traditional approach of offering software applications that are installed on user company premises and have introduced cloud applications.

Enterprises are renaming their services and operations with the prefix "C" to replace the "e" in what formerly used to be refered under the umbrella term "e-systems". Major developments have occurred in the manufacturing related enterprises which have realised new terminologies such as cloud manufacturing (CM), cloud based maintenance, cloud based design (CBD), cloud based design and manufacture (CBDM). Wu, et al. (2015) defines CB and CBDM as follows;

Cloud-Based Design (CBD) refers to a networked design model that leverages cloud computing, service-oriented architecture (SOA), Web 2.0 (e.g., social network sites), and semantic web technologies to support cloud-based engineering design services in distributed and collaborative environments. Cloud-based design and manufacturing (CBDM) refers to a service-oriented networked product development model in which service consumers are enabled to configure, select, and utilize customized product realization resources and services ranging from CAE software to reconfigurable manufacturing systems.

Bohu, et al. (2010) define Cloud manufacturing as a new, networked and intelligent manufacturing model that is service-oriented, knowledge based, high performance, and energy efficient. State-of-the-art technologies such as digital manufacturing, cloud computing (CC), Internet of Things (IoT), semantic Web, and high-performance computing are integrated in order to provide secure, reliable, and high quality on-demand services at low prices across the whole manufacturing lifecycle.

Cloud computing is enabled by various technologies as listed below (Hassan, 2011);

- **Broadband:** High-speed Internet access enabled systems and data to reside in one continent while users access them from different continents,

- **Grid Computing:** A distributed computing model that tends to gather underutilized computing resources available in organizations to process computing-intensive tasks faster,
- **ASP:** In the mid 1990s, ASP came to the surface as a business model that enabled organizations to access applications hosted by third parties, freeing them to focus on their business instead of being distracted by IT complexities,
- **Service-Oriented Architecture (SOA):** The idea of SOA is to turn functionalities of both existing and new applications into a set of granular components. SOA has encouraged software vendors to offer their products as services that clients can use/reuse and compose together to fulfill business requirements in an agile manner,
- **Web 2.0:** Innovations in web technologies help organizations to offer their information as sets of services that allow others to easily access them to mix and match underlying functionalities in their own websites/applications,
- **Multi-Tenancy:** A software architecture that allows software vendors to offer a single instance of one or more of their systems to different tenants (clients). Multi-tenancy represents an evolution of the ASP model that offers similar services at lower costs. The savings mainly come from sharing the same software instance and underlying infrastructures by clients rather than dedicating resources to each single one,
- **Virtualization:** Virtualization is creating a temporarily simulated or extended version of computing resources (software or hardware) such as processors, operating systems, storages, and network resources. The simulated or extended version (virtual machine) will look like a real resource. Virtualization creates an abstraction layer between computing resources and the application that use them (Hamdaqa and Tahvildari, 2012).

CC has not yet reached high maturity and its dynamic nature may pose risks to enterprises. Issues such as vendor lock-in and cyber security are of concern. Research is being carried out to seek ways of solving such challenges. However the advantages of the technology far outweigh the setbacks.

INTEROPERABILITY AND STRATEGIC ADVANTAGES

Interoperability is the ability of computer-based systems to exchange information seamlessly. ISO16100-1 defines interoperability as the ability to share and exchange information using common syntax and semantics to meet an application-specific functional relationship across a common interface (Mourad et al., 2016). Three levels of connectivity between various systems which allow seamless integration are:

unified systems where essentially all information is transferred with no requirement for alterations, integrated systems where information is exchanged in accordance to predefined standards, and interoperable systems where dynamic information exchange rules are created as required. Interoperability is a strategic advantage to enterprises, because of its ability to bring flexibility, efficiency, responsiveness and collaboration with other enterprises. Modern supply chain networks are getting more complex, and interoperability comes in as a solution to solve the maze.

Products achieve interoperability with other products using either or both of two approaches:

- By adhering to published interface standards,
- By making use of a "broker" of services that can convert one product's interface into another product's interface "on the fly".

Software systems achieve the goal of information integration used by following a common standard of communicating with each other in such a way to eliminate the possibility of any type of information or data loss. The nature of data to be exchanged continue to be more complex as more enterprises adopt latest technology in almost all their business processes. Manufacturing enterprises are the most affected because of a large number of value adding processes involved in the PLM phases (inception and conceptualisation, design, manufacture, marketing, useful life, maintenance and disposal).

Efforts to establish standardised information representation of manufacturing resources and capabilities within an integrated manufacturing system, aiming for seamless data transfer and exchange include; STEP/STEP-NC, WSDL, ontology techniques and XML. These technologies are deployed for the identification and application of standardised data models and structures for manufacturing resources and capabilities utilised through the product life cycle processes (Mourad et al., 2016).

CYBER SERCURITY

Advanced internet based technologies can be major risk concern to organisations, both private and public. Confidential information is kept in the cloud or on in-house servers. This calls for organisations and individuals to be very careful when manoeuvring in cyber space. Risks stem from both physical and cyber threats and hazards. Cyber attackers identify vulnerabilities to steal information and money and are always developing capabilities to disrupt, destroy, or threaten the delivery of essential services. Integral requirements for cyber space include the following:

- **Privacy:** Information exchanged must be kept from unauthorized parties,
- **Integrity:** The exchanged information must not be altered or tampered with,
- **Authentication:** Both sender and recipient must prove their identities to each other, and
- **Non-Repudiation:** Proof is required that the exchanged information was indeed received.

Obregon (2015) discusses security patterns in four core information security domains: access control, log management, network security and remote access. Other information security domains include; host security, vulnerability management and wireless security that apply to industrial control systems (ICS) environments. Deploying appropriate security measures around these domains can greatly reduce an organization's attack surface while increasing its security posture.

Enterprises must adopt policies, technologies, and controls deployed to protect data, applications, and the associated infrastructure as part of their strategy for e-systems. Service providers must ensure that their infrastructure is secure and that their clients' data and applications are protected, while the user must take measures to fortify their application and use strong passwords and authentication measures. Use of virtualization in cyber space infrastructure brings unique security concerns for customers or tenants of public cloud services.

CONCLUSION

A wide range of e-systems concepts have been introduced. Current trends have also been highlighted, which comprise of voice commerce, IoT voice commerce, artificial intelligence (AI), machine learning and cloud computing (CC). Although these new technologies have not yet reached maturity, they present superior opportunities to enterprises. The dynamic nature of current market trends favors early adopters. The wait and see attitude has resulted in many traditional enterprises being booted out of business due to highly competitive counterparts that are equipped with modern technology, giving them superior advantage. Subsequent chapters of the book explore practical ways of integration of the cloud in the health, banking, manufacturing, and product life cycle management (PLM) sectors. Technology acceptance, quality and cyber security issues are also covered in-depth in order to guide organisations on safeguarding themselves as well as their customers against cybercrime.

REFERENCES

Bohu, L., Lin, Z., & Xudong, C. (2010). *Introduction to cloud manufacturing*. Retrieved from: http://wwwen.zte.com.cn/endata/magazine/ztecommunications/2010Year/no4/articles/201012/t20101219_196752.html

Cheng, K., & Bateman, R. J. (2008). e-Manufacturing: Characteristics, applications and potentials. *Progress in Natural Science*, *18*(11), 1323–1328. doi:10.1016/j.pnsc.2008.03.027

Coye, C., & Loftesness, B. S. (2012). *Interoperability in electronic payments: Lessons and opportunities*. Retrieved from: http://www.cgap.org/sites/default/files/Interoperability_in_Electronic_Payments.pdf

European Commission. (2003). *E-health*. Retrieved from http://ec.europea.eu/information_society/eeurope/ehealth/conference/2003/doc/min_dec_22_may_03.pdf

Hamdaqa, M., & Tahvildari, L. (2012). Cloud computing uncovered: A research landscape. *Advances in Computers*, *86*, 41–84. doi:10.1016/B978-0-12-396535-6.00002-8

Hassan, Q. F. (2011, January). Demystifying Cloud Computing. *The Journal of Defense Software Engineering. Crosstalk*, *2011*, 16–21. Retrieved from http://www.crosstalkonline.org/storage/issue-archives/2011/201101/201101-Hassan.pdf

Koç, M., & Lee, J. (2003). E-Manufacturing-fundamentals, requirements and expected impacts. *International Journal of Advanced Manufacturing Systems*, *6*(1), 29–46.

Mell, P., & Grance, T. (2011). *The NIST Definition of Cloud Computing*. National Institute of Standards and Technology. doi:10.6028/NIST.SP.800-145

Micheal, M. (2009). *Cloud computing: Web- based applications that change the way you work and collaborate online*. Que Publishing.

Mourad, M., Nassehi, A., & Schaefer, D. (2016). Interoperability as a key enabler for manufacturing in the cloud. *Procedia CIRP, 52,* 30-34. Retrieved from: http://www.sciencedirect.com/science/article/pii/S2212827116308010

Obregon, L. (2015). Secure architecture for industrial control systems. *Technet Magazine.*

Openclinical. (2005). Retrieved from: http://www.openclinical.org/e-Health.html

Ramarao, A. (2016). Analysis of e-manufacturing capabilities and characteristics. *International Research Journal of Engineering and Technology, 3*(9).

Saha, R., & Grover, S. (2011). Identifying enablers of e-manufacturing. *ISRN Mechanical Engineering*, *2011*, 1–6. doi:10.5402/2011/193124

Shivanand, H. K., Nanjundaradhya, N. V., Kammar, P., Divya, S. S., & Keshavamurthy, Y. C. (2008). E-Manufacturing: A technology review. *Proceedings of the World Congress on Engineering 2008*, *2*.

Wikipedia. (2017). *Cloud computing*. Retrieved from: https://en.wikipedia.org/wiki/Cloud_computing

Wu, D., Rosen, D. W., Wang, L., & Schaefer, D. (2015). Cloud-based design and manufacturing: A new paradigm in digital manufacturing and design innovation. *Computer Aided Design*, *59*, 1–14. doi:10.1016/j.cad.2014.07.006

Chapter 1
IoT and Smart Manufacturing

Aqeel ur Rehman
Hamdard University, Pakistan

Iqbal Uddin Khan
Hamdard University, Pakistan

Ahmar Murtaza
Hamdard University, Pakistan

Uzma Naz
Hamdard University, Pakistan

ABSTRACT

Internet of things (IoT) is a concept of providing uniquely identifiable objects connectivity to the internet. Under the roof of IoT, it is predicted that above 28 billion devices will be connected to the internet by the year 2020. When billions of things connect, it will be difficult to manage and analyze huge amount of data as each object will send and retrieve data. Smart manufacturing is an emerging concept where the manufacturing process is supported by technology and the required information is made available during the manufacturing process to get the flexibility and the product as per customer changing needs. Internet of things (IoT) may provide a good platform to enhance the manufacturing process into smart manufacturing. The advantages of smart manufacturing include the higher quality of a product, improved productivity, increased energy efficiency, enhanced scalability in manufacturing process, etc. This chapter presents in depth the IoT and smart manufacturing concepts, their requirements, relevance, and available solutions.

DOI: 10.4018/978-1-5225-3628-4.ch001

THE CONCEPT AND VISIONS OF IoT

There are various definitions that are available in the literature. The core concept was proposed by Kevin Ashton that was based on well-known Ubiquitous Computing concept proposed by Mark Weiser in 1989 (Weiser, 1991). The vision of IoT based on core concept progressed with the time that leads to several visions based on advancement in technologies.

The Concept of IoT

The term "Internet of Things" was coined by Kevin Ashton in 1999. IoT concept is dreadfully simple (refer to figure 1) and it will be realized when all objects used in daily life will be uniquely identified and will have the ability to communicate with each other and react on the occurrence of any event.

Ashton presented IoT concept in an article of RFID journal (Ashton, 2009). As per the concept, the technologies like RFID and sensors will enable computers to observe, identify and understand the world of their own without the intervention of humans. Such computers will be able to provide their own status like when to repair, replace and discard. IoT concept is also presented by other.

As per dictionary definition of Oxford:

Figure 1. The Internet of things for Everyone, Every Where, Any Thing

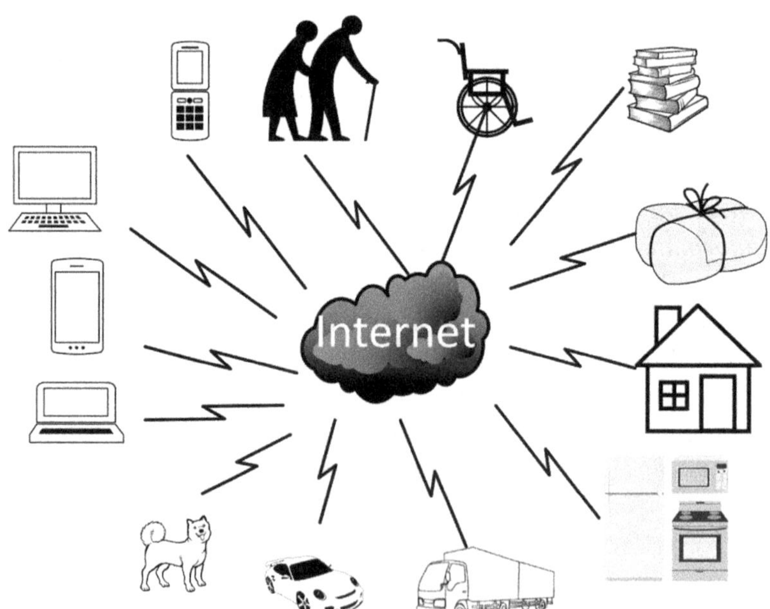

The interconnection via the Internet of computing devices embedded in everyday objects, enabling them to send and receive data.

The Internet Architecture Board (IAB) describes IoT as (Rose, Eldridge & Chapin, 2015):

The term Internet of Things (IoT) denotes a trend where a large number of embedded devices employ communication services offered by the Internet protocols. Many of these devices, often called "smart objects", are not directly operated by humans, but exist as components in buildings or vehicles, or are spread out in the environment.

Above concepts' definitions entail a much different variety of hardware, software and networking solutions in terms of cost, performance, power consumption, availability, bandwidth, and other key attributes.

The Visions of IoT

A variety of visions has been presented by researchers since the inception of the IoT concept. The vision of IoT in terms of things perspective and network-oriented perspective is substantial for stakeholders, business alliances, and research and standardization bodies (Sundmaeker, Guillemin, Friess, & Woelfflé, 2010; Ericsson Mobility Report, 2016; ITU Internet Reports, 2005).

There are three visions of IoT that are:

- Things Oriented
- Internet Oriented
- Semantic Oriented

Things Oriented Vision

Things Oriented vision that provided the foundations to connected devices concept, things are identified with unique IDs and RFID Tags were considered as the first identification technology. Later on, NFC (Near Field Communication) and WSN (Wireless Sensor Networks) are also considered for identification under this vision.

Internet Oriented Vision

The Internet as a global network is the core of the Internet-oriented vision. According to the Vision, all uniquely identified objects will be connected to the global network i.e. Internet. ITU described this vision as (ITU Internet Reports, 2005): "from anytime, anyplace connectivity for anyone, we will now have connectivity for anything".

Semantic Oriented Vision

It is predicted that there will be 28 billion connected devices by 2021 (Ericsson Mobility Report, 2016). This huge number of devices will need to connect and communicate autonomously from anywhere by anyone leading to the consequences of scalability and interoperability. As per the researchers, semantic technologies may provide the solutions to the issues of storage, connectivity, representation, and organization of information generated by the plethora of connected devices. The Semantic oriented vision is based on semantic technologies that will be helpful in dealing with billions of devices expected to be connected to the global network in future.

IoT ARCHITECTURES

Architecture is the first step towards the realization of any technological concept and that is the case with IoT. Due to the utilization of various heterogeneous technologies, a single reference architecture for IoT is not appropriate that is why different IoT architectures are being used as per requirements. Several architectures have been proposed since the inception of the IoT concept that is evolved from the 3- layered to the 5- layered architecture, service oriented IoT-A architecture to the special purpose BeTaaS, Open IoT and IoT@Work architectures (Laeeq & Shamsi, 2015). A brief description of architectures is offered in the following section:

THREE LAYERS AND FIVE LAYER ARCHITECTURES

Some key parameters like security, privacy, QoS, and reliability should be focused while planning to create IoT architecture. Some of these concerns have been involved in very generic architecture models which were proposed and based on layered architecture. These models are called 3-Layer and 5-Layer architecture (Said & Masud, 2013).

3-Layer architecture model (refer to figure 2) is considered as a primary model that has been approved for IoT work, is based on the application layer, network layer, and perception layer. The task of Perception layer is to identify objects along with the collection of information. Some of the key devices that operate on this layer are GPS, Camera, and RFID tags. The second layer of the architecture is Network layer which is responsible for processing received data from the connected layer

Figure 2. The Three Layers of IoT

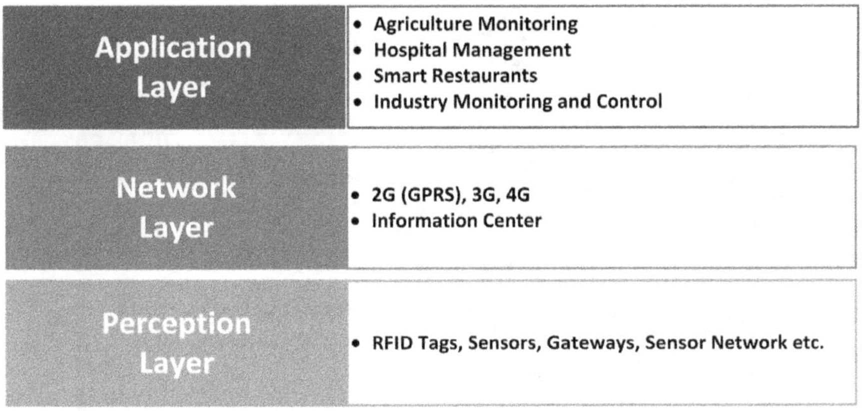

namely the perception layer and also to transmit the collected data. The top most layer, known as Application layer is liable to provide informational services.

5-layer architecture is more well-defined architecture and highly recommended for the IoT use. Most experts believe that the 5-layer architecture model (refer to figure 3) will cover all the IoT features as it is consist of 2 additional layers as compared to 3-layer architecture that are Processing and Business layers.

Processing layer in this architecture is used to store all the data received from transport layer (often called network layer) and process it. Technologies like Cloud computing, Ubiquitous computing etc. are used on this specific layer. In Business layer, management of task (applications management and business model) is performed that is realization and charge management of different applications.

IoT-A

IoT-A is a business-driven architecture which generates various reference architectures that are dependent on domain-specific requirements including users, services, physical/virtual objects, resources and devices (sensors and tags). IoT-A outlines rules for the design of protocols and algorithms for IoT. Special importance is given in IoT-A towards the security aspects of IoT that includes integrity, confidentiality, privacy, security, availability, and trust (Alexander & Saied, 2013).

BETAAS

For the requirement of communication between machine to machine, BeTaaS architecture is purposed. BeTaaS is founded on the reference model named as Things

Figure 3. The Five Layers of IoT

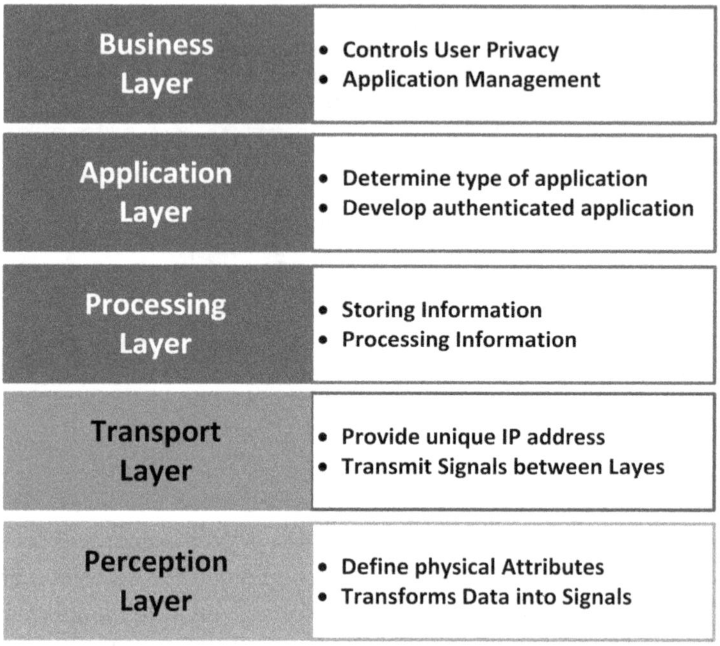

of Service (TaaS) (BETaaS Consortium, 2013). In this architecture, four layers are presented that is the physical layer, adaptation layer, TaaS layer and Service layer. BeTaas architecture emphasizes on identity management and privacy (BETaaS Consortium, 2012).

OPENIoT

This architectural model (OpenIoT Consortium, 2013) is based on the reference model defined in IoT-A. The major objective of OpenIoT is to provide the infrastructure that is cloud-based middleware. There are two security modules presented in OpenIoT architecture specification, known as security and privacy module and trust module.

IoT@WORK

IoT@work is proposed (IoT@Work Consortium, 2013) to harness IoT technologies in industrial and automation domain to achieve reliable communication, connectivity, and security in a highly complex and demanding arena.

COMMUNICATION TECHNOLOGIES OF IoT

The major communication technologies that can be utilized by IoT devices are summarized below. These technologies are categorized into three mediums:

1. Short Range
2. Medium Range
3. Long Range

SHORT RANGE IoT COMMUNICATION TECHNOLOGIES

Li-Fi

Light Fidelity (Li-Fi) is a wireless communication technology which travels at 224 Gbps. It is a visible light communication system that uses LED lights to transfer data. Professor Harald Haas belongs to the University of Edinburgh introduced the term Li-Fi. It is quite similar to Wi-Fi but instead of radio waves to transfer data, Li-Fi uses visible light spectrum. It covers short range because Li-Fi signals cannot cross walls. So, it is secured from hackers. It requires LED lights as ON state every time to provide smooth and fast internet connectivity (Mercer, 2016).

ZigBee

ZigBee is the wireless technology mostly used for home and industry automation. It consumes low power to provide long battery life (months to years). It operates on three unlicensed bands that are 2.4 GHz, 900 MHz, and 868 MHz. It transfers data at 250 kbps data rate. It targets Radio-Frequency (RF) applications and provides secure networking with low power usage ranging from 20 to 60 mW.

Radio Frequency

Radio Frequency technology is used for tracking and identifying objects by using small electronic chips that are known as *Tags*. RFID tags are characterized into passive, active and battery assisted passive (BAP) tags. The *passive tags* cannot transmit information without power supply because they don't have a battery. *Active tags* can broadcast data continuously because they have their own battery. *BAP tags* are sometimes known as semi-passive or semi-active. These tags have an internal battery. They respond after receiving signals from RFID then it returns the result to the RFID reader (Atzori, Iera & Morabito, 2010) It is low cost and efficient technology to track and communicate objects over the range of 100m to 1km. It is mostly used in superstores and library.

Bluetooth

Bluetooth is an IEEE 802.15.1 standard. The radio technology used by Bluetooth is known as *Frequency-hopping spread spectrum.* It is low cost, short range and cheap devices of wireless radio technology. It uses ISM band short range from 2400-2480 MHz. It can create peer networks automatically and connect small devices. We can connect few computers, mobiles and some other small devices to interchange files. It also has some drawbacks like slow speed, less security and shortened battery life (Barett, 2016).

Bluetooth 4.0

It is a short-range wireless radio technology f communication between devices uses low power. Bluetooth 4.0 is more energy efficient. It collects information from devices very quickly. It is a packet-based protocol which transfers data by dividing it into small chunks.

It is specially designed for longer battery life. It can transmit or receive data over the short range up to 200 feet. Its maximum data rate is 25 Mbps. As per Foley (Bennett, 2012), it is an extremely optimized version of Bluetooth and it enables an entirely new class of product into the Bluetooth world.

Z-Wave

Z-Wave is developed by ZenSys and promoted by Z-Wave Alliance for home automation and a light commercial environment. It is a low-cost wireless technology for communication. All communications through Z-Wave are secure and reliable with low consumption of power. Its main purpose is to transmit short messages to one or more nodes in the network from a control unit (Barett, 2016). It operates on 900 MHz and 2.4 GHz. It transfers data at the speed of 9.6 kbps, 40kbps and 200 kbps (Gonzalez, Organero & Kloos, 2008; Suresh, 2015).

Wi-Fi

Wireless fidelity is known as (Wi-Fi) the IEEE 802.11x standards are used to connect the devices (that are in the range of Wi-Fi) to the internet without wires. It works on the unlicensed spectrum of 2.4 GHz (ISM) band. These networks can be established easily and thus IoT devices connect directly to the internet via Wi-Fi module. It consumes more power. Wired connection is more secure than Wi-Fi. It supports devices indoor as well as outdoor.

6LoWPAN

6LoWPAN is an acronym of Low Power Personal Area Network, where '6' represents the use of IPv6. The thought behind 6LoWPAN is that, to enable the use of TCP/IP protocol suite by a device having very low computational and battery power, which will allow these small computational devices to connect and enable internet connectivity for IoT applications. This platform will enable the user to connect small devices to the internet by micro controllers' platforms. IEEE 802.15.4 is the network standard intended to use in 6LoWPAN for encapsulation and header comparison mechanism.

MEDIUM RANGE IoT COMMUNICATION TECHNOLOGIES

WiMAX

World Interoperability for Microwave Access (WiMAX) (Gonzalez, Organero & Kloos, 2008) is known as IEEE 802.16 standard for Wireless metropolitan area network (WMAN). It can operate in both licensed and unlicensed spectrum. Its speed is faster than Wi-Fi and covers larger distance up to 50 Km. It can support hundreds of users with the quality of service. WiMAX transfers data at 70 Mb/s speed over the range of 50 Km approximately. It will provide asymmetric (up and down) bit rates of 70Mbps. operate in the frequency range between 2-6 GHz.

Weightless

Weightless technology was introduced in 2014 (Weightless.org, 2016; Namiot, 2016) to deliver connectivity for low power wide area network specifically designed for IoT. It is open standard communication technology and uses the unlicensed band for Television (TV) broadcast which enables long battery life, long range, and low-cost IoT products for rapidly and economically deployment in the entire world. It transfers data at the data rate of 2.5Kbps to 16Kbps. It covers distance up to 5Km. This technology is most applicable in transferring messages from machine to machine because its devices remain in standby mode, waking up every 15 minutes acting on any message.

LONG RANGE IoT COMMUNICATION TECHNOLOGIES

2G / GPRS

It is the technology of the second generation follows Time-Division Multiple-Access (TDMA) spectrum sharing, introduced by the European Telecommunications Standards Institute (ETSI). ETSI proposed 900 MHZ spectrum band reservation for GSM in 1986.In 1991 work began to expand GSM standard expanded up to 1800 MHz frequency band and being operational in the UK in 1993. It described a digital circuit-switched network for full duplex voice telephony and later on it started packet data transportation via GPRS to include data communication (Wikipedia. 2016).

3G and 4G

The first cellular network was 3G, it meets the requirements of IMT-2000 standards which is define by International Telecommunication units (ITU) to increase growth, bandwidth and support more diverse applications. It delivers data at 14.4 Kbps but the device having 3G facility must require a constant power source and ability to recharge. 4G follows 3G and mostly used by people for mobile cellular data and IoT devices. 4G set its goals to increase data rate up to 100Mbps, streaming multimedia, IP based packet switched delivery and use of OFDMA transmission (SearchTelecom, 2009; Link Labs., 2016).

LTE

LTE stands for Long Term Evaluation. It is the latest technology for wireless mobile communication. User demand for high speed and quality of services leads to the development of LTE. Its goals are to apply new DSP techniques and redesign network architecture into the simplest form to reduce data transfer latency. It is developed to support Time Division Duplex Technology (TDD) and Frequency Division Duplex (FDD) as well as it covers the spectrum range from 1.4 MHz to 20 MHz. It provides only 5ms transfer delay in radio access network, with the data rate of 300Mbps (3gpp.org, 2016).

LoRaWAN

LoRaWAN is a Low Power Wide Area Network (LPWAN) for operating batteries wirelessly. It fulfills the requirement of IoT by providing mobility, bi-directional communication, and localization services. It transfers data at the speed of 0.3 Kbps to 50 kbps. It facilitates the range of 10mile. Maximum battery life of LoRaWAN is about 10 years. It eliminates the infrastructure cost (Anon, 2016).

Wi-Fi HaLow

It is IEEE 802.11ah technology which offers low power connectivity but covers the longer range. Its frequency band is below 1 GHz. Wi-Fi HaLow range is double than Wi-Fi. It extends Wi-Fi into the 900 MHz band. Many Wi-Fi Halow supported devices operate in 2.4 GHz and 5 GHz as well as in 900 MHz. It will support new power saving system like a connected car, digital healthcare, smart home, agriculture etc. It also has the ability to penetrate walls and other barriers in challenging environment (Las Vegas, 2016).

CHALLENGES ASSOCIATED WITH IoT

Architectures which support the Internet of Things are entailed to deal with an estimated growth of billions of objects, which would interact with each other and with other entities, e.g. human beings or things (European Union Report, 2013; Mendez, Papapanagiotou and Yang, 2015). Following are some important challenges that should be taken into account for the promotion of Internet of Things in the actual world:

HETEROGENEITY

Billions of devices are connected that create a huge network of networks which lead towards one of the major challenges that are heterogeneity. It requires specialized protocols that can provide seamless connectivity and the capability for optimal key management for better security.

SECURITY AND PRIVACY

Security concerns had a great impact on IoT. Billions of devices with trillions of users create major security threats like eavesdropping, data integrity, authentication, sniffing, data loss, authorization, etc. The security-related issues must be taken into account for the provision of trusted secure connectivity. Privacy is also a big challenge like confidentiality in universal recognition of omnipresent attachment from side to side IoT for trillions of objects.

STANDARDIZATION

Multi-vendor environment leads towards monopoly, which limits the growth of economy so; standards must be developed as an infrastructure of IoT. These standards will improve rapid growth of IoT paradigm and will open the doors of new innovations for future implementations. Standardization is required for deployment of IoT worldwide and also for consistency that is extremely essential.

IDENTIFICATION

A Huge quantity of connected devices and their heterogeneity generate the problem related to device identification. The use of IPv6 may overcome this problem to some extent however new routing algorithms and protocols are required to solve this issue.

POWER MANAGEMENT

Processing power is a critical problem in all phases of deployment of the IoT environment. There is a need to focus on an appropriate solution for deployment of this technology which is a combination of various technologies including; Nano electronics, semiconductor, sensor technology, micro systems integration and they have to serve as ultra-low power devices. Using "More Than Moore's" technologies for IoT network that promise to provide the integrated system with efficient power management may deliver an effective power solution to the upcoming technology.

INTELLIGENCE

Some of the superlative features including self and/or context awareness and inter-machine communication are the core features of IoT object with a combination of memory and processing power, and the ability to configure the harsh environments. At this point, the security at the physical layer explores the characteristics of wireless channels and addresses the scalability problems. The large-scale deployments of smart "things" may raise issues for available physical layer security solutions. Considering it an important research challenge there is a need to propose "Intelligent" techniques to discover and control the device.

CONNECTIVITY

Communication among the devices and their infrastructure is a big challenge for researchers and developer for better connectivity as devices will be of different types and will be at different locations over a wireless channel using new smart antennas (fractal antennas, adaptive antennas, receptive directional antennas, plasma antennas). Furthermore, the extensions of different Application Programming Interfaces (APIs) that will deal with different applications like modulation schemes, transmission rates, and transmission speed, are the main development issues in IoT for connectivity.

SEMANTIC TECHNOLOGIES AND IoT

IoT networks have a capability to connect and interchange information easily within a network to another network. This will need to involve devices and services to enhance capabilities in the proper direction (IoT Semantic Interoperability, 2015). To allow the interoperability in the IoT network semantic technologies may present themselves as a good candidate. There are many challenges for semantic technologies like, distribute in large-scale ontologies and have to support web services in a semantic way, define new rule engines and methodologies for reasoning over a large heterogeneous data and facts, the semantic-based discovery of devices and semantically driven code generation for device interfaces etc.

REGULATION

There are three different way for regulation; i) regulation via government, ii) regulation via international agreements, and iii) self-regulation. Government regulation is restricted to protective limitations and not feasible for comprehensive IoT (Internet of Thing) organization. However, using international agreements can be extended as global IoT structure. The self-regulation is cost effective and more feasible however it requires self-motivation.

THE CONCEPT OF SMART MANUFACTURING

The term Smart Manufacturing is an approach that requires internet connectivity and Nano scale computational devices to monitor end to end production process.

The primary goal of this concept is to improve and enhance manufacturing process by the automation of maximum operations and in-depth use of data analytics. In simple words, Smart Manufacturing is the use of technology and real-time data as per requirement of manufacturing processes and its associated entities.

The Smart Manufacturing is the implementation of the concept known as Industrial Internet of Things. Which is the practical approach to deploy Nano scaled computational devices along with sensors and direct or controller based internet connectivity, not only to collect operational stats but also to control from any location. In current practice, the data produced are kept in local databases and are only utilized when needed in case of any failure or renovation of system and processes. In Smart Manufacturing environment data streams are available for all types of data analytics and engineers to analyze and overlook the signs of any process failure or any malfunctioning equipment or component. It will enable the opportunity to take pre-failure measures to avoid and reduce down time in any stage of manufacturing. This will also help manufacturers to overlook the trends of data streams from various processes to point out the causes of over timed manufacturing. Furthermore, data scientists as well as many other analysts can utilize the generated data stream to run their simulations processes of different manufacturing stages to identify the most efficient way of manufacturing in various circumstances.

In addition, Smart Manufacturing has the ability to integrate data and information, synchronizing of maximum multiple open and paid applications and products to compile new required solutions. It can be implemented on a single line of manufacturing as well as on entire plant.

APPROACHES FOR SMART MANUFACTURING

Approach One

The manufacturing systems which are able to adopt and respond to the changes required by the demands, supply network, customer's needs and in the environment of manufacturing (Yan, Morris, and Frechette, 2016).

Approach Two

The ability of manufacturing process to solve existing problems and to overcome the problems that might come in future by analyzing past data and circumstances (CMTC's Guide, 2015).

RELEVANCE OF IoT IN MANUFACTURING

With the passage of time manufacturers have adopted various options for the effective and efficient factory management and manufacturing operations, which includes changing in manufacturing processes and timely monitoring of each spot of manufacturing throughout the factory. Previously, monitoring requires additional resources including human resource which also increases the probability of human error. Now, where industries are adopting recent and advanced technologies are reducing the chances, risks of human error and improving productivity within least possible time and cost. The term Smart Manufacturing was introduced a few years back. Smart manufacturing encompasses a wide range of manufacturing industry, aiming to optimize manufacturing and production. Manufacturing can be understood as a set of multiple phases to create a finished product from a variety of raw materials, while Smart Manufacturing is a term to employ computer controlled equipment and assembly lines in the phases of manufacturing to avoid any error and loss and also to increase production. In a current and upcoming era, the term smart manufacturing will deliver the advanced manufacturing and information technology to enable flexibility in physical processes to address the challenges in dynamics of the global market.

As IoT has been discussed earlier, the benefits of IoT can optimize the manufacturing process so the industry can deliver its maximum. IoT based data analytics can be utilized to improve asset and inventory with efficiency. By combining recent and previous data of inventories will give manufacturers and their workers capability to make the decision to deal with the constantly increasing complexity of the market and with the inconsistent variability of markets' demand (IDC Manufacturing Insight, 2015). It was stated that in 2015, 65% of the companies possessing and running more than 10 plants will authorize their workers on the floor of the factory to make their decision by investing in operational intelligence.

Internet of things is making itself the number one of all technologies enabling remote access to the devices and equipment via internet from anywhere and everywhere. IoT can be enabled in any equipment which is used in manufacturing domain no matter they are mounted in storage silo or on any assembly line. This property of IoT allows IT platform to inject itself in manufacturing systems. IoT opens new windows of connectivity to empower factory with a wide variety of applications to run, monitor and control production. This variety encompasses the possibility of production facility sharing as a service, also allowing flexibility and agility among the production systems (River Publishers Series in Communication, 2013) and this also includes the connectivity of factory to the smart grid of power as well as supply chain of various vendors. Considering discussed vision, IoT can be considered one of smarter and efficient ecosystem of production and manufacturing in near future.

Talking about the industry in today's world, works of any manufacturing plant are buried under the huge and continuously increasing amount of inventory related data which is making difficult to sort and analyze data that what is present and what is not, which makes decision making difficult. For smart manufacturing, it is necessary to place an intelligent layer that will be responsible for providing connectivity between executive's layer and shop floor which will enable to provide real-time stats for decision making (LOPEZ Research, 2014). In this regard, IoT enabled platform having operational intelligence can bridge updated information receiving from machine level with the higher level of intelligence residing over the top of enterprise applications as illustrated in figure 4. One of the advantages is having track and keep an eye on assets, IoT can easily eliminate the need of a person dedicated to acquire and enter the data collected from the manufacturing site, last but not least IoT can make any manufacturing site or plant to run on its own providing all required information on a single desktop or palmtop computing device used by any manager.

IoT RELEVANT AREAS IN SMART MANUFACTURING

This era is the commencing age of IoT and this is the age of innovation in the products and processes related to manufacturing equipped with IoT. To create innovative, adaptive and reliable IoT oriented processes and products, manufacturers may depend on the platforms which are built upon technologies like cloud, big data analytics, social and mobile technologies. Fundamentally, IoT is itself acting as an innovation accelerator.

Figure 4. Illustration of IoT Optimized Manufacturing

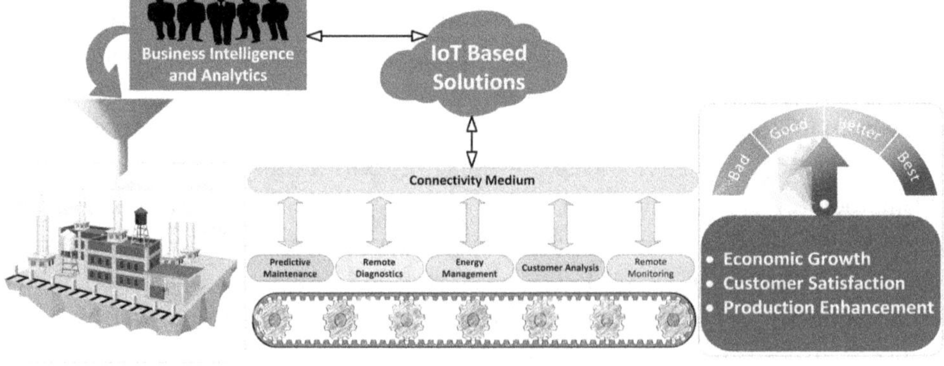

The platform that encompasses all aforesaid properties is known as 3rd Platform (IDC White Paper, 2013). The 3rd platform resides over the connectivity layer within IoT. The said platform might become core skill to transform their products, processes and business models and by reviewing all in delivered technologies may provide an improvement in developed models.

There might be two vital roles that can IoT play very effectively in manufacturing. As a first role, it can easily connect all equipment and machines to make manufacturing processes more even and hassle-free and it can easily be appended on the low level along with sensors, actuators, and ADCs. The second one might be to deal with the data that is gathered or generated by smart products allowing utilizing it for benefit of business. Manufacturers are now using palmtop devices equipped with applications to read Radio Frequency Identification Tags, Bar Codes or QR Codes for recording transactions on the manufacturing floor or in their warehouses. Tablets are the most popular handheld devices to replace clipboard carried by inspectors or supervisors, those handheld devices are connected directly to the enterprises' applications like ERPs. This is just a simple example of IoT enabled smart manufacturing.

DATA

Bringing IoT within any company will change the devices used to connect into its systems. The IoT enabled device will produce data which will be entirely different from the devices without IoT functionality. As every IoT enabled device will be able to connect to the internet and with the devices around it. These devices include different type of sensors, actuators, display units, video cameras, motion detectors and RFID readers to provide instantaneous reading over the internet and to their connected ERP. Due to there continues collection of data; big data processing and analytics will be required either on site, within company's premises or on a connected cloud service. These types of data collection and analysis will turn the collected data into a context which can further be used to make immediate, valuable and relevant decisions.

SUPERVISORS AND MANAGERS

Mobility in IoT enable devices and their associated sensors will be able to provide matchless visibility over manufacturing operations and supply chain's flow in more wide range as compared to conventional process monitoring as it will provide factual information in real time, which will improve business due to the availability of right information at right time from the right point or location. The received information can be shared with the associated suppliers, vendors, distribution chains and even retailers. Mobile-based applications will allow company managers to access in deep data of every equipment including line efficiency, equipment efficiency, data processing and visualizing tools from anywhere much easier as compared to the currently used accessing tools and techniques.

PROCESSES

Deployment of IoT will enable manufacturers to seek supply problems along with outsourced solutions. After some time manufacturers will be able to rely and trust on IoT based platforms as well as IoT will also get more pervasive which will allow faster information flow, quicker decisions and better market awareness and customer trends by interconnecting both business software process as well as manufacturing operational processes. Inter process communication that is also known as Machine to Machine (M2M) Communication, will enable all new aspects of industrial automation, just for example a painting unit checks humidity condition to paint an automobile and finds non-suitable, it will route it to another unit for some other manufacturing process till the conditions are suitable for painting. Such type of M2M communication can save time and increase production within minimum uptime of any plant.

FACTORY VISIBILITY

IoT enabled devices utilizes IP based networks to connect enterprise-based systems and executive decision makers to the floor of the factory which also includes production line information connectivity for real-time data visualization and making fair decisions to improve factory's deliverability. For understanding, we can state that any production line supervisor or manager walking on the floor can also use IoT and visualizing tools to access any machine to check its throughput, monitor production from any location at any time reducing time to access data and stats required for any decision to be made for necessary action. Considering one more example, if any plant is using Mobile Enabled Supervisory Control and Data Acquisition (SCADA) applications allowing handheld devices to display the performance data and status updated which are normally available to a PC placed on a factory floor. IoT will be the reason that manager and production supervisors will not be tied up to their rooms to access real-time information and can collaborate more effectively. Using IoT, suppliers of capital equipment will be able to offer new business models which might involve production oriented revenue rather than revenue based on the sale of capital equipment as equipment can be monitored sufficiently for output and as well as for maintenance. IoT enabled maintenance and repair operations empowers suppliers of parts, consumables, and services to manage and monitor their distributed inventory along with the storage level, parts condition and current status of the plant, helping to create a close business coordination and the relation among suppliers and manufacturers.

AUTOMATION

Networks within any manufacturing plant are mostly private networks and isolated from remote offices or sale suites. IoT and IP networks provide connectivity and information sharing among and across various locations and as well as with business networks simultaneously. As machinery and its related systems are assembled and connected within the manufacturing plant, its manufacturers can use coming information to automate workflow to optimize and keep plant maintained without the involvement of any human being. For example, of an automation, consider a manufacturing company used IoT based solution to keep track, updated records and real-time stats of equipment like monitoring a high-speed fan at painting block, a system measures temperature and humidity constantly as the environment deviates from threshold values the system will adjust the speed of fan as per requirement to maintain the values required by painting block.

ENERGY MANAGEMENT

Energy is the most vital component in the industry as it affects the cost of product per manufacturing cycle. Due to the non-efficient management of energy cost cannot come under expectation and make it hard for the company to run in the competition. This is due to lack of performance measuring and modeling tools in the company, especially within operations' zones. There are various ways that IoT and environment control systems like Heating Ventilation and Air-conditioning (HVAC) and electricity can control cost. IoT enabled HVAC can maintain the temperature of production line or factory according to the weather predicting servers over the internet, saving the cost of running unnecessarily until it is manually set as per demand, which can save the cost of energy.

MAINTENANCE CONTROL

In the current era, manufacturers have adapted the preventative and monitoring solutions. These solutions include low-cost sensors, wireless connectivity along with big data analytical and processing soft tools makes it cost effective and easier to collect real-time stats to analyses equipment's health. For example, if any manufacturer equipment is supposed to run between guided temperature range, the running company can deploy IoT based sensors to monitor it actively and company's system will maintain the temperature of equipment according to the values generated and transmitted by IoT based sensors to prevent any hazard or malfunction. Equipment mostly malfunctions and damaged due to vibrations generated while operated, detection and measuring of vibration is also an example of IoT in maintenance control. IoT enabled vibration detectors can alert and keep the system as well as manufacturer updated about overall equipment effectiveness, saving downtime due to failure by maintaining it in time or as per requirement.

CONNECTIVITY WITH SUPPLY CHAIN

IoT based IP networks and data analytics can help manufacturers to collect updated information about supply chain to keep the mark of Just in Time (JIT). Interconnecting plant equipment suppliers and a production line with real-time data stats availability all parties can understand and synchronize the availability and flow of material with the cycle of manufacturing. These IoT enabled devices can be configured and utilized for real-time location tracking, remote health monitoring of material, inventory and equipment. It also involves reporting of parts and material mobilized for a destination in the supply chain. These IoT enable systems are able to update Inventory in any company's ERP after collection of info of deliveries in real time, providing updated information for accounts and billing section. These real-time updates keep manufacturers up to date so they can identify and rectify the issues before they can create panic, reducing cost over capital requirements.

AVAILABLE PLATFORMS – SMART MANUFACTURING

The base to build a smart manufacturing plant along with its applications is the ability to connect various kinds of equipment with each other and to the main server. Types of equipment could be like machines, Sensors, Programmable Logic Controllers (PLC) itself and their Human Machine interface (HMI), assembly line robots, large-scale printers or plotters and various controllers and other IT equipment which run within business networks. The platform components could be a User Interface (UI) to enter required instructions or information displaying unit to display real-time parameters of working unit. The IoT based smart manufacturing pat form can be deployed to monitor, control and configure any equipment for analyzing purpose or for improvement depending on the information gathered by platform generated by IoT enable devices.

THINGWORX

ThingWorx delivers connectivity and intelligence for factory operations, plant's management, supply chain updating and fleets using IoT with their business systems. It provides a platform (refer to figure 5) for device management capabilities made easy to use for business analysts and other daily use business management without obtaining or accruing any high development skills (ThingWorx, 2016).

ThingWorx Foundation

Foundation module facilitates all the relevant developers to get connected, start creating the assigned object, and launch revolutionary solutions, their proper applications. It also allows sharing the experiences in the benefits of smart, connected world (ThingWorx, 2016). ThingWorx Foundation module is a comprehensive, higher end to lower end's technological platform to incorporate Internet of Thing successfully and in a long run. ThnigWorx is further split into three sections, illustrated in figure 6:

Figure 5. ThingWorx Platform

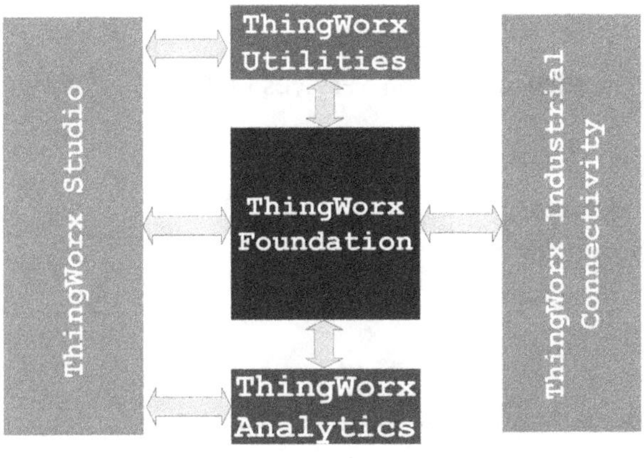

Figure 6. ThingWorx Foundation split

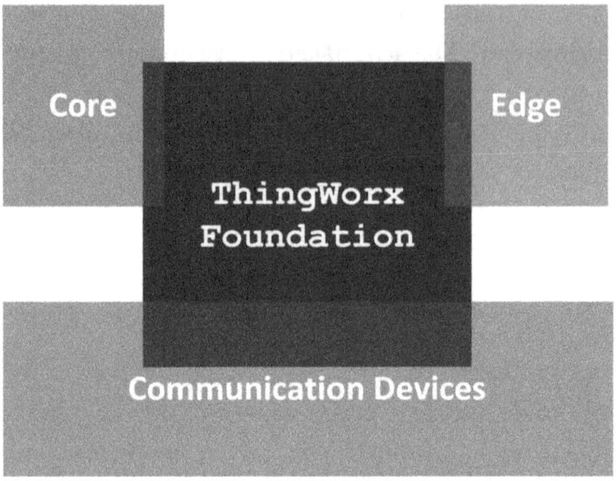

1. Core
2. Connection Services
3. Edge

Core

The Foundation Core/module includes the following:

- **Application Enablement Platform (AEP):** A Composer, Model Runner, Integration, Mashup Builder
- **Platform Services:** Security, DevOps

Connection Services

The Foundation Module Connection Services include the following:

- Connection Servers
- Device/Cloud Adapters
- Tunneling Servers

Edge

The Foundation Edge includes the following:

- Edge MicroServer
- Edge "Always On" SDK

ThingWorx Utilities

Utility module has capabilities to develop IoT based applications that empower business handlers to effectively monitor and manage the routine and performance of all interconnected devices or even goods. IoT developers create the innovative connected devices and IoT applications in the smart interconnected environment. Nonetheless, as it comes to the more frequent management, business analysts and users it is necessary that they should be able to outline, manage, monitor, and able to optimize the routines and performances of their concerned devices and goods. The developers of the concerned area may require specific hard and soft tools as well as skills that should enable them to customize IoT based applications to realigned or customized where it is necessary.

As discussed above, ThingWorx Utilities provides a broad spectrum of devices' management, delivering capabilities to business analysts and line-of-business users for very frequent management of connected devices, without the need for developer skills. Utility module is further split into three blocks (refer to figure 7):

Figure 7. ThingWorx Utilities split

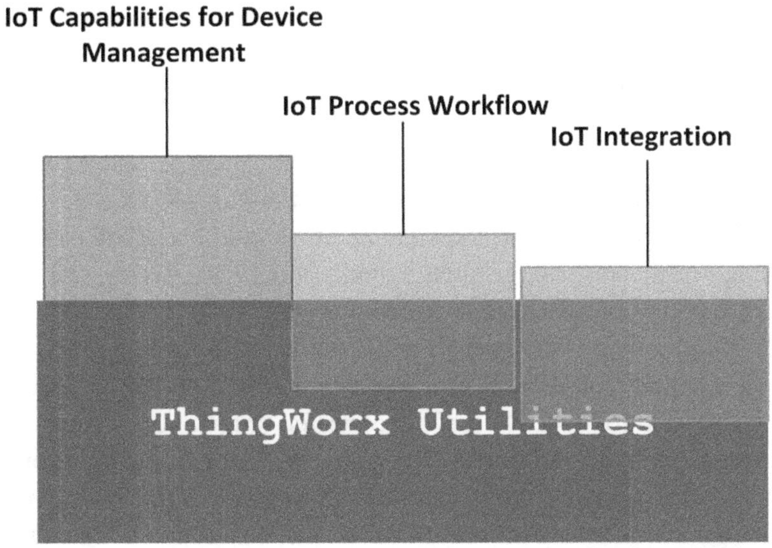

IoT Aptitudes Intended for Device Management

- Boundary less options and pre-defined capacity to optimize, monitor and manage the connected devices.
- It includes remote monitoring, provision for utilities and real-time connectivity as well as updating of assets per session.

Procedure of Workflow for IoT

- To deliver and accommodate all requirements for non-developer level, so a layman can organize and customize his process and products.
- Procedures may be fashioned in the direction of implementation across various available systems, originated on warnings and monitored events, and directed by a pre-defined set of rules.

Integration of IoT

- Allows connecting all sort of business-driven processes and working persons with deployed systems to run communal as well as dependable IoT related information across the enterprise.
- This provision also provides a standardized basis for developers to create and implement available connectors which may be pre-built and reusable and integrating available and previously deployed business systems with emerging IoT based applications.

ThingWorx Analytics

Analytics module is responsible for delivering influential and operationalized analytics to the solutions which are developed on the ThingWorx's stage. Analytics module (refer to figure 8) is considered to be the core of every IoT based solution, each "Thing" in the IoT can generate countless data sets within the span of 24 hours. It is already being expected that by the year 2020, the number of devices having ability to connect with the internet will be 30 billion. So, one can easily imagine the amount continuously growing data will be a problem to deal with. As a result, the demand for data by users of connected "Thing" or even manufacturers can be outpaced by the rapid growth of the data itself. So, at that time conventional algorithms will not work according to the demand and expectations of deployed IoT based system, understanding the real-time volume, generation of data with greater velocity as compared to today's data generation speed and challenge of variety of

Figure 8. ThingWorx Analytics split

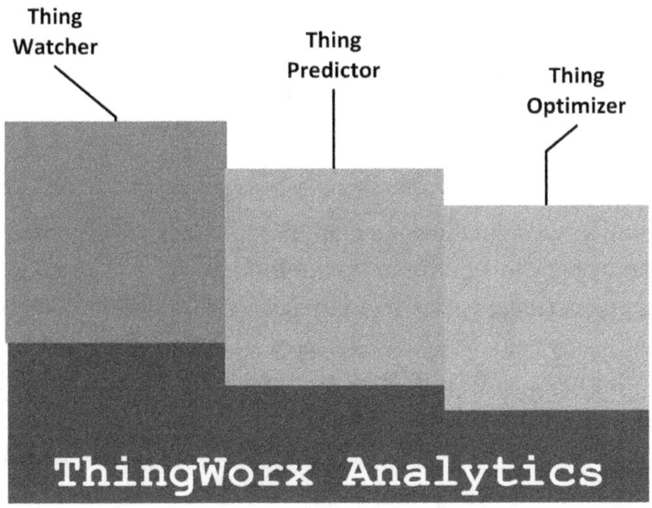

required information from the data of Internet of Things are keys to deliver even computationally more powerful IoT solutions for users and enterprises.

Analytics module empowers enterprises to extract factual values in their available IoT's data – enabling to machine learning from previous data, develop future predictions and upon past experience and future prediction make decisions to enhance outcomes. Following are four analytical properties that are applied by ThingWorx Analytics module:

To Watch

It keeps monitoring of connected devices and as per threshold, does anomaly detection on real-time incoming data streams.

To Predict

It is to provide autonomous predictive modeling and operationalization for a large variety of different outcomes and requirements.

To Adapt

It has the ability to deliver a prescriptive and simulative intelligence, so that system identifies factors that might contribute to an outcome and must have provision to change a predicted outcome of user's will.

To Optimize

The system must automatically operationalize and also maintain predictive and simulative intelligence to deliver optimized performance to end-users.

How Does It Work?

ThingWorx Analytics module provided an integrated capability of on the platform that enables developers to rapidly and straightforwardly add real-time pattern as well as anomaly detection, predictive analytics and simulation to the solutions they had developed for any requirement. The three major working modules are shown in figure 5. Furthermore, for ThingPredictor and ThingOptimizer an extra block is required called a ThnigWorx Analytic Server.

Challenges of the Growing IoT

As already mentioned earlier the IoT based devices will be crossing 30 billion by 2020 and each of them will be running a purpose-built application to monitor and control it. Nevertheless, all concerned enterprises and developers are observing their outcomes for solutions that:

- Improve manageability.
- Create more compelling user experiences.
- Better contextualize information from smart, connected devices.

The Solution

The Augmented Reality (AR) is shifting the paradox for enterprise and developers to more connected and composite IoT solutions. Rather having of stand-alone, limited applications as currently in practice, augmented reality has the ability to create a dynamic user experience through single but a powerful tool. AR offers a unique opportunity for swift scale and adoption in any enterprise. By incorporating AR into their IoT related strategy, any enterprise can improve the service, customer's experiences, related operations, and even the engineering and manufacturing of their products.

The ThingWorx Studio

ThingWorx Studio is a powerful tool as well as a platform for developing and also publishing AR related experiences in the related enterprise. As it is designed not only for developers only, it does not require such programming skills. Any content creators can build experiences in a matter of minutes and can experience as well as enjoy the advantage of existing 3D assets which are created with top CAD tools. As it is the one of a major component of the industry-leading ThingWorx® IoT platform, it delivers inbuilt AR experiences for all connected products within any enterprise.

How Does It Work?

The Studio has the ability to connect everyone to the creation of AR experiences for any specific object. The studio is a very effectual and easy-to-use developing tool for producing use-case oriented detailed experiences, which are delivered over ThingWorx's View. The experience built upon ThingWorx Studio enables a merged view of virtual and physical product's data, dashboards and most demanded, alerts are presented in 2D, 3D, and AR - without writing its code.

ThingWorx Industrial Connectivity

ThingWorx connects all relevant, required and pre-deployed devices and their applications. It provides a single source for all manufacturing automation data in IoT based solutions. Industrial Internet of Things controls all current standing technologies in industrial settings, including machine-to-machine connectivity and their communication and automation technologies. It also integrates upcoming big data and machine learning technologies to enable smart manufacturing via connected machines.

Smart, Connected Products

Smart, all connected machines constantly and precisely capture, communicate, and interpret data to update their inventory and raw material data. The reliable collection and reporting of data enable companies to detect and identify inefficiencies or problems in operations earlier to continue their production stream. As all data is collected in real time, companies are better able to improve processes, increase productivity, and maintain an inexpensive edge.

Market-Leading Industrial Connectivity

To take gain of a complete solution till now for device-to-cloud interoperability, KEPServerEX exhibits a unit solution for industrial automation data for various applications. Users can connect, easily manage, continuously monitor and control diverse automation devices and software applications through one inbuilt user interface.

The KEPServerEX is continually evolving to support platform independence. These proficiencies are vital to enable and improve operational productivity through various applications like preventative maintenance, energy monitoring, product lifecycle insights and global production tracking.

E-F@CTORY, AN ADAPTABLE SMART MANUFACTURING PLATFORM

The Mitsubishi Electric Corporation making available an IoT based Smart Manufacturing platform as the vision discussed earlier. The offered platform is known as *e-F@ctoryAlliance*. Mitsubishi Electric is an international and assorted industrial entity, encompassing sectors like transportation, automotive, semiconductors, electrical goods, aerospace, energy management and industrial automation. The immense industrial experience of Mitsubishi Electric along with union of a huge number of the equipment manufacturer and other service providers in a wide range of electrical and automation offers *e-F@ctory Alliance* to provide automation and control solution using the Internet of Things in smart manufacturing.

e-F@ctory Alliance, joining a large count of equipment and service providers in a wide span of automation as well as electrical engineering, automation and motion control systems, ICT, HMI, SCADA and MES, vision systems, industrial networking, and much more (ARC White Paper, 2015). The Alliance's contribution and experience are available to the market via the platform of alliance founder Mitsubishi Electric Factory Automation.

As support architecture is illustrated in Figure 9, at the core of any solution which can be built from the large range of products is the MELSEC Q-series controller (Programmable Controller by Mitsubishi Electric) and the CClink fieldbus. The Q-series controller has the ability to provide a variable and comfort platform to develop IoT based solutions. For solution and hardware's flexibility specimen consider a model has any CPU variant that runs compiled C-code, which is not vulnerable to any cyber world threats, this type model can easily be configured directly or through any alliance partner's available applications which might act as the gateway which permits direct connections with Manufacturing and Executing System (MES) and other IT applications.

The interconnectivity of devices is provided via CC-Link or CC-Link-IE fieldbuses, CC-Link is an open industrial network which allows various devices from many different manufacturers of devices to communicate. The recent version of the fieldbus with around 1 GB/s bandwidth that has collected data, controls, motion coordinated, and safety profiles and can allows to communicated all devices with different wired mediums and data rates. The CC-Link fieldbus utilizes the presumed

Figure 9. e-F@ctory Core Architecture (ARC White Paper, 2015)

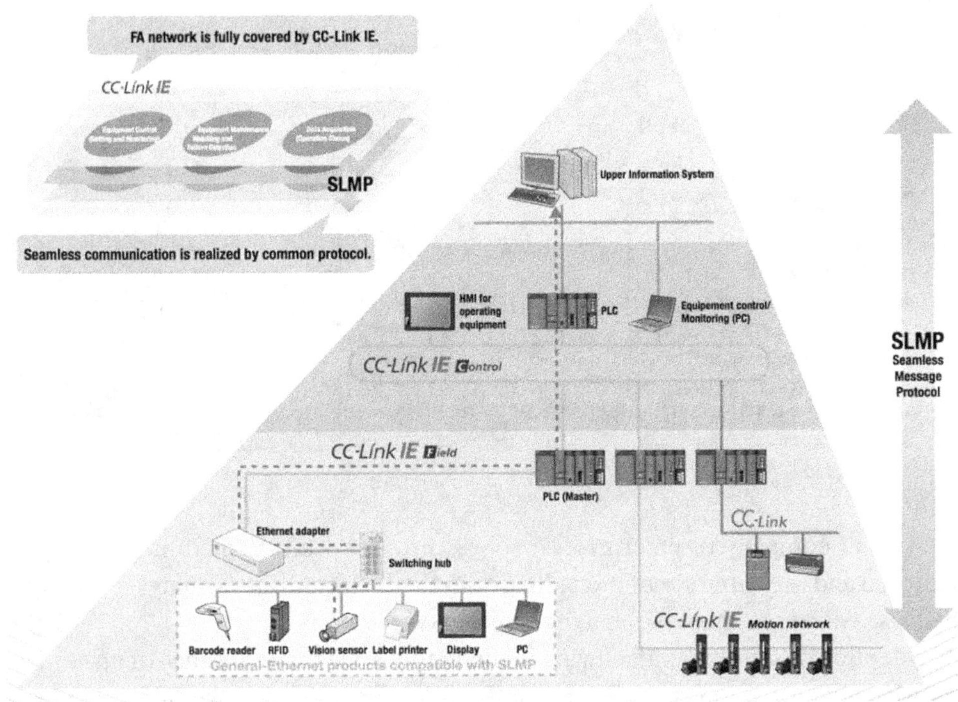

SeamLess Message Protocol (SLMP) which has the certain advantage of providing a continuous connection from the lower level i.e. Sensor level, via a gateway control system to the external IT ecosphere. Data volumes generated in the field are expected to continue to rise as disused earlier, as soon as bulky amounts of innovative and low-cost sensors will be available in the market. The broad bandwidth enables all small and Big Data analytics applications, which can be beneficial, when sufficient data will be available, the smaller the latency to analytics results, thereby leaving more options to take remedial decisions.

SMART MANUFACTURING LEADERSHIP COALITION

Smart Manufacturing Leadership Coalition (SMLC) provides a unique platform (refer to Figure 10) to drive any real-time applications of high level for manufacturing industries to augment their line of production and to enhance quality, their gross quantity, innovation and the plant productivity (Swink, 2014). SMLC presents a well-defined business model no very complex and costly to deploy and to adapt for

Figure 10. SMLC's Architecture (Swink, 2014)

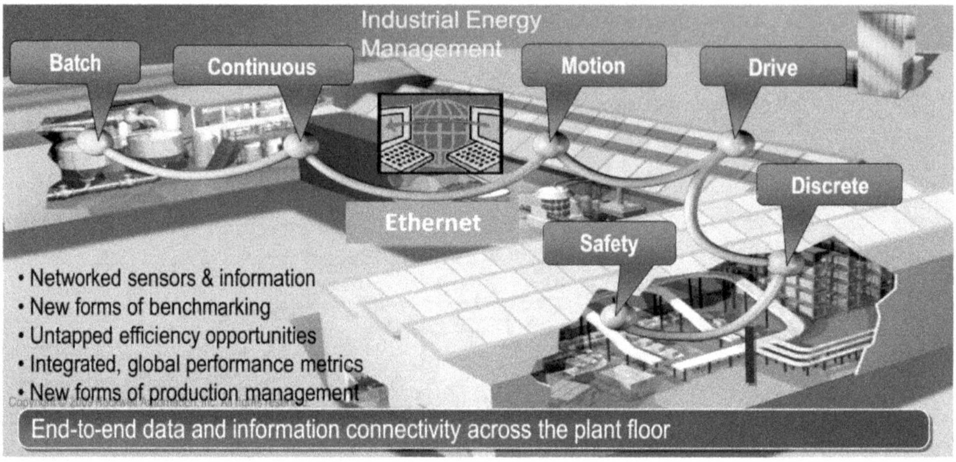

any size of company or enterprise. It solely provides well-defined modeling, pre-deployed and real-time simulation along with the collected data's analysis, enabling business entities to meet the market requirement.

Adoption of smart manufacturing is challenging due to the absence of up-to-date knowledge and awareness, over calculated risks and quick Return of Investment's (ROI) barricades. The accompanying costs over deployment make it almost unbearable for maximum companies to adopt all the necessary capabilities to deploy the Smart Manufacturing. In contrast to aforesaid, SMLC is providing a truly Open Smart Manufacturing Platform for highly accessible, industry-driven, enabling an infrastructure that is missing in current time. Significantly, over its open architecture design, the platform will interface with commercial and open source technologies providing boundless opportunities for innovation, sustainability, and affordability to all manufacturing stakeholders. This infrastructure will provide real-time, data-based application development, deployment, performance, and reuse implemented in "as needed services" available through the cloud. Few of available "Open Platforms" offered by SMLC:

- **Open Architecture:** Non-popular standards-based integration and ability to interface with commercial and open source platform technologies.
- **Open Access:** Low-cost access to Smart Manufacturing Platform for technologies.
- **Open Marketplace:** Open access (for community contributors and users) to market-driven commercial and open source application libraries inclusive

of deployment, data management, modeling and analytics and metrics applications and associated nonproprietary deployment data, certifications, services, experiences and quick solutions.

BRIDGING PLATFORMS: IoT AND SMART MANUFACTURING

There are various alliances and companies which are delivering platforms to bridge deployed systems and IoT based smart manufacturing solutions. These bridging solutions are not so costly to deploy or adopt in as compared to deploy complete new infrastructure. Two of the platforms having major solutions are discussed below:

INTEL® IoT SOLUTION ALLIANCE

Intel along with its more than 400-member companies around the globe provides interoperable, scalable solution to accelerate the use of smart devices and end to end analysis solution (Intel® Internet of Things Solutions Alliance, 2016). The close collaboration of all member organizations under *Intel® IoT Solution Alliance* provides innovative IoT based solution for smart and efficient manufacturing. The Intel® IoT Solution Alliance provides a wide variety of solution on a single web, considering Smart manufacturing and IoT domain it provides solutions as:

- **Smart Manufacturing Systems:** Systems from various companies like LEX Computech Co., Ltd, Neousys Technology Inc. and MSC Technologies GmbH etc.
- **Smart Manufacturing Hardware Components:** Provides Interfacing Cards and components and communication models from Ennoconn Corporation, ADLink, Critical Link LLC and AXIOMTEK etc.
- **Smart Manufacturing Software Components:** Striim, Inc., Softweb Solutions, Kainos Technology Sdn Bhd, azeti Networks AG, azeti Networks AG and infiswift etc.
- **Smart Manufacturing Services:** Services from Kainos Technology Sdn Bhd, Sophic Automation Sdn Bhd and Wind River Systems, Inc etc.

INDUSTRIAL IoT: KAA MIDDLEWARE

Kaa middleware empowers data managing among connected "things" like sensors, automated manufacturing plants and back-end systems like servers, analytical application by providing the server and endpoint SDK components (Driving New Business Economies with the New Triple Play, 2016). The Kaa's server offers some back-end functionality which is scalable according to enterprise-level deployments, warranting data consistency and security, devices' interoperability, and consistent connectivity.

Kaa offers a variety of IoT features from various perspectives that have the ability to empower smart devices and the upcoming Telecom infrastructure along with cloud-driven services and applications. As a hardware- and transport-skeptical

Figure 11. Kaa's middleware application's illustration

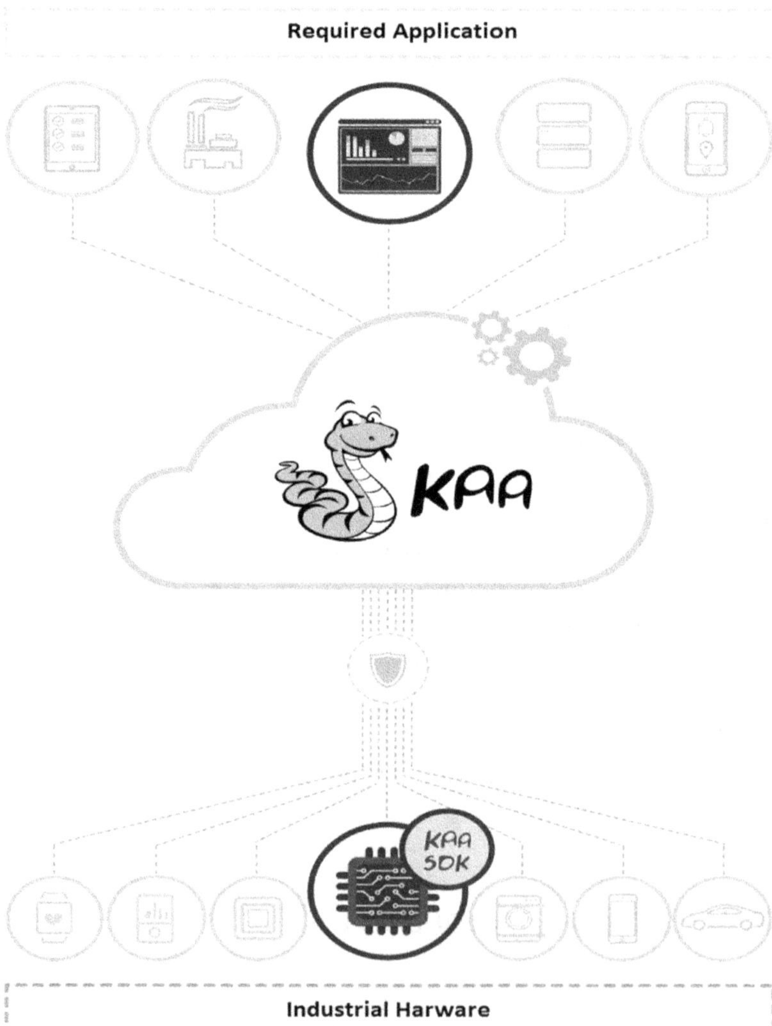

platform, Kaa is well-suited with virtually any type of connected device and can handle all popular protocols till used in the IoT. This is the Kaa's unique facility for IoT platform - a top-of-the-range feature set delivered as 100% open-source. Figure 11 illustrates the discussed concept of Kaa.

INDUSTRY 4.0: THE UPCOMING TREND

The fourth industrial revolution which is expected is known as industry 4.0. It will integrate and digitize all processes across the entire adopting organization from the purchase of raw material to manufacturing till logistics and its service (Industry 4.0, 2016). It will eliminate the various thoughts about smart manufacturing. In figure 12, the contributing technologies and proposed framework are illustrated, which encompasses all IT related technologies and analytic core to analyze raw data generated by industry and its components. Still, some challenges are yet to address in upcoming technology and platforms like technological transfer, adoption rate, cost factor, and implementation time frame factor (Thomas Hahn, 2017). The platforms like SmartFactoryOWL from Fraunhofer Society and OWL University of Applied Sciences is a manufacturer-independent Industry 4.0 application and demonstration center and also a testing area for the Industry 4.0 research and development.

Figure 12. Illustration of Industry 4.0 architecture (Industry 4.0, 2016)

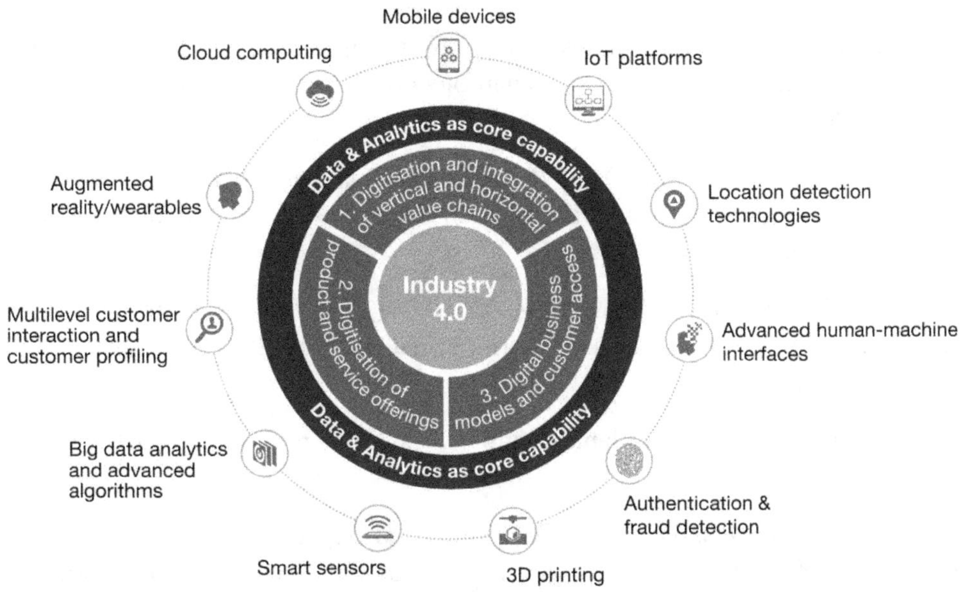

REFERENCES

ARC White Paper. (2015). *The Industrial Revolution Continues*. Available at: http://www.treasurebox.com.sg/meap/nl_201605/pdf/ARC_White_Paper_e-Factory_final_2015-12.pdf

Ashton, K. (2009). That 'Internet of Things' thing. *RFID Journal*, *22*(7), 97–114.

Atzori, L., Iera, A., & Morabito, G. (2010). The internet of things: A survey. *Computer Networks*, *54*(15), 2787–2805. doi:10.1016/j.comnet.2010.05.010

Barett, J. (2016). *The Disadvantages of Bluetooth Technology*. Available at: https://www.techwalla.com/articles/the-disadvantages-of-bluetooth-technology

Bennett, B. (2012). *The power of Bluetooth 4.0: It'll change your life*. Available at: http://www.cnet.com/news/the-power-of-bluetooth-4-0-itll-change-your-life

BETaaS Consortium. (2012). *Building the environment for the things as a service*. Retrieved from http://www.betaas.eu/

BETaaS Consortium. (2013). *D1.4.2 – TaaS Reference Model*. Retrieved from http://www.betaas.eu/docs/deliverables/BETaaS%20-%20D1.4.2%20%20TaaS%20Reference%20Model%20 v1.0.pdf

CMTC's Guide to Smart Manufacturing. (2015). Available at: https://cdn2.hubspot.net/hubfs/103829/Ebooks_Whitepapers/CMTC-Smart-Manufacturing_Ebook_Final.pdf?t=1509064707903

Driving New Business Economies with the New Triple Play, IoT, RTC, and OTT. (2016). Available at: https://www.kaaproject.org/wp-content/uploads/2016/07/telecom-whitepaper.pdf

European Union Report. (2013). *IoT Privacy, Data Protection, Information Security*. Available at: http://ec.europa.eu/information_society/newsroom/cf/dae/document.cfm?doc_id=1753

Gonzalez, G. R., Organero, M. M., & Kloos, C. D. (2008, July). Early Infrastructure of an Internet of Things in Spaces for Learning. In ICALT (pp. 381-383). doi:10.1109/ICALT.2008.210

3. gpp.org. (2016). *LTE*. Available at: http://www.3gpp.org/technologies/keywords-acronyms/98-lte

IDC Manufacturing Insight. (2015). *Transforming Manufacturing with the Internet of Things*. Available at https://www.cognizant.com/InsightsWhitepapers/transforming-manufacturing-with-the-Internet-of-Things.pdf

IDC White Paper. (2013). *The 3rd Platform: Enable Digital Transformation*. Available at: http://achievabledigitaltransformation.com/TCS-DSS/tcs-white-paper_244515.pdf

Intel® Internet of Things Solutions Alliance for Premier Members (2016). Available at: https://www.intel.com/content/dam/m/enrollment/isalegalpdfs/us/en/IOTPremier_Lug_093015.pdf

IoT Semantic Interoperability: Research Challenges, Best Practices, Solutions and Next Steps (2015). Available at http://www.internet-of-things-research.eu/pdf/IERC_Position_Paper_IoT_Semantic_Interoperability_Final.pdf

IoT@Work Consortium. (2013). *D1.2 – Final framework architecture specification*. Retrieved from https://www.iot-at-work.eu/data/D1.3 IoT@Work Architecture final v1.0-submitted.pdf

ITU Internet Reports. (2005, Nov.). *The Internet of Things*. Available at https://www.itu.int/net/wsis/tunis/newsroom/stats/The-Internet-of-Things-2005.pdf

Laeeq, K., & Shamsi, J. A. (2015). *A Study of Security Issues, Vulnerabilities and Challenges in Internet of Things*. Securing Cyber-Physical Systems. doi:10.1201/b19311-9

Las Vegas. (2016). *Wi-Fi Alliance® introduces low power, long range Wi-Fi HaLow™ | Wi-Fi Alliance*. Available at: http://www.wi-fi.org/news-events/newsroom/wi-fi-alliance-introduces-low-power-long-range-wi-fi-halow

Link Labs. (2016). *The Complete List of Wireless IoT Network Protocols*. Available at: http://www.link-labs.com/complete-list-iot-network-protocols

LOPEZ Research. (2014). *Part 2. of "The IoT Series", "Building Smarter Manufacturing with The Internet of Things (IoT)"*. Available at: http://cdn.iotwf.com/resources/6/iot_in_manufacturing_january.pdf

Lu, Y., Morris, K. C., & Frechette, S. (2016). Current Standards Landscape for Smart Manufacturing Systems. *NISTIR 8017*. Available at: 10.6028/NIST.IR.8107

Mendez, D., Papapanagiotou, I., & Yang, B. (2015). *Internet of Things: Survey on Security and Privacy*. Available at: https://arxiv.org/pdf/1707.01879.pdf

Mercer, C. (2016). Li-Fi explained | What is it, how does it work and why Wi-Fi might be a thing of the past. *Techworld*. Available at: http://www.techworld.com/big-data/what-is-li-fi-everything-you-need-know-3632764

Namiot, D. (2016). On Internet of Things and Smart Cities educational courses. *International Journal of Open Information Technologies*, *4*(5), 26–38.

OpenIoT Consortium. (2013). *OPENIoT project description*. Available at: http://www.openiot.eu/

Ericsson Mobility Report. (2016). *Ericsson Mobility Report: On the Pulse of the Network Society*. Available at https://www.ericsson.com/res/docs/2016/ericsson-mobility-report-2016.pdf

River Publishers Series in Communication. (2013). *Internet of Things: Converging Technologies for Smart Environments and Integrated Ecosystems*. Available at: http://www.internet-of-things-research.eu/pdf/Converging_Technologies_for_Smart_Environments_and_Integrated_Ecosystems_IERC_Book_Open_Access_2013.pdf

Rose, K., Eldridge, S., & Chapin, L. (2015). *The internet of things: An overview. The Internet Society*. ISOC.

Said, O., & Masud, M. (2013). Towards Internet of Things: Survey and future vision. *International Journal of Computer Networks*, *5*(1), 1–17.

Salinas & Ben Saied. (2013). Dissemination Level. Internet of Things Architecture Concepts and Solutions for Privacy and Security in the Resolution Infrastructure. (257521).

SearchTelecom. (2009). *What is 3G (third generation of mobile telephony)?* Available at: http://searchtelecom.techtarget.com/definition/3G

Sundmaeker, H., Guillemin, P., Friess, P., & Woelfflé, S. (2010). *Vision and challenges for realizing the Internet of Things. In Cluster of European Research Projects on the Internet of Things*. European Commission.

Suresh, K. (2015). *Past, Present & Future of IoT. Internet of Things*. Retrieved December 3, 2015, from http://www.tips2secure.com

Swink, D. (2014). Smart Manufacturing Leadership Coalition (SMLC). Available at http://cdn.osisoft.com/corp/en/media/presentations/2014/UsersConference2014/PDF/UC2014_SMLC_Swink_AVisionforInfusingIntelligenceintoManufacturingEnterprises.pdf

ThingWorx. (2016). Retrieved from https://www.thingworx.com/

Weightless.org. (2016). *What is Weightless - Weightless*. Available at: http://www.weightless.org/about/what-is-weightless

Weiser, M. (1991). The computer for the 21st century. *Scientific American*, *265*(3), 94–104. doi:10.1038/scientificamerican0991-94 PMID:1675486

Wikipedia. (2016). *GSM*. Available at: https://en.wikipedia.org/wiki/GSM#GSM_carrier_frequencies

ADDITIONAL READING

Aggarwal, C. C., Ashish, N., & Sheth, A. (2013). *The internet of things: A survey from the data-centric perspective Managing and mining sensor data*. Springer.

Athreya, A. P., & Tague, P. (2013). Network self-organization in the internet of things. Paper presented at the 2013 10th Annual IEEE Communications Society Conference on Sensor, Mesh and Ad Hoc Communications and Networks (SECON).

Bandyopadhyay, D., & Sen, J. (2011). Internet of things: Applications and challenges in technology and standardization. *Wireless Personal Communications*, *58*(1), 49–69. doi:10.1007/s11277-011-0288-5

Coalition, S. M. L. (2011, June). Implementing 21st century smart manufacturing. In *Workshop Summary Report*.

Costa-Pérez, X., Festag, A., Kolbe, H.-J., Quittek, J., Schmid, S., Stiemerling, M., & Van Der Veen, H. (2013). Latest trends in telecommunication standards. *Computer Communication Review*, *43*(2), 64–71. doi:10.1145/2479957.2479968

Kusiak, A. (2017). Smart manufacturing must embrace big data. *Nature*, *544*(7648), 23–25. doi:10.1038/544023a PMID:28383012

Kusiak, A. (2017). Smart manufacturing. *International Journal of Production Research*, 1–10. doi:10.1080/00207543.2017.1351644 PMID:28383012

Mainetti, L., Patrono, L., & Vilei, A. (2011). Evolution of wireless sensor networks towards the internet of things: A survey. Paper presented at the 2011 19th International Conference on Software, Telecommunications and Computer Networks (SoftCOM).

Wang, S., Wan, J., Li, D., & Zhang, C. (2016). Implementing smart factory of industrie 4.0: An outlook. *International Journal of Distributed Sensor Networks*, *12*(1), 3159805. doi:10.1155/2016/3159805

Wang, S., Zhang, C., & Li, D. (2016, March). A Big Data Centric Integrated Framework and Typical System Configurations for Smart Factory. In *International Conference on Industrial IoT Technologies and Applications* (pp. 12-23). Springer International Publishing. doi:10.1007/978-3-319-44350-8_2

Whitmore, A., Agarwal, A., & Da Xu, L. (2014). The Internet of Things—A survey of topics and trends. *Information Systems Frontiers*, 1–14.

KEY TERM AND DEFINITIONS

Internet of Things: Global network of smart devices.

IoT: Standard abbreviation of internet of things.

OWL: Standard abbreviation of web ontology language.

Semantic Technology: A way to process data to discover relationships within varied data sets.

Smart Manufacturing: A technology-driven approach that utilizes internet-connected machinery to monitor the production process.

Ubiquitous Computing: Computing everywhere for everyone.

Wireless Sensor Networks (WSN): Wireless network consisting of spatially distributed autonomous devices using sensors.

Chapter 2
Importance of Big Data and Hadoop in E-Servicing

Karthiga Shankar
Thiagarajar College of Engineering, India

Suganya R.
Thiagarajar College of Engineering, India

ABSTRACT

Consumers are spending more and more time on the web to search information and receive e-services. E-commerce, e-government, e-business, e-learning, e-science, etc. reflect the growing importance of the web in all aspects of our lives. Along with the tremendous growth of online information, the use of big data has become a vital force in growing revenues. Consumers are today shopping multiple products across multiple channels online. This transformation is substantial and many of the e-commerce companies have now turned to big data analytics for focused customer group targeting using opinion mining for evaluating campaign strategies and maintaining a competitive advantage, especially during the festive shopping season. So, the role of intelligent techniques in e-servicing is massive. This chapter focuses on the importance of big data (since there is a large volume of data online) and big data analytics in the field of e-servicing and explains the various applications, platforms to implement the big data applications, and security issues in the era of big data and e-servicing.

DOI: 10.4018/978-1-5225-3628-4.ch002

INTRODUCTION

E-Services

Typically the term e-services is used to describe a variety of internet based electronic interactions ranging from basic services, such as the delivery of news and stock quotes, to smart services, Products, clothes etc. These electronic-services involve various types, delivery systems, advanced information technologies, methodologies and applications of online services and they are provided by e-government, e-business, ecommerce, e-market, e-finance, and e-learning systems.

There services produce a large volume of data online. Where is this data stored and how this data is processed and analyzed to produce the useful information. The key is Big Data Analytics where the huge data data generated by this various applications over online is processed and analyzed. There is a need of some intelligent approaches to evaluate e-service systems, conduct web user classification, help users online trading and support users online decision making.

Big Data

Big Data is transforming healthcare, business, and ultimately society itself, as e-services becomes one of key driving features during the modernization process. Many business cases utilizing big data have been grasped in recent years; Facebook Twitter & LinkedIn are examples of companies in the social networking realm. Other big data use cases have focused on capturing of value from streaming of video movies, monitoring of network traffic or upgrading of processes in the manufacturing industry. With the e-services initiatives across the world especially in India, the transactional data is developing exponentially and it has become difficult to analyze such huge volume of data stored at multiple locations with conventional data mining algorithms. The alternative to analyze large data is to use Big Data analytical tools. The tools having capabilities of mining large data sets in distributed environment help in investigating and analyzing e-services projects.

The progresses in Information Technology (IT) services and hardware have led to the generation of huge amount of data referred to as Big Data. There are projects/ systems in e-service where this big data needs to be processed and analyzed for better and effective decision making. Due to the fact that the data generated has unstructured and formats, it is difficult to analyze such huge data using traditional analyses tools. Big data analytics can manage this dynamic nature of Big data, keeping it secure by applying the correct analytical technique to use the information in an effective manner. It has the capacity to interact with huge Volume, Velocity, Veracity and Variety of data. With its scalable nature, it can expand the technologies that can correlate data and produce exploitable results. Big data analytics can help the government in providing its services directly to its citizens. It has the ability to recognize patterns in a set and make predictions regarding past experiences and provide results for taking future actions. Big Data Analytics refers to the use of advanced analytic techniques against very large and different data sets that include structured/semi-structured/unstructured data and of different sizes. Big data is the term applied to data sets whose size and type is beyond the ability of traditional relational databases to capture, manage and process the data. Big data is defined in terms of 3 V's i.e. Volume, Velocity and Variety [Gartner]. Analyzing Big data allows analysts, researchers, government and business users to make better and faster decisions using data that was previously inaccessible or unusable. Using advanced analytics techniques, government or business agencies can analyze previously unexploited data sources autonomous or together with their existing enterprise data to gain new imminent resulting in significantly better and faster decisions.

Quite simply, the Big data era is in full force today as the world is changing through instrumentation, we are clever to sense more objects, and if we can sense it, we tend to try and store it. Through advances in communications technology, people and things are becoming increasingly connected –and not just some of the time, but all of the time. This intersecting rate is a escape train. Generally referred to as machine-to-machine (M2M), interconnectivity is responsible for double digit year over year data growth rates. Finally, because small integrated circuits are now so cheap, we are able to add intelligence to almost everything.

Experiencing the Art of Big Data

Each day, 2.5 quintillion bytes of data are produced. These data come from digital pictures, videos, posts to social media websites, intelligent sensors, purchase transaction records, cell phones GPS signals etc. This is known as Big Data. In small, the term Big Data applies to information that can't be processed or analyzed using traditional processes or tools. Increasingly, organizations nowadays are facing more and more Big Data challenges. They have access to prosperity of information, but they don't know how to acquire value out of it because it is sitting in its most raw form or in a semi structured or unstructured format; and as a result, they don't even recognize whether it's merit keeping. There is no doubt that Big Data and particularly what we do with it has the probable to become a powerful force for innovation and value creation.

Data now stream from daily life: from phones and credit cards and televisions and computers; from the infrastructure of smart cities; from sensor-equipped buildings, trains, buses, planes, bridges, and factories.

Big Data Analytics

This outline of big data analytics will help you to know big data analytics, the e-service worth it gets to many corporate industries and government sector in the world, and how organizations across various industries are applying it to tackle their unique business requirements. Every big data source has different characteristics, including the occurrence, volume, velocity, type, and veracity of the data. When big data is developed and stored, extra dimensions come into play, such as governance, security, and policies. Choosing an structural design and constructing an appropriate big data solution is challenging because so many issues have to be considered

Big data analytics is the procedure of gathering and analyzing large sets of data (called big data) to find out patterns and other useful information. Big data analytics can assist organizations to better appreciate the information contained within the data and will also help recognize the data that is most important to the business and future business decisions. For most organizations, big data analysis is a challenge. Consider the huge volume of data and the different unstructured formats of the data that is collected across the entire organization and the many different ways diverse types of data can be joined, contrasted and analyzed to find patterns and other useful business information.

Big data, data is broken down into multiple data sets and accumulated on various nodes of a cluster environment. One of the nodes (Master node in Hadoop) keeps track of the data stored at various nodes (Slave node in Hadoop) by creating index of the data. When a request for data search appears, first index is checked and the based on that required data is found and fetched. This helps in making the search process faster even if data is too bulky and large. For example, twitter and facebook are using the concept of Big data. Data from any database management system like Oracle and MySql can be imported into Big data. Through data mining activities, data can be collected from various servers. Then data analytics techniques are applied on the data retrieved and reports are generated. After that reports are analyzed to find out the outcome or behavior in regard to a particular condition and the future action plans are prepared. According to various studies done, it has been observed that in the year 2008 enterprise servers had processed about 9.57 * 1021 bytes and this number is doubled every couple of years. So, managing such huge volume of data using simple techniques is quite difficult. But if this big data is examined using Big data analysis techniques, the results produced would be much reliable and efficient and hence help in taking effective decisions. e.g. Walmart saves near about 2.5 peta byte of data every hour of its purchase history to do analysis for future using Big data analysis techniques.

Doug Cutting was the first developer who designed the structure for Nutch, a venture by Apache Software foundation. The program used MapReduce to split big data files into controllable areas which could be prepared by computer systems linked in similar.

A lot of people comprehend the importance of programming when they are to work with some applications. This is the result of the fact that they acknowledge the importance of knowing how to run the codes for the application. Apart from this, the code listings can furthermore trigger questions that deal with the possibility and operation of various business and games software. Hence, the latest hadoop framework serves as good business tools for each business operation to be a success. For leading search engines like Google, hadoop performance utilizes Map Reduce for indexing purposes. It comprises of the dynamic application that can improve the task of searching in at a faster rate than it had been done before.

Hadoop could run on current devices, which made it very practical. These days this is one of the most extensive alternatives or solutions to handle big data that are used by such significant organizations as Yahoo!, Facebook or Myspace. The achievements of the structure designed a big requirement for professionals who are dedicated to modifying and establishing up the several elements of the structure. A person operating in this area may be termed as a Hadoop developer or he may be called as a Hadoop consultant.

In addition to this, the usage of mobile phones has tremendously increased much beyond just sending or receiving the calls or text. Ranging from watching movies to booking movie tickets, listening to music to downloading every new song, drafting to sending email, and much more, mobile phones can carry out day-to-day activities easily and conveniently. All, this has become possible with the advent on internet on mobile phones. Due to this, windows and apps development took place. Looking all this, there are many companies which have introduced latest mobile windows & apps development, services Development, Android Application Development, gaming applications which includes iPhone Application Development, BlackBerry Applications Development, iPad Application development, Office/ Business Applications, Mobile Website Development, Productivity Applications, Networking applications and much more.

The growing role of Big data analytics in business is obvious. To survive in the frames of constant competition most organizations will need to support their business by having their own Big data analytics. If you already have an analytic idea you are welcome to visit Enterprise Mobility Service Provider Emorphis Technologies website and fill in the form with your requirements to help Emorphis team better understand and build your Enterprise Analytics solution for your business. We Enterprise mobility services provider company Emorphis Technologies have in-depth experience for creating engaging mobile apps for businesses. We are one of the premier Enterprise mobility Solution development company and have a long list of credentials and portfolio ranging from m-commerce solutions for retail services, Hadoop & Big data analytics, e-commerce web development, Cloud computing services, and outsourced web development. If you are on the lookout for a partner to help you leverage your online business through Mobile Applications, then do connect with Emorphis Technologies. Our proficient developers experienced in free source mobility and cloud computing applications will help you get the best solution for your company.

Benefits of Using Big Data Analytics for E-Servicing Projects

The major profits of introducing Big data platform in e-Governance projects are to assist government. These benefits are:

- Enhancement in online information and service delivery by government for business Analytics.
- Making practices open and apparent to endusers.
- Documenting and Visualizing government presentation through predictive policing.
- Producing insight for new business projects and improving business growth.
- Providing better Customer Services feedback

CHALLENGES OF USING BIG DATA ANALYTICS WITH E-SERVICING

Though analyzing Big Data using various analytical techniques has proved to be useful for e-Services projects but still there are some confronts that need to be handled to attain desirable results. Some of them are:

Privacy

Big data analytics refers to examining historical data. Each e-service project consists of a group of private information related to various end users. This data can be regarding UIDs of individuals, sales/purchase information of a business firm, information of clients of an organization or records of patients of hospital. Such organizations may oppose in revealing their private information. If provided, such information should be preserved properly so that it is safe and beneficial.

Security

Big data generated from e-Governance projects is a combination of large data which is stored on various servers. Storage and security of this big data is one of the major challenges. Processing of this data using data analytics further produces more data in the form of reports. Managing this data and protecting it from unauthorized access and usage should be carefully done.

Lack of Technical Knowledge

Still it is very difficult to find skilled and qualified people who are proficient in using big data analytics techniques with e-Governance projects.

Unavailability of Appropriate Software Tools

Data generated by different e-Governance projects during analysis is of different types and different formats. Sometimes application of relevant software tools is not possible because of its unavailability or that technique/software is still under development.

Lack of Common Standards

No proper standards are defined of how data is to be represented using analytical techniques.

APPLICATIONS OF BIG DATA IN E-SERVICING

E-Health Services

In the context of e-Health, variety of flows have generated from various sources such as electronic medical records (EMR) systems, mobilized health records (MHR), personal health records, mobile health care monitors and predictive analytics as well as a large array of biomedical sensors and smart devices including IoT. The electronic medical record (EMR) initiative has resulted data streams from all types of patients at the hospital, insurance companies and doctor's office. A single patient stay generates thousands of data elements, including diagnoses, procedures, medications, medical supplies, digital image, lab results and billing. These need to be validated, processed and integrated into a large data pools to enable meaningful analysis. Multiplying this by all the patient-stays across the health processing systems and combining it with the large number of points where data is generated and stored and the scope of the big data challenge begins to emerge.

E- Government

Big data Applications have the potential to transform the digital Government services, communication between the Government, citizens and business sectors. The Government initiatives such as digital India and Smart City are some of the examples for the contribution of big data applications in the era of E-Government. Today there are lot of Government services available through online such as ration shop stocks (status of the ration items), e-certificates (community, birth certificate, etc), Applying aadhar cards online, etc.

E- Shopping

Internet shopping transactions have recently raised big concerns. But the problem on making businesses through internet is that it is impossible for the retailers to check whether the customer is the genuine or not who accept online payment mode transaction. Using Machine learning, a method of data analytics that iteratively learn from data and allows computers to find hidden insights without being any explicit programmed. Also there are algorithms to analyze the users previous transactions and purchases and it can predict what the customer's next purchase would be. It will be very useful for the retailers to promote the particular items with discounts so that they can increase the customer's buying.

PLATFORMS TO IMPLEMENT BIG DATA APPLICATIONS

Hadoop

Open source Hadoop is a cutting-edge device for harvesting big data. One of the most highly grade, and used, platforms, Hadoop's impact in Retail is as exciting as frontline. Hadoop has established itself as the go-to technology for Big Data management and data pipeline. What Hadoop does is extraordinary: it takes the dice-rolling out of sell business and allows the hub to shift to the individual consumer by providing a customized retail experience. The attractiveness of Hadoop is that there are tangible practical advantages to boast about. Enterprises are steadily moving to deploy Hadoop to increase returns.

Metascale is one such Hadoop solution which also has Big Data training as a service and helps enterprises speed up Hadoop implementation. MetaScale's key impact factor is in solving capacity and cost problems that accompany swelling data demands. With retail, viable storage and efficient processing capabilities are the top demands. Hadoop and NoSQL provide Big Data value to retail enterprises particularly when used to refine analytics along with brand monitoring, product perception, sentiment analysis and inventory management. Mitigating latency of voluminous data is another aspect where MetaScale outscores other Hadoop solutions. Processing data overnight for real-time application is a great Hadoop property which retailers use. Extract, Transform and Load (ETL) refers to a process in database usage. Software cost and management of ETL invariably are so large that projects often fail at launch. Moreover, with conventional ETL techniques it can be weeks before data created is utilized effectively.

Hadoop on HDFS

HDFS, the Hadoop Distributed File System, is a distributed file system designed to hold very large amounts of data (terabytes or even zetabytes), and provide high-throughput access to this information. Files are stored in a redundant fashion across multiple machines to ensure their durability to failure and high availability to very parallel applications. Hadoop Map/ Reduce is a software framework that process large amount of data, in parallel, in a fault tolerant manner. Map Reduce is highly efficient and scalable, and thus can be used to process huge datasets. HDFS has a master/ slave architecture.

An HDFS cluster consists of single Name Node, a master server that manages the file system namespace and regulates access to files by clients. In addition, there are a number of DataNodes, usually one per node in the cluster, which manages storage attached to the nodes that they run on. MapReduce programming consists of writing two function, a map function and a reduce function. The map function takes a key, value pair and outputs a list of intermediate values with the key. The map function takes a key, value pair and outputs a list of intermediate values with the key. The map function is written in such a way that multiple map functions can be executed at once, so it's the part of the program that divides up tasks. The reduce function then takes the output of the map functions, and does some process on them, usually combining values, to generate the desired result in an output file.

Since real-time data intelligence and processing is core to retail data's success, another great Hadoop use-case is the 'market basket analysis'. Popularly called MBA, market basket analysis, uses data mining algorithms to find similar patterns in consumer behavior on-site. This is much like predicting what a consumer will do in a current session bases on historic data patterns. The metrics here involve 'support' and 'confidence', which predict degree of interest a consumer has in buying in his, or her, active session. MBA thus is a useful indicator that can be used to devise new promotional initiatives and identifying optimal schemes. The fun part is that the MBA interface can be integrated with other standard business intelligence (BI) tools and can work independently for users as well as disparate data sets.

Spark

Spark is another platform to implement the Big data applications. It is the much advance cluster computing engine to process the online requests. It is mainly designed for fast computation. Spark is one of the favorite choices of data scientist and thus Apache Spark is growing very quickly and replacing Hadoop Mapreduce.

USE OF BIG DATA METHODOLOGIES AND ALGORITHMS IN E-SERVICING

Recommender Systems

Recommender system is very much helpful in the online websites for the users to obtain their exact search in the net surfing. The system learns the user's interests from their history and recommend desired items to purchase. For example in the e-shopping website, a user may buy a book and logs out. When the next time the same user comes in, the system recommends some of the books of same author or same topic of previously purchased book. Here the system learns and stores the users interests and accordingly recommending information. There are lot of algorithms available to enable recommender systems, some of the popular algorithms are,

- Collaborative filtering
- Content-based filtering
- Hybrid recommender systems
- Clustering
- Multi Criteria Recommender Systems

Sentiment Analysis

Sentiment analysis or Opinion Mining is the process of computationally identifying and categorizing opinions online that is expressed in a piece of text, especially in order to determine the user's attitude towards a particular topic or product, etc. and check whether it is a positive comment, negative or neutral comment. In many social networking services or e-commerce websites, users can provide text review, comment or feedback to the items. This will be helpful for the online retailers or distributors to check their product ratings and reviews and they can improve the product quality as well they can take up different solutions to make up the product better with the help of suggestions from the user's perspective. The most widely used Opinion Mining algorithms are

- Social Sentiment Analysis
- Sentiment by Term Algorithm
- Natural Language Processing
- Text Analytics Algorithms

FUTURE RESEARCH DIRECTIONS

The future research could be of comparative nature of the identified factors as barriers and drivers for ICT based e-services and the users satisfaction to use e-Service so as to indicate the aspects that need extra effort to enhance the usage of e-Services for all the societal needs. E-Services have better scope in all the domains like Medical where the patient can be monitored remotely by the doctors say E-Health, E-Certification, where the students write exams online and they can get certificates, E- Governance Project, etc. There is a wide range of challenges in all the fields and researchers can take up different problems and solve it with Big data technologies to experience the user friendly, interactive and intelligent E-Services.

CONCLUSION

The world is moving towards digitization. In India central and state governments are steadily moving towards digitizing all the government departments, scheme and services. Due to the awareness of people and availability of network access that emerges, use of e-servicing application has increased. Now a days all the basic requirements from Mobile recharge to Electricity bill payment, Shopping, Passport application and everything is seeking the help of E-Servicing. As a result data is getting more complex in size and its volume. Handling big data with huge volume is also a major challenge but however there are lot of research is being taken by various scientist to solve the issues with the Big Data Handling. Since Hadoop's prime functional advantage is making Big Data more viable, knowing how to select appropriate hardware is critical.

This research received no specific grant from any funding agency in the public, commercial, or not-for-profit sectors.

REFERENCES

Bertot & Choi. (n.d.). Big data and e-government: Issues, policies, and recommendations. *University of Maryland College Park.*

Chen, C. P., & Zhang, C. (2014). Data-intensive applications, challenges, techniques and technologies: A survey on big data. *Inf. Sci.*, *275*, 314–347. doi:10.1016/j.ins.2014.01.015

Clark, B. Y., Brudney, J. L., & Jang, S.-G. (2013). Coproduction of government services and the new information technology: Investigating the distributional biases. *Public Administration Review*, *73*(5), 687–701. doi:10.1111/puar.12092

Coronel, C., Morris, S., & Rob, P. (2016). Database Systems: Design, Implementation, and Management (11th ed.). Cengage Learning.

Delena, D., & Demirkanb, H. (2017). Data, information and analytics as services. *Decision Support Systems*, *55*(1), 359–363. doi:10.1016/j.dss.2012.05.044

Fan, S., Lau, R. Y., & Zhao, J. L. (2015). Demystifying big data analytics for business intelligence through the lens of marketing mix. *Big Data Res.*, *2*(1), 28–32. doi:10.1016/j.bdr.2015.02.006

Gandomi, A., & Haider, M. (2015). Beyond the hype: Big Data Concepts, Methods, and Analytics. *International Journal of Information Management*, *35*(2), 137–144. doi:10.1016/j.ijinfomgt.2014.10.007

Hamed, T., Samshul, S., & Neda, J. (2015). *A Review paper on e-service technologies and concepts. 8th International Conference Interdisciplinary in Engineering*, Romania.

Holsapplea, C., Lee-Postb, A., & Pakath, R. (2014). A unified foundation for business analytics. *Decision Support Systems*, *64*, 130–141. doi:10.1016/j.dss.2014.05.013

Inder Jit Singh, M., Vinod, K., Hanuv, M., & Uma, K. (2015). *Scope of City E-Government Initiative*. Sprott School of Business, Carleton University.

Mahesh Kumar, S., & Kunwar Singh, V. (2016). E-health for Rural Areas of Uttarakhand under Governance Service Delivery Model. *IEEE 1st International Conference on Recent Advances in Information Technology.*

Mahmud & Sattar. (2016). Deployment of Contextual E-healthcare System: A prospective e-service based on context aware conceptual framework and ICTization framework model. *IEEE 11th Conference on Industrial Electronics and Applications (ICIEA).*

Masatsugu, T. (2015). Analysis of the Long-run Effect of e-Health Intervention on Chronic Diseases: A DID-PSM Approach. *17th International Conference on E-health Networking, Applications& Services.*

Mashaqbeh. (2016). Computers and e-Health: Roles and New Applications. *IEEE Transactions.*

Minelli, M. Chambers, M., & Dhiraj, A. (2013). Big Data, Big Analytics: Emerging Business Intelligence and Analytic Trends for Today's Businesses. Wiley & Sons.

Nishant Srivastava, S., & Hashank, S. (2016). E- Business Scope and Challenges in India. *International Journal of Business and Management Invention.*

Poonam, M., & Priyanka, D. (2016). Challenges and Future Prospects for E-Governance in India. International Journal of Science, Engineering and Technology Research, 3(7).

Rina, D. (n.d.). Challenges and Future Scope of E-commerce in India. *International Journal of Emerging Trends & Technology in Computer Science.*

Sun, Z., Firmin, S., & Yearwood, J. (2012). Integrating online social networking with e-commerce based on CBR. *International Proceedings of the 23rd ACIS 2012,* 3–5.

Sun, Z., Strang, K., & Yearwood, J. (2014). Analytics service oriented architecture for enterprise information systems. International Proceedings of iiWAS2014, 506-518. doi:10.1145/2684200.2684358

Tan, O., Ng, J., Wong, A., & Koh, W. K. (2012). Bridging gaps between three generation family's Needs and Attitudes towards e-health technologies. *IEEE 14th International Conference on e-Health Networking, Applications and Services.*

Van der Meulen, R., & Rivera, J. (2014). *Gartner Says Worldwide Business Intelligence and Analytics Software Market Grew 8.* Academic Press.

Vesset, D., McDonough, B., Schubmehl, D., & Wardley, M. (2013). Worldwide Business Analytics Software. 2013–2017 Forecast and 2012 Vendor Shares.

KEY TERMS AND DEFINITIONS

Big Data: Large volume of data with vital characteristics such as volume, variety, and velocity.

Big Data Analytics: Techniques and algorithms to analyze large amount of data that comes from different sources over online.

E-Services: A variety of internet-based electronic interactions.

Hadoop: An open source platform for processing big data in a distributed environment.

HDFS: HDFS is a Hadoop-distributed file system to store data, gives high performance access to data across Hadoop clusters, and it is a data management layer in a Hadoop framework.

Intelligent E-Services: Various intelligent technologies like neural networks, fuzzy logic, expert systems, machine learning, data mining, etc. are being applied in various e-services approaches, so that these e-services will provide much higher quality of online information searching, presentation, etc.

Spark: Spark is the advanced technology in processing big data. It overcomes the drawbacks of Hadoop map reduce system, and it is mainly designed for fast computation.

Chapter 3
Enterprise Information Systems and Digital Marketing:
Advanced Issues and Implications

Kijpokin Kasemsap
Suan Sunandha Rajabhat University, Thailand

ABSTRACT

With the support of modern technologies, enterprise information systems (EISs) and digital marketing are the significant approaches in modern business and can lead to the establishment of a consolidated business system toward improving the business performance. EISs and digital marketing are related to enterprise resource planning (ERP), electronic commerce (e-commerce), cloud computing, and social media platforms. EISs and digital marketing help modern businesses gain the rapid access to the mass market at an affordable price, increase business profit, and attract new customers in a timely and effective manner. EISs and digital marketing allow managers and executives to establish their profitable strategies and to find various business opportunities in the competitive business environments.

INTRODUCTION

With global operations in place, there is a requirement for suitable enterprise information systems (EISs), such as enterprise resource planning (ERP) and electronic commerce (e-commerce) for the integration of extended enterprises along the supply chain in order to gain the flexibility and responsiveness (Tavana, 2012). ERP includes the advanced software system that can organize the resource requirements

DOI: 10.4018/978-1-5225-3628-4.ch003

of an enterprise (Kasemsap, 2015a). Concerning EISs, e-commerce involves the procedure of purchasing and selling products and services through computerized systems (Kasemsap, 2016a). Business managers are a powerful source of influence toward shaping the post-adoption attitudes and behaviors of users and the success of EISs (Rezvani, Dong, & Khosravi, 2017).

The continual development of EISs is essential for organizational advancement (Cruz-Cunha & Varajao, 2011). As information technology (IT) is applied in the modern enterprise management (e.g., financing, marketing, and ERP), the high availability of EISs is more important (Fan, Tiejun, & Liuyi, 2012). EISs change the way organizations execute their business information (Sun & Mouakket, 2015) and improve the automated information exchange in networked enterprise scenarios (Agostinho et al., 2016). The application of EISs offers a high level of customization of delivered services and data (El Kadiri et al., 2016). The key to successful implementation of EISs lies in achieving an alignment among strategy, technology, organization, and people, and also in achieving commitment to the new technologies (Kidd, 2011).

Digital marketing is a proliferating field that has opened new challenges for marketers (Jain, Ahuja, & Medury, 2015) since they can be performed through the Internet (Şengül & Eren, 2017). The platforms of digital marketing are focused on being highly interactive toward providing many unique ways to create and distribute the marking contents (Rathore & Panwar, 2016). The development of web technologies has enhanced the availability of online business opportunities for entrepreneurs (Capatina & Rancati, 2017). From a customer perspective, the Internet has given power to get up-to-date information, to compare products and services more easily, and to get in touch with marketers (Merisavo, 2009).

To achieve effective digital marketing, integrating information and communication technology (ICT) in marketing functions requires marketers to take an active managerial role far beyond their traditional areas of marketing competence (Millman & El-Gohary, 2013). As ICT rapidly develops, a new enterprise model must be created to face the digital enterprise requirement (Weichhart, Molina, Chen, Whitman, & Vernadat, 2016). To survive in today's competitive business environment, marketing professionals must look to develop the innovative methods of reaching their customers and stakeholders (Hajli, 2015). Thus, it is imperative that organizations have a presence in the Internet spaces that their customers and potential customers are exploring on a daily basis (Dickey & Lewis, 2012).

This chapter is based on a literature review of EISs and digital marketing. The extensive literature of EISs and digital marketing provides a contribution to practitioners and researchers by explaining their advanced issues and implications in global business.

BACKGROUND

Most business enterprises are open in a competitive market, and are dependent on distributed information infrastructure across various geospatial location (Manna, Sengupta, & Mazumdar, 2016). Many organizations are looking for methods, tools, and solutions which enable them to improve business compliance, and reduce the likelihood of situations that may jeopardize their operational performance (Marques, 2017). EISs remain one of the last levers to increase the performance and competitiveness of the enterprise (Zacharewicz et al., 2017), and can provide an advanced platform that enables the distant collections of organizations to effectively integrate their operations (Tung et al., 2010).

The business capabilities of modern enterprises rely on EISs and IT infrastructure (Farwick, Schweda, Breu, & Hanschke, 2016). Enterprise architecture indicates the components of company employed in its business operations regarding the business structure and information flows (Kasemsap, 2015b). Concerning enterprise architecture, enterprise performance can be enhanced with the support of sophisticated technology (Lezoche, Yahia, Aubry, Panetto, & Zdravković, 2012). Continuous improvements in EISs have created the profitable opportunities for modern businesses to develop more advanced processes (Tavana, 2017).

Businesses in the 21st century are reliant on the Internet to enhance the growth of their organizations (Mekonnen, 2016). As no company can be profitable in the long run without satisfying customers' needs, marketing aims to serve the demand by creating superior value for customers (Lahuerta-Otero & Cordero-Gutiérrez, 2016). Online social networks have become the fastest growing phenomenon on the Internet, and firms begin to take advantage of them as a marketing tool (Saavedra, Criado, & Andreu, 2013). Advanced technology enhances communication between marketers and customers in digital marketing, and alters its methods to improve the interaction between them in the marketplace (Kaltenrieder, D'Onofrio, & Portmann, 2016).

Modern web-based marketing applications are pertinent for businesses, as they often encourage their core competencies and capabilities (Tarnanidis, Vlachopoulou, & Papathanasiou, 2017). In the digital age, as financial systems migrate to an online environment, both business leaders and investors must adapt to changes in the market (Khosrow-Pour, 2015). The modern business landscape demands that organizations maintain an online presence to network with their customers and investors (Li, Loh, Evans, & Lorenzi, 2013).

ASPECTS OF ENTERPRISE INFORMATION SYSTEMS AND DIGITAL MARKETING

This section provides the overview of EISs; EISs and cloud computing; the overview of digital marketing; and the relationship between digital marketing and social media in modern business.

Overview of Enterprise Information Systems

The complexity of the modern enterprise poses an array of challenges in the fast changing environment in which business enterprise evolves (Yahia, Lezoche, Aubry, & Panetto, 2012) among them the cooperation between autonomous and heterogeneous information system in distributed networks (Yahia, Aubry, & Panetto, 2012). Regarding EISs, enterprise information management can help company effectively find the strategic methods for the application of business information within organizations (Kasemsap, 2017a).

Involving an integrated technology platform, EISs enable organizations to integrate and coordinate all business processes, thus offering a strategic area of business perspectives (Cruz-Cunha, 2010). EISs present a wide variety of roles, ranging from managerial strategy to operational support of the company (Trigo et al., 2013). The highly innovative information system is essential for enterprises to apply information toward gaining competitive advantage (Xue & Xu, 2017).

The presence of information system-related mindfulness increases the information system unit's ability to identify and interpret opportunities to assimilate enterprise-wide IT within a company (Mu, Kirsch, & Butler, 2015). Modern EISs are the distributed systems typically built on multitiered client server architectures, which can be modeled using well-known frameworks, such as Zachman enterprise architecture (Nikolaidou & Alexopoulou, 2009).

The implementation of EISs is one of the most challenging parts of IT strategy for an organization, since implementation brings in efficiency in the system, and justifies the investments made (Raman & Goyal, 2017). EISs implemented in the organizations are the critical assets to provide competitive advantage in changing the continuity of business processes and the management of enterprise resources (Eroğlu & Çakmak, 2016). Thus, the disruption of EISs, even for a short period of time, can cause large problems for business operations (Parthasarathy, 2010).

Enterprise Information Systems and Cloud Computing

The increasing usage of Internet-based services makes design of software system complex and their power unlimited in scale (Venkatesan & Sridhar, 2017). With the rapid development of computer technology, the concept of cloud computing is greatly applied in EISs and digital marketing (Xue, 2014). Cloud computing is the delivery of hosted services over the Internet (Kasemsap, 2018a). The advances in cloud computing establish the distribution of the EISs-related functional processes and storage capabilities as the default approaches to their designs (Zdravković & Panetto, 2017).

In order to save costs, many enterprises and their employees have hired the enterprise cloud services, and put much important information in the cloud computing applications (Niu, Liu, Zhang, Lü, & Li, 2016). Cloud computing makes it possible to deliver EISs where each party can contribute to the establishment of the advanced system, while delivering superior integration capabilities with any party's legacy systems and web-based intranets applications (Hug, 2008).

Overview of Digital Marketing

In recent years, technology has been the impetus for progress in various processes, systems, and businesses (Bowen & Ozuem, 2015). Digital technologies and digital media have changed the way marketers communicate with their target customers, as well as the way in which customers interact with each other (Montecchi & Nobbs, 2017). Digital media professionals face many changes within mass media that will continue to impact the creation and delivery of persuasive messages toward improving digital marketing in the digital marketplace (Eastin, Daugherty, & Burns, 2011).

Digital marketing has emerged as an aid in allowing businesses to reach a broader audience (El-Gohary & Eid, 2013), toward establishing the exchange of products and services in the marketplace (Kahraman, Yazıcı, & Karaşan, 2016). Regarding the widespread adoption of advanced technology, it is imperative for modern businesses to manage the social technologies to stay competitive in the global economy (Lee, 2014). Concerning digital-marketing area, e-mail, browsers, and web pages become universal in various conversations across the globe (Gonzalez-Fuentes, 2017).

The ubiquity of online retail stores and availability of hard-to-find products promotes a growing interest in research on the best procedures and strategies in the e-commerce era (Lee, 2016). E-commerce provides better connectivity for its potential customer as their respective website can be virtually accessed from anywhere through the Internet (Kasemsap, 2016b), and increases the commercial ability toward customer satisfaction, thus maintaining competitive costs and pricing in modern business (Kasemsap, 2015c).

In the digital age, customer relationship management (CRM) is one of the most significant business strategies, and includes the capability of organizing business interactions with customers toward effective e-commerce (Kasemsap, 2015d). With the support of CRM, the company's goal can be best achieved through the indication of the customer's requirement toward gaining customer satisfaction (Kasemsap, 2018b). Customer satisfaction presents the degree of satisfaction provided by the products or services of a company (Kasemsap, 2018c). Satisfied customers are loyal, and make repeat orders in order to apply a wide range of services offered by service companies (Kasemsap, 2017b). Regarding e-commerce perspectives, customer satisfaction, customer value, and brand loyalty are positively associated with CRM performance (Kasemsap, 2014a).

The growing importance of digital branding and advertising is demonstrated by the investment that many organizations are making in branding and advertising through digital marketing (Rowley & Edmundson-Bird, 2013). As marketing professionals look for ever more effective ways to promote their products and services to customers, an understanding of customer needs and the ability to predict a target audience's reaction to the advertising campaigns is essential (Management Association, 2015).

Relationship Between Digital Marketing and Social Media in Modern Business

Social media provides an essential shift in how buyers and sellers interact, and its interactivity can facilitate the opportunity for developing brands in international settings (Kasemsap, 2016c). Social media can promote the development of knowledge value chain toward customizing information and delivery for business growth (Kasemsap, 2014b). Social media can allow marketers to create the valuable brand with larger business communities than ever before, and can enhance the effective interactions and relationships in international advertising (Kasemsap, 2015e).

Social media and brand communities provide virtual places where the consumers can publicize their opinions about products, and play an important role as marketers, advertisers, and value creators through electronic word-of-mouth in the digital marketplace (Costanza, 2017). Value creation is a pivotal aspect of the modern business industry (Rozenes & Cohen, 2017). Social media marketing programs usually center on efforts to create the content that attracts attention and encourages readers to share it across their social networks toward competition in the global marketplace (Kasemsap, 2016d).

The utilization of social media develops effective communication, increases thought leadership, enhances sales, and provides better customer service (Kasemsap, 2016e). Social media offers a chance to redefine the delivery of service to customers, thus changing the way they think about a company's brands while reducing service costs (Kasemsap, 2017c). Depending on the number and type of social media sites, executives and social media users should be educated and trained to have digital literacy and IT competency to effectively manage social media platforms (Kasemsap, 2017d). Digital literacy includes the ability to find, assess, and create the content with the application of IT and the Internet (Kasemsap, 2017e).

FUTURE RESEARCH DIRECTIONS

The classification of the extensive literature in the domains of EISs and digital marketing will provide the potential opportunities for future research. Utilizing EISs and digital marketing can lead to the improved interdepartmental communication and enhanced business performance. Business analytics can help business managers explore the causal relationships with the consideration of business input, process, and output (Kasemsap, 2015f). Electronic procurement (e-procurement) allows procurement professionals to communicate information, organizes the global procurement process, and reduces both time and costs (Kasemsap, 2017f). Investigating the correlations among EISs, digital marketing, business analytics, and e-procurement should be further studied.

CONCLUSION

This chapter presented the overview of EISs; EISs and cloud computing; the overview of digital marketing; and the relationship between digital marketing and social media in modern business. With the support of modern technologies, EISs and digital marketing are the significant approaches in modern business, and can lead to the establishment of a consolidated business system toward improving the business performance. EISs and digital marketing are related to ERP, e-commerce, cloud computing, and social media platforms. EISs and digital marketing help modern businesses gain the rapid access to the mass market at an affordable price, increase business profit, and attract new customers in a timely and effective manner. EISs and digital marketing allow managers and executives to establish their profitable strategies and to find various business opportunities in the competitive business environments.

REFERENCES

Agostinho, C., Ducq, Y., Zacharewicz, G., Sarraipa, J., Lampathaki, F., Poler, R., & Jardim-Goncalves, R. (2016). Towards a sustainable interoperability in networked enterprise information systems: Trends of knowledge and model-driven technology. *Computers in Industry*, *79*, 64–76. doi:10.1016/j.compind.2015.07.001

Bowen, G., & Ozuem, W. (2015). *Computer-mediated marketing strategies: Social media and online brand communities*. Hershey, PA: IGI Global. doi:10.4018/978-1-4666-6595-8

Capatina, A., & Rancati, E. (2017). *Key challenges and opportunities in web entrepreneurship*. Hershey, PA: IGI Global. doi:10.4018/978-1-5225-2466-3

Costanza, F. (2017). Social media marketing and value co-creation: A system dynamics approach. In S. Rozenes & Y. Cohen (Eds.), *Handbook of research on strategic alliances and value co-creation in the service industry* (pp. 205–230). Hershey, PA: IGI Global. doi:10.4018/978-1-5225-2084-9.ch011

Cruz-Cunha, M. M. (2010). *Enterprise information systems for business integration in SMEs: Technological, organizational, and social dimensions*. Hershey, PA: IGI Global. doi:10.4018/978-1-60566-892-5

Cruz-Cunha, M. M., & Varajao, J. (2011). *Enterprise information systems design, implementation and management: Organizational applications*. Hershey, PA: IGI Global. doi:10.4018/978-1-61692-020-3

Dickey, I. J., & Lewis, W. F. (2012). An overview of digital media and advertising. In I. Management Association (Ed.), E-marketing: Concepts, methodologies, tools, and applications (pp. 31–61). Hershey, PA: IGI Global. doi:10.4018/978-1-4666-1598-4.ch003

Eastin, M. S., Daugherty, T., & Burns, N. M. (2011). *Handbook of research on digital media and advertising: User generated content consumption*. Hershey, PA: IGI Global. doi:10.4018/978-1-60566-792-8

El-Gohary, H., & Eid, R. (2013). *E-marketing in developed and developing countries: Emerging practices*. Hershey, PA: IGI Global. doi:10.4018/978-1-4666-3954-6

El Kadiri, S., Grabot, B., Thoben, K. D., Hribernik, K., Emmanouilidis, C., von Cieminski, G., & Kiritsis, D. (2016). Current trends on ICT technologies for enterprise information systems. *Computers in Industry*, *79*, 14–33. doi:10.1016/j.compind.2015.06.008

Eroğlu, S., & Çakmak, T. (2016). Enterprise information systems within the context of information security: A risk assessment for a health organization in Turkey. *Procedia Computer Science*, *100*, 979–986. doi:10.1016/j.procs.2016.09.262

Fan, L., Tiejun, W., & Liuyi, W. (2012). Research and implementation on model for high availability of enterprise information system. *IERI Procedia*, *3*, 181–185. doi:10.1016/j.ieri.2012.09.030

Farwick, M., Schweda, C. M., Breu, R., & Hanschke, I. (2016). A situational method for semi-automated enterprise architecture documentation. *Software & Systems Modeling*, *15*(2), 397–426. doi:10.1007/s10270-014-0407-3

Gonzalez-Fuentes, M. (2017). The organization of the future and the marketing function: Marketers' competencies in the era of information technology. In R. Batko & A. Szopa (Eds.), *Strategic imperatives and core competencies in the era of robotics and artificial intelligence* (pp. 126–145). Hershey, PA: IGI Global. doi:10.4018/978-1-5225-1656-9.ch007

Hajli, N. (2015). *Handbook of research on integrating social media into strategic marketing*. Hershey, PA: IGI Global. doi:10.4018/978-1-4666-8353-2

Hug, M. (2008). *Will cloud-based multi-enterprise information systems replace extranets?* Retrieved from https://www.infoq.com/articles/will-meis-replace-extranets

Jain, N., Ahuja, V., & Medury, Y. (2015). Digital marketing optimization. In M. Khosrow-Pour (Ed.), *Strategic e-commerce systems and tools for competing in the digital marketplace* (pp. 162–170). Hershey, PA: IGI Global. doi:10.4018/978-1-4666-8133-0.ch008

Kahraman, C., Yazıcı, İ., & Karaşan, A. (2016). A literature survey on the usage of fuzzy MCDM methods for digital marketing. In A. Kumar & M. Dash (Eds.), *Fuzzy optimization and multi-criteria decision making in digital marketing* (pp. 1–19). Hershey, PA: IGI Global. doi:10.4018/978-1-4666-8808-7.ch001

Kaltenrieder, P., D'Onofrio, S., & Portmann, E. (2016). Applying the fuzzy analytical network process in digital marketing. In A. Kumar & M. Dash (Eds.), *Fuzzy optimization and multi-criteria decision making in digital marketing* (pp. 202–232). Hershey, PA: IGI Global. doi:10.4018/978-1-4666-8808-7.ch010

Kasemsap, K. (2014a). The role of brand loyalty on CRM performance: An innovative framework for smart manufacturing. In Z. Luo (Ed.), *Smart manufacturing innovation and transformation: Interconnection and intelligence* (pp. 252–284). Hershey, PA: IGI Global. doi:10.4018/978-1-4666-5836-3.ch010

Kasemsap, K. (2014b). The role of social media in the knowledge-based organizations. In I. Lee (Ed.), *Integrating social media into business practice, applications, management, and models* (pp. 254–275). Hershey, PA: IGI Global. doi:10.4018/978-1-4666-6182-0.ch013

Kasemsap, K. (2015a). Implementing enterprise resource planning. In M. Khosrow-Pour (Ed.), *Encyclopedia of information science and technology* (3rd ed.; pp. 798–807). Hershey, PA: IGI Global. doi:10.4018/978-1-4666-5888-2.ch076

Kasemsap, K. (2015b). The role of information system within enterprise architecture and their impact on business performance. In M. Wadhwa & A. Harper (Eds.), *Technology, innovation, and enterprise transformation* (pp. 262–284). Hershey, PA: IGI Global. doi:10.4018/978-1-4666-6473-9.ch012

Kasemsap, K. (2015c). The role of electronic commerce in the global business environments. In F. Cipolla-Ficarra (Ed.), *Handbook of research on interactive information quality in expanding social network communications* (pp. 304–324). Hershey, PA: IGI Global. doi:10.4018/978-1-4666-7377-9.ch019

Kasemsap, K. (2015d). The role of customer relationship management in the global business environments. In T. Tsiakis (Ed.), *Trends and innovations in marketing information systems* (pp. 130–156). Hershey, PA: IGI Global. doi:10.4018/978-1-4666-8459-1.ch007

Kasemsap, K. (2015e). The role of social media in international advertising. In N. Taşkıran & R. Yılmaz (Eds.), *Handbook of research on effective advertising strategies in the social media age* (pp. 171–196). Hershey, PA: IGI Global. doi:10.4018/978-1-4666-8125-5.ch010

Kasemsap, K. (2015f). The role of business analytics in performance management. In M. Tavana & K. Puranam (Eds.), *Handbook of research on organizational transformations through big data analytics* (pp. 126–145). Hershey, PA: IGI Global. doi:10.4018/978-1-4666-7272-7.ch010

Kasemsap, K. (2016a). Implementing electronic commerce in global marketing. In I. Lee (Ed.), *Encyclopedia of e-commerce development, implementation, and management* (pp. 591–602). Hershey, PA: IGI Global. doi:10.4018/978-1-4666-9787-4.ch043

Kasemsap, K. (2016b). Advocating electronic business and electronic commerce in the global marketplace. In S. Dixit & A. Sinha (Eds.), *E-retailing challenges and opportunities in the global marketplace* (pp. 1–24). Hershey, PA: IGI Global. doi:10.4018/978-1-4666-9921-2.ch001

Kasemsap, K. (2016c). Role of social media in brand promotion: An international marketing perspective. In A. Singh & P. Duhan (Eds.), *Managing public relations and brand image through social media* (pp. 62–88). Hershey, PA: IGI Global. doi:10.4018/978-1-5225-0332-3.ch005

Kasemsap, K. (2016d). The roles of social media marketing and brand management in global marketing. In W. Ozuem & G. Bowen (Eds.), *Competitive social media marketing strategies* (pp. 173–200). Hershey, PA: IGI Global. doi:10.4018/978-1-4666-9776-8.ch009

Kasemsap, K. (2016e). Role of social media in brand promotion: An international marketing perspective. In A. Singh & P. Duhan (Eds.), *Managing public relations and brand image through social media* (pp. 62–88). Hershey, PA: IGI Global. doi:10.4018/978-1-5225-0332-3.ch005

Kasemsap, K. (2017a). Mastering intelligent decision support systems in enterprise information management. In G. Sreedhar (Ed.), *Web data mining and the development of knowledge-based decision support systems* (pp. 35–56). Hershey, PA: IGI Global. doi:10.4018/978-1-5225-1877-8.ch004

Kasemsap, K. (2017b). The importance of customer satisfaction and customer loyalty in the service sector. In S. Rozenes & Y. Cohen (Eds.), *Handbook of research on strategic alliances and value co-creation in the service industry* (pp. 59–82). Hershey, PA: IGI Global. doi:10.4018/978-1-5225-2084-9.ch004

Kasemsap, K. (2017c). Mastering social media in the modern business world. In N. Rao (Ed.), *Social media listening and monitoring for business applications* (pp. 18–44). Hershey, PA: IGI Global. doi:10.4018/978-1-5225-0846-5.ch002

Kasemsap, K. (2017d). Professional and business applications of social media platforms. In V. Benson, R. Tuninga, & G. Saridakis (Eds.), *Analyzing the strategic role of social networking in firm growth and productivity* (pp. 427–450). Hershey, PA: IGI Global. doi:10.4018/978-1-5225-0559-4.ch021

Kasemsap, K. (2017e). Digital storytelling and digital literacy: Advanced issues and prospects. In D. Loveless, P. Sullivan, K. Dredger, & J. Burns (Eds.), *Deconstructing the education-industrial complex in the digital age* (pp. 151–171). Hershey, PA: IGI Global. doi:10.4018/978-1-5225-2101-3.ch009

Kasemsap, K. (2017f). Mastering electronic procurement, green public procurement, and public procurement for innovation. In R. Shakya (Ed.), *Digital governance and e-government principles applied to public procurement* (pp. 29–55). Hershey, PA: IGI Global. doi:10.4018/978-1-5225-2203-4.ch002

Kasemsap, K. (2018a). Cloud computing, green computing, and green ICT. In K. Munir (Ed.), *Cloud computing technologies for green enterprises* (pp. 28–50). Hershey, PA: IGI Global. doi:10.4018/978-1-5225-3038-1.ch002

Kasemsap, K. (2018b). Facilitating customer relationship management in modern business. In M. Khosrow-Pour (Ed.), *Encyclopedia of information science and technology* (4th ed.; pp. 1594–1604). Hershey, PA: IGI Global. doi:10.4018/978-1-5225-2255-3.ch138

Kasemsap, K. (2018c). Customer lifetime value. In M. Khosrow-Pour (Ed.), Encyclopedia of information science and technology (4th ed.; pp. 1584–1593). Hershey, PA: IGI Global. doi:10.4018/978-1-5225-2255-3.ch137

Khosrow-Pour, M. (2015). *Strategic e-commerce systems and tools for competing in the digital marketplace*. Hershey, PA: IGI Global. doi:10.4018/978-1-4666-8133-0

Kidd, P. T. (2011). Enterprise information systems: Aligning and integrating strategy, technology, organization and people. In I. Management Association (Ed.), Enterprise information systems: Concepts, methodologies, tools and applications (pp. 314–328). Hershey, PA: IGI Global. doi:10.4018/978-1-61692-852-0.ch201

Lahuerta-Otero, E., & Cordero-Gutiérrez, R. (2016). How to promote teaching centers by using online advertising: A case study. In I. Lee (Ed.), *Encyclopedia of e-commerce development, implementation, and management* (pp. 262–271). Hershey, PA: IGI Global. doi:10.4018/978-1-4666-9787-4.ch020

Lee, I. (2014). *Integrating social media into business practice, applications, management, and models*. Hershey, PA: IGI Global. doi:10.4018/978-1-4666-6182-0

Lee, I. (2016). *Encyclopedia of e-commerce development, implementation, and management*. Hershey, PA: IGI Global. doi:10.4018/978-1-4666-9787-4

Lezoche, M., Yahia, E., Aubry, A., Panetto, H., & Zdravković, M. (2012). Conceptualising and structuring semantics in cooperative enterprise information systems models. *Computers in Industry*, *63*(8), 775–787. doi:10.1016/j.compind.2012.08.006

Li, E. Y., Loh, S., Evans, C., & Lorenzi, F. (2013). *Organizations and social networking: Utilizing social media to engage consumers*. Hershey, PA: IGI Global. doi:10.4018/978-1-4666-4026-9

Manna, A., Sengupta, A., & Mazumdar, C. (2016). A survey of trust models for enterprise information systems. *Procedia Computer Science*, *85*, 527–534. doi:10.1016/j.procs.2016.05.212

Marques, R. P. (2017). Continuous assurance and business compliance in enterprise information systems. In M. Tavana (Ed.), *Enterprise information systems and the digitalization of business functions* (pp. 99–119). Hershey, PA: IGI Global. doi:10.4018/978-1-5225-2382-6.ch005

Mekonnen, A. (2016). Digital marketing strategy for affinity marketing: Utilising the new marketing arena. In W. Ozuem & G. Bowen (Eds.), *Competitive social media marketing strategies* (pp. 1–19). Hershey, PA: IGI Global. doi:10.4018/978-1-4666-9776-8.ch001

Merisavo, M. (2009). The effects of digital marketing on customer relationships. In I. Lee (Ed.), *Electronic business: Concepts, methodologies, tools, and applications* (pp. 1430–1441). Hershey, PA: IGI Global. doi:10.4018/978-1-60566-056-1.ch088

Millman, C., & El-Gohary, H. (2013). New digital media marketing and micro business: A UK perspective. In H. El-Gohary (Ed.), *Transdisciplinary marketing concepts and emergent methods for virtual environments* (pp. 59–81). Hershey, PA: IGI Global. doi:10.4018/978-1-4666-1861-9.ch005

Montecchi, M., & Nobbs, K. (2017). Let it go: Consumer empowerment and user-generated content – An exploratory study of contemporary fashion marketing practices in the digital age. In A. Vecchi (Ed.), *Advanced fashion technology and operations management* (pp. 294–317). Hershey, PA: IGI Global. doi:10.4018/978-1-5225-1865-5.ch013

Mu, E., Kirsch, L. J., & Butler, B. S. (2015). The assimilation of enterprise information system: An interpretation systems perspective. *Information & Management*, *52*(3), 359–370. doi:10.1016/j.im.2015.01.004

Nikolaidou, M., & Alexopoulou, N. (2009). Frameworks for model-based design of enterprise information systems. In M. Syed & S. Syed (Eds.), *Handbook of research on modern systems analysis and design technologies and applications* (pp. 326–356). Hershey, PA: IGI Global. doi:10.4018/978-1-59904-887-1.ch021

Niu, D. D., Liu, L., Zhang, X., Lü, S., & Li, Z. (2016). Security analysis model, system architecture and relational model of enterprise cloud services. *International Journal of Automation and Computing*, *13*(6), 574–584. doi:10.1007/s11633-016-1014-2

Parthasarathy, S. (2010). *Enterprise information systems and implementing IT infrastructures: Challenges and issues*. Hershey, PA: IGI Global. doi:10.4018/978-1-61520-625-4

Raman, A., & Goyal, D. P. (2017). Extending IMPLEMENT framework for enterprise information systems implementation to information system innovation. In M. Tavana (Ed.), *Enterprise information systems and the digitalization of business functions* (pp. 137–177). Hershey, PA: IGI Global. doi:10.4018/978-1-5225-2382-6.ch007

Rathore, S., & Panwar, A. (2016). *Capturing, analyzing, and managing word-of-mouth in the digital marketplace*. Hershey, PA: IGI Global. doi:10.4018/978-1-4666-9449-1

Rezvani, A., Dong, L., & Khosravi, P. (2017). Promoting the continuing usage of strategic information systems: The role of supervisory leadership in the successful implementation of enterprise systems. *International Journal of Information Management*, *37*(5), 417–430. doi:10.1016/j.ijinfomgt.2017.04.008

Rowley, J., & Edmundson-Bird, D. (2013). Brand presence in digital space. *Journal of Electronic Commerce in Organizations*, *11*(1), 63–78. doi:10.4018/jeco.2013010104

Rozenes, S., & Cohen, Y. (2017). *Handbook of research on strategic alliances and value co-creation in the service industry*. Hershey, PA: IGI Global. doi:10.4018/978-1-5225-2084-9

Saavedra, F. U., Criado, J. R., & Andreu, J. L. (2013). Is social media marketing really working?: Its impact on the relationships among market orientation, entrepreneurial orientation, and business performance. In H. El-Gohary & R. Eid (Eds.), *E-marketing in developed and developing countries: Emerging practices* (pp. 174–193). Hershey, PA: IGI Global. doi:10.4018/978-1-4666-3954-6.ch011

Şengül, Ü., & Eren, M. (2017). Selection of digital marketing tools using fuzzy AHP-fuzzy TOPSIS. In I. Management Association (Ed.), Advertising and branding: Concepts, methodologies, tools, and applications (pp. 587–616). Hershey, PA: IGI Global. doi:10.4018/978-1-5225-1793-1.ch027

Sun, Y., & Mouakket, S. (2015). Assessing the impact of enterprise systems technological characteristics on user continuance behavior: An empirical study in China. *Computers in Industry*, *70*, 153–167. doi:10.1016/j.compind.2015.01.003

Tarnanidis, T., Vlachopoulou, M., & Papathanasiou, J. (2017). *Driving agribusiness with technology innovations*. Hershey, PA: IGI Global. doi:10.4018/978-1-5225-2107-5

Tavana, M. (2012). *Enterprise information systems and advancing business solutions: Emerging models*. Hershey, PA: IGI Global. doi:10.4018/978-1-4666-1761-2

Tavana, M. (2017). *Enterprise information systems and the digitalization of business functions*. Hershey, PA: IGI Global. doi:10.4018/978-1-5225-2382-6

Trigo, A., Varajão, J., Barroso, J., Soto-Acosta, P., Molina-Castillo, F. J., & Gonzalvez-Gallego, N. (2013). Enterprise information systems adoption in Iberian large companies: Motivations and trends. In I. Management Association (Ed.), Enterprise resource planning: Concepts, methodologies, tools, and applications (pp. 1035–1052). Hershey, PA: IGI Global. doi:10.4018/978-1-4666-4153-2.ch060

Tung, H., Marshall-Bradley, T., Wood, J., Sofge, D. A., Grayson, J., Bergman, M., & Lawless, W. (2010). Enterprise information systems: Two case studies. In M. Cruz-Cunha (Ed.), *Social, managerial, and organizational dimensions of enterprise information systems* (pp. 434–458). Hershey, PA: IGI Global. doi:10.4018/978-1-60566-856-7.ch022

Venkatesan, D., & Sridhar, S. (2017). A novel programming framework for architecting next generation enterprise scale information systems. *Information Systems and e-Business Management*, *15*(2), 489–534. doi:10.1007/s10257-016-0330-y

Weichhart, G., Molina, A., Chen, D., Whitman, L. E., & Vernadat, F. (2016). Challenges and current developments for sensing, smart and sustainable enterprise systems. *Computers in Industry*, *79*, 34–46. doi:10.1016/j.compind.2015.07.002

Xue, C., & Xu, Y. (2017). Influence factor analysis of enterprise IT innovation capacity based on system dynamics. *Procedia Engineering*, *174*, 232–239. doi:10.1016/j.proeng.2017.01.125

Xue, J. (2014). Research on ERP enterprise management information system in the cloud computing model. *BioTechnology: An Indian Journal*, *10*(20), 12556–12562.

Yahia, E., Aubry, A., & Panetto, H. (2012). Formal measures for semantic interoperability assessment in cooperative enterprise information systems. *Computers in Industry*, *63*(5), 443–457. doi:10.1016/j.compind.2012.01.010

Yahia, E., Lezoche, M., Aubry, A., & Panetto, H. (2012). Semantics enactment for interoperability assessment in enterprise information systems. *Annual Reviews in Control*, *36*(1), 101–117. doi:10.1016/j.arcontrol.2012.03.008

Zacharewicz, G., Diallo, S., Ducq, Y., Agostinho, C., Jardim-Goncalves, R., Bazoun, H., & Doumeingts, G. et al. (2017). Model-based approaches for interoperability of next generation enterprise information systems: State of the art and future challenges. *Information Systems and e-Business Management*, *15*(2), 229–256. doi:10.1007/s10257-016-0317-8

Zdravković, M., & Panetto, H. (2017). The challenges of model-based systems engineering for the next generation enterprise information systems. *Information Systems and e-Business Management, 15*(2), 225–227. doi:10.1007/s10257-017-0353-z

ADDITIONAL READING

Appelbaum, D., Kogan, A., Vasarhelyi, M., & Yan, Z. (2017). Impact of business analytics and enterprise systems on managerial accounting. *International Journal of Accounting Information Systems, 25*, 29–44. doi:10.1016/j.accinf.2017.03.003

Bianchi, C., & Andrews, L. (2015). Investigating marketing managers' perspectives on social media in Chile. *Journal of Business Research, 68*(12), 2552–2559. doi:10.1016/j.jbusres.2015.06.026

Boell, S. K. (2017). Information: Fundamental positions and their implications for information systems research, education and practice. *Information and Organization, 27*(1), 1–16. doi:10.1016/j.infoandorg.2016.11.002

Boiko, A., & Shendryk, V. (2017). System integration and security of information systems. *Procedia Computer Science, 104*, 35–42. doi:10.1016/j.procs.2017.01.053

Brink, T. (2017). SME routes for innovation collaboration with larger enterprises. *Industrial Marketing Management, 64*, 122–134. doi:10.1016/j.indmarman.2017.01.010

Chuang, S. H., & Lin, H. N. (2017). Performance implications of information-value offering in e-service systems: Examining the resource-based perspective and innovation strategy. *The Journal of Strategic Information Systems, 26*(1), 22–38. doi:10.1016/j.jsis.2016.09.001

Fayoumi, A., & Loucopoulos, P. (2016). Conceptual modeling for the design of intelligent and emergent information systems. *Expert Systems with Applications, 59*, 174–194. doi:10.1016/j.eswa.2016.04.019

Galster, M., & Avgeriou, P. (2015). An industrial case study on variability handling in large enterprise software systems. *Information and Software Technology, 60*, 16–31. doi:10.1016/j.infsof.2014.12.003

Geerts, G. L., & O'Leary, D. E. (2015). A note on an architecture for integrating cloud computing and enterprise systems using REA. *International Journal of Accounting Information Systems, 19*, 59–67. doi:10.1016/j.accinf.2015.11.001

Grewal, D., Roggeveen, A. L., & Nordfält, J. (2016). Roles of retailer tactics and customer-specific factors in shopper marketing: Substantive, methodological, and conceptual issues. *Journal of Business Research, 69*(3), 1009–1013. doi:10.1016/j.jbusres.2015.08.012

Järvinen, J., & Karjaluoto, H. (2015). The use of web analytics for digital marketing performance measurement. *Industrial Marketing Management, 50*, 117–127. doi:10.1016/j.indmarman.2015.04.009

Killian, G., & McManus, K. (2015). A marketing communications approach for the digital era: Managerial guidelines for social media integration. *Business Horizons, 58*(5), 539–549. doi:10.1016/j.bushor.2015.05.006

Kirikova, M. (2017). Variable contents of enterprise models. *Procedia Computer Science, 104*, 89–96. doi:10.1016/j.procs.2017.01.077

Kladou, S., & Mavragani, E. (2015). Assessing destination image: An online marketing approach and the case of TripAdvisor. *Journal of Destination Marketing & Management, 4*(3), 187–193. doi:10.1016/j.jdmm.2015.04.003

Krafft, M., Goetz, O., Mantrala, M., Sotgiu, F., & Tillmanns, S. (2015). The evolution of marketing channel research domains and methodologies: An integrative review and future directions. *Journal of Retailing, 91*(4), 569–585. doi:10.1016/j.jretai.2015.05.001

Lee, C., & Lee, C. (2017). Method to reduce the gap between construction and IT companies to improve suitability before selecting an enterprise system. *Computers in Industry, 85*, 23–30. doi:10.1016/j.compind.2016.12.005

Li, X., Wang, Y., & Yu, Y. (2015). Present and future hotel website marketing activities: Change propensity analysis. *International Journal of Hospitality Management, 47*, 131–139. doi:10.1016/j.ijhm.2015.02.007

Mathews, S., Bianchi, C., Perks, K. J., Healy, M., & Wickramasekera, R. (2016). Internet marketing capabilities and international market growth. *International Business Review, 25*(4), 820–830. doi:10.1016/j.ibusrev.2015.10.007

Mostaghel, R., Oghazi, P., Beheshti, H. M., & Hultman, M. (2017). Strategic use of enterprise systems among service firms: Antecedents and consequences. *Journal of Business Research, 68*(7), 1544–1549. doi:10.1016/j.jbusres.2015.01.049

Panetto, H., Zdravkovic, M., Jardim-Goncalves, R., Romero, D., Cecil, J., & Mezgár, I. (2016). New perspectives for the future interoperable enterprise systems. *Computers in Industry, 79*, 47–63. doi:10.1016/j.compind.2015.08.001

Pazos, V., & Chalmeta, R. (2017). Reusing enterprise models to build platform independent computer models. *Information Systems and e-Business Management, 15*(2), 377–421. doi:10.1007/s10257-016-0307-x

Romero, D., & Vernadat, F. (2016). Enterprise information systems state of the art: Past, present and future trends. *Computers in Industry, 79*, 3–13. doi:10.1016/j.compind.2016.03.001

Roumani, Y., Nwankpa, J. K., & Roumani, Y. F. (2017). Adopters' trust in enterprise open source vendors: An empirical examination. *Journal of Systems and Software, 125*, 256–270. doi:10.1016/j.jss.2016.12.006

Sedera, D., Lokuge, S., Grover, V., Sarker, S., & Sarker, S. (2016). Innovating with enterprise systems and digital platforms: A contingent resource-based theory view. *Information & Management, 53*(3), 366–379. doi:10.1016/j.im.2016.01.001

Semerádová, T., & Vávrová, J. N. (2016). Using a systemic approach to assess Internet marketing communication within hospitality industry. *Tourism Management Perspectives, 20*, 276–289. doi:10.1016/j.tmp.2016.09.007

Stephen, A. T. (2016). The role of digital and social media marketing in consumer behavior. *Current Opinion in Psychology, 10*, 17–21. doi:10.1016/j.copsyc.2015.10.016

Tsadimas, A., Kapos, G. D., Dalakas, V., Nikolaidou, M., & Anagnostopoulos, D. (2016). Simulating simulation-agnostic SysML models for enterprise information systems via DEVS. *Simulation Modelling Practice and Theory, 66*, 243–259. doi:10.1016/j.simpat.2016.04.001

Vernuccio, M., & Ceccotti, F. (2015). Strategic and organisational challenges in the integrated marketing communication paradigm shift: A holistic vision. *European Management Journal, 33*(6), 438–449. doi:10.1016/j.emj.2015.09.001

Weichhart, G., Guédria, W., & Naudet, Y. (2016). Supporting interoperability in complex adaptive enterprise systems: A domain specific language approach. *Data & Knowledge Engineering, 105*, 90–106. doi:10.1016/j.datak.2016.04.001

Yadav, M., Joshi, Y., & Rahman, Z. (2015). Mobile social media: The new hybrid element of digital marketing communications. *Procedia: Social and Behavioral Sciences, 189*, 335–343. doi:10.1016/j.sbspro.2015.03.229

KEY TERMS AND DEFINITIONS

Cloud Computing: The computing resource that can be delivered as a service through the internet.

Digital Marketing: The business activity of promoting products and brands through electronic media.

Electronic Commerce: The exchange of business information and transaction through the internet.

Enterprise Information Systems: The systems that integrate many business processes and technologies toward increasing business performance.

Enterprise Resource Planning: The software system that can help individuals organize both data and operations.

Information Technology: The utilization of advanced technologies and computerized devices.

Internet: The online network that can communicate among computerized devices.

Marketing: The business operation that includes products, services, advertising, branding, and sales.

Social Media: The online platform that can share information among users through digital devices.

Chapter 4
Manufacturing Information and Database Systems Adoption and Usage Trends in Developing Countries

Norman Gwangwava
Botswana International University of Science and Technology, Botswana

ABSTRACT

Data and information has become a central productive and strategic asset, and the success of the organization depends on its ability to gather, produce, maintain, and disseminate this information for its benefit and those in its value chain. The chapter details the results from a study concerning different types of software packages used by manufacturing and engineering firms in developing countries. The participants were categorized into six industries: academic, power generation, manufacturing chemicals, manufacturing engineering, mining, and transportation. The study revealed that 97% of the companies had a software package installed for their operations. The "big six" software packages found to be in use are Syspro, Sage, SAP, ELLIPSE, Navision, and SCADA DELTA. In the manufacturing sector, Syspro has the largest market share: 86%. ELLIPSE is largely used in the mining sectors. Companies spend about $20,000 as initial investment on the software and about $2,000 annually on license fees. More than 50% of the software users are satisfied with the performance of their software packages.

DOI: 10.4018/978-1-5225-3628-4.ch004

INTRODUCTION

Manufacturing Information and Database Systems (MIDS) are systems designed to collect and present the data which managers need in order to plan and direct operations within the company. MIDS are used in many manufacturing industries, they vary in type but they help the industries in their manufacturing endeavours. MIDS is a management information system used anywhere in the production line. A firm needs to determine its competitive strategy before deciding how to use information technology. That is, it must first understand its competitive position, the competitive forces affecting it, and its overall business strategy. The software package selected by an organization should support its competitive strategy and enable it to deal effectively with the competition forces in its industry. Industrial applications of Manufacturing Information and Database Systems are largely in ERP/CRM, PDM/PLM, SCM, and MOM/MES where the bulky of transactional and operational processes are executed.

The purpose of this chapter is to analyse adoption and usage trends of MIDS in developing countries. By assessing industry case studies, conceptual and literature review articles, a set of recommendations are outlined to assist companies make competitive strategies for ICT adoption.

The rest of the chapter is organized as follows: The next section provides a background to the operating environment for enterprises in the developing world. This is followed by literature review on various categories of MIDS, such as ERP and PLM systems. The research methodology applied in this study follows after the literature review. The chapter ends with recommendations to enterprises in order to be competitive in the modern business environment where business is heavily dependent upon the internet.

BACKGROUND

The modern day competitive global market place is characterized by stringent regulations, higher operating costs, scarcity of resources and demands from increasingly informed customers. In a changing competitive environment, there is need to develop systems and products that are significantly more flexible and responsive than existing ones. Companies in developing countries, with no exception need instant feedback, traceability, uniformity and transparency as some factors to make key decisions hence eliminating any of the loop holes like result forging and bias will hinder fast and effective decision making. Many organisations rely on systems that are disconnected from each other i.e. planning and execution processes. This situation often hinders manufacturers to fully explore market opportunities because production plans and planning plans cannot be adjusted to changes in each other quickly. There is need to allow for open communication within and between all company functions. There is need for instant feedback to achieve high Customer Case fill on Time (CCFOT).

Research has attributed closure of many companies to decision making on inadequate, incomplete, inaccurate and obsolete information. As a result, the most popular trend in corporations nowadays, is to rely more on Enterprise Resource Planning (ERP) systems as they move towards increased integration and team-based work processes. These ERP systems allow companies to work as a whole instead of relying on islands of information. However, most companies have not been able to fully benefit from their very expensive ERP implementations. The failure of ERP projects remains a major concern for all organisations thinking of adopting ERP systems.

LITERATURE REVIEW

This section covers recent, historical and empirical reviews laying the foundation for the present study.

Manufacturing Operations Management (MOM) and Manufacturing Execution Systems (MES)

MOM is a holistic solution to improve manufacturing operations performance. The systems are built to consolidate the management of several production processes, such as quality management, sequencing, production capacity analysis, Work-in-Process (WIP), inventory turns, standard lead times, non-conformance management, asset management, and many other processes within one system. MOM systems expand focus from a single facility to the entire supply network, monitoring a variety of aspects of the manufacturing process. Managing different aspects of the whole manufacturing organisation is associated with various challenges in trying to automate the tasks. Different software tools are needed to collect and analyse real-time data and translate it into valuable knowledge that can be used to inform decision making. Traditionally manufacturers had their own tailor made software applications because of lack of standard platforms to integrate systems at the shop floor level. In order to resolve the challenge, software providers started to package multiple execution management components into single, integrated solutions called manufacturing execution systems (MES) (Saenz, et al, 2009). An MES keeps track of all manufacturing information in real time, receiving up-to-the-minute data from robots, machine monitors and employees. MES solutions now include integrated components such as Computer Aided Process Planning (CAPP), CAM, Product Data Acquisition (PDA), Machine Data Acquisition (MDA), and Personnel Time Recording, as well as Time and Attendance (PTR/T&A).

Generally MES is a process-oriented manufacturing management system which acts as the comprehensive driving force for the organization and execution of the production process. Its major tasks can be summarised as follows:

- Organisation and support of all activities related to the production process.
- Implementation of the closed loop of all actions related to the execution of the production processes.
- Exchange of information with other levels such as corporate management (Enterprise Resource Planning, ERP) and the manufacturing / process levels, as well as operational support systems, and Supply Chain Management (SCM).

In order to achieve integration and interoperability in MES, common standards are required in modeling manufacturing information systems from the shop floor to the business logistics level. The Instrumentation, Systems, and Automation Society (ISA) developed a standard to address the integration issues. ISA-95 is a multi-part standard that defines the interfaces between enterprise activities and control activities (Gifford, 2013). The ISA-95 standard aims to enhance the development of MES applications and integrate them to other information systems of manufacturing companies, particularly ERP systems. Another standard for MES is the IEC 62264 "Enterprise-Control System Integration" set of standards, which defines the functional hierarchy levels of an enterprise, in which decisions with differing timescales and varying levels of detail must be made (IEC®, 2007). The hierarchy of levels are as listed below;

- **Level 4:** Business planning and logistics.
- **Level 3:** Manufacturing operations management.
- **Level 2:** Monitoring as well as supervisory and automated control of the production process (Batch control, Continuous control and Discrete control).
- **Level 1:** Control of process or machinery and data acquisition.
- **Level 0:** Production process.

Enterprise Resource Planning (ERP) Systems

In recent years there has been an increase in using Enterprise Resource Planning (ERP) systems in large companies and government corporations mainly in developed countries. While there is wide adoption of ERP systems in Western economies, developing countries lag far behind. However, due to recent economic growth, developing countries are increasingly becoming major targets of ERP vendors. There is an urgent need for understanding ERP implementation issues in developing countries, as ERP systems are still in their early stages in these countries (Otieno, 2008). Developing countries face additional challenges related to economic, cultural and basic infrastructure issues. This research investigates the organisational and national context within which ERP is adopted and used, and how the context and ERP influence each other.

According to Nah et al (2001), the most important attributes of an ERP system are its abilities to:

- Automate and integrate business processes across organizational functions and locations.
- Enable implementation of all variations of best business practices with a view towards enhancing productivity.
- Share common data and practices across the entire enterprise in order to reduce errors, produce and access information in a real-time environment to facilitate rapid and better decisions and cost reductions.

Edward and Stefan (2001) discussed the ERP selection in small-scale and large-scale companies in Austria. In particular they addressed the fields of software packages considered and chosen by companies. When purchasing a software package weights are assigned to different selection criteria, the size and structure of the team responsible for the decision, the methods employed and the effort used. In most developing countries, there is not much documentation of processes involved prior to purchasing software package and usually the software is not purchased by the people who use it on a daily basis neither are they involved in selection process for the packages. Gwangwava et al (2013) developed a technology transfer decision support system (TTDSS). The system can be used to assist companies in decision making when purchasing a software package.

Gargeya & Brady (2005) state that studies, mostly conducted in developed countries, show that organizations often run into costly and sometimes fatal difficulties with implementation and subsequent maintenance of ERP systems. For example, Akkermans & Van Helden (2002) and Monk & Wagner (2006) observe that a typical ERP implementation initiative takes anywhere between one and three years and typical budgets are in tens to hundreds of millions of dollars. The difficulty in ERP enactment in developing countries may be aggravated by the claim that ERP embodies established ways of doing business thereby requiring organizations adopting ERP systems to change their business processes to conform to business practices inbuilt in ERP packages. Vendors argue that the adoption of these best practices makes the configuring of the software less costly and brings about improvement in the organization's processes. Consequently, organizations and their members often experience pressure to adopt these practices (Gosain, 2004).

Product Data Management (PDM)/ Product Lifecycle Management (PLM)

Product Lifecycle Management can be taken as a strategic business approach that applies a consistent set of business solutions that support the collaborative creation, management, dissemination, and use of product definition information. Product Lifecycle Management (PLM) systems support the management of a portfolio of products, processes and services from initial concept, through design, launch, production and use to final disposal. The product life cycle covers the lifespan of a product from the idea through development up to disposal or recycling. The product's life cycle - period usually consists of five major steps or phases: product development, product introduction, product growth, product maturity and finally product decline. These phases exist and are applicable to all products or services (Ioannis, 2002). The phases can be split up into smaller ones depending on the product and must be considered when a new product is to be introduced into a market since they dictate the product's sales performance. Product Data Management (PDM) systems provide the tools to control access to and manage all product definition data. It does this by maintaining information (meta-data) about product information. (Npd-Solutions, 2011). PDM systems are file repositories or "vaults" that hold mechanical CAD files, including parts and assembly models as well as drawing files. In 3D CAD systems these files rely heavily on each other - they have relationships and contain dependencies that drive details like feature size and placement in other models. If engineers are not careful, small changes in one file can break an entire product assembly. The PDM vault allows mechanical engineers to better manage the complex interrelationships between the part, assembly and drawing files. They can share files with other team members and keep each other up to date on design modifications through a file check-in/check-out process.

RESEARCH METHODOLOGY

A survey research was used to tap data on the software packages used by firms in the case study developing country. The study used stratified random sampling to ensure the survey truly represents all the firms. In this study the questionnaire was used as the main research instrument and was complimented by interviews to extract key data for the research. Sampling enabled the researchers to draw conclusions about the MIDS packages used in industry and the challenges faced in using these systems as organizations endeavour to meet global standards. The sampling process also enabled the researchers to come up with researcher findings within a short space of time.

Data Collection

Data was collected from a sample of organizations. The participants were categorized into six classes of industries academic, chemical manufacture, engineering manufacturing, mining, and transportation industries. A random sampling approach was used in this study. This approach was chosen because it is unbiased and covered the greater proportion of companies in the case study country. A simple random sample is meant to be an unbiased representation of a group. Hence, the size of the sample was a trade-off between rising cost of data collection and the diminishing cost of sampling error. The research instruments used are detailed in the next sub-section.

Questionnaires

Thirty four (34) questionnaires were sent to manufacturing organizations. The researchers made sure that the respondents were guaranteed about the confidentiality of the information to be elicited thus the confidentiality clause. Open-ended questions were used. These questions allowed the respondents to reply freely without selecting one of the several provided responses. In addition open ended questions gave room for respondents to freely state their views on the particular aspect of the research. Close-ended questions were also used. These questions provided respondents with guided answers that enabled systematic analysis of data. It was easier to analyse close-ended questions. Close-ended questions also make it easy and quick to answer questions for the respondents. Questionnaires were used because of their ability to save time and money. The survey covered the whole of the case study country and the use of questionnaires was ideal because they were emailed to the respondents and that was a lot cheaper than travelling.

Structured Interviews

An interview session was held with engineers and operations managers of different manufacturing industries. Clear and concise interview guides were prepared and given to the interviewees a day before the interview so that they could study and understand the guides before the interview. This enhanced the preparedness of both the researchers and the respondents. Only one question was asked at a time, making sure that it was concise and clear. No attempt was made by the researchers to put words into the respondents' mouths; it was made clear from the onset that the purpose of the interview was to get the views of the respondents, not those of the researchers. The researchers avoided reacting to the respondents' replies by expressing approval, disapproval, surprise, or shock; the researchers were always appreciative of what the respondents provided. The researchers were neutral in recording responses so that the collected data would be accurate and objective and a run-down of the major recorded points was read to the interviewees for them to confirm if the points had been captured correctly. Structured interviews were used because they allowed for the creation of rapport between the researcher and the interviewees. Doan (2009) mentioned that structured interviews had an advantage of providing structural relationships of concepts and one concept can lead to the definition of other unknown related concepts. Structured interviews were also used because they maintained focus on the given issues and provided information on the issues raised.

Research Findings

This section outlines the survey results of the software packages used by various organisations in the case study country. The respondents were categorized into specialized industry types that encompass academic, power generation, manufacturing, mining and transport. Major investigations targeted by the research include the types of ERP packages used by each industry category, the age of the installed packages, and implementation challenges, as well as the level of integration of the information systems in companies.

Response Rate

Of the thirty four (34) questionnaires distributed, twenty eight (28) were responded to, which amounts to 82% response rate. Table1 shows that the manufacturing engineering, manufacturing chemicals and power generation industries had the highest response rate of 100%. The transportation industry was the lowest with 0% response.

Market Share of Various Software Applications

The pie chart in Figure 1 shows the market share of various software applications used by industries in the case study country. Syspro has the greatest market share overally, followed by Sage, Sap, Navision, Elipse and other softwares like Pastel, Oracle, Eagle and Hansa world.

Figure 1. Software applications market share in the case study country

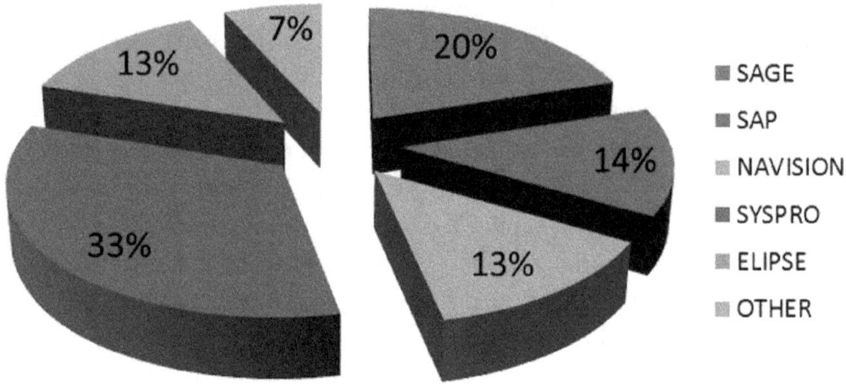

Average Initial Installation Costs of a Multi-User Software Package

The average initial cost of installing a multi-user software system in the case study country is between $US 10 000 and $US 20 000. It was however noted that the most software users are not aware of the investments made by their company towards the purchase of such systems as shown in Figure 2.

Figure 2. Installation costs of a multi-user software package

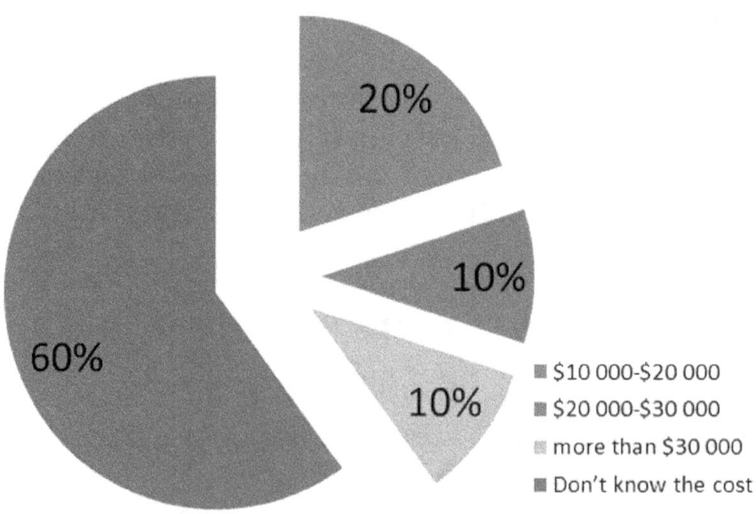

Software Package Usage Analysis by Industry Category

Figure 3 shows the packages used by the mining industry. These packages include ELLIPSE, SAP and PASTEL. SAP is the mostly used with 34%.

Figure 4 shows the software packages used in the manufacturing chemicals industry and SAGE 500 is the mostly used system.

In the manufacturing engineering industry, SCADA DELTA V as shown in Figure 5 is the mostly used software package at 50% with other softwares being the lowest at 17% and 16% respectively.

Each academic institution seems to be using its own system. This is shown in Fig 6 where all other systems have equal percentages of twenty percent (20%) respectively.

Table 1. Response rate

Industry	Number of Questioners Sent	Number of Respondents	Percentage Response
Academic	8	5	63%
Power generation	2	2	100%
Manufacturing chemicals	6	6	100%
Manufacturing Engineering	6	6	100%
Mining	10	9	90%
Transportation	2	0	0%
Total	34	28	82%

Figure 3. Software packages used by the mining industry

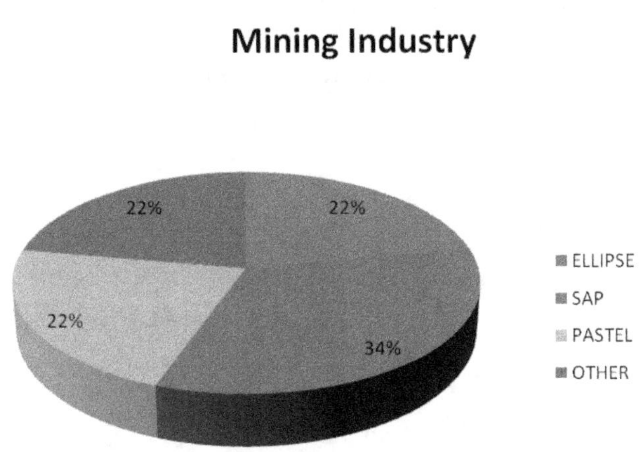

Figure 4. Software packages used by the Manufacturing Chemicals Industry

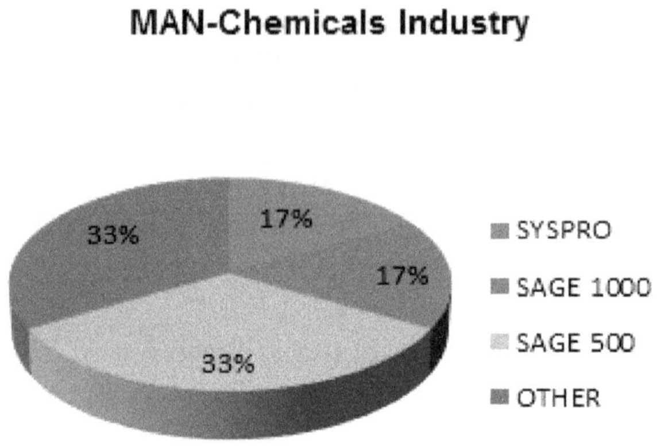

Figure 5. Software packages used by the manufacturing engineering industry

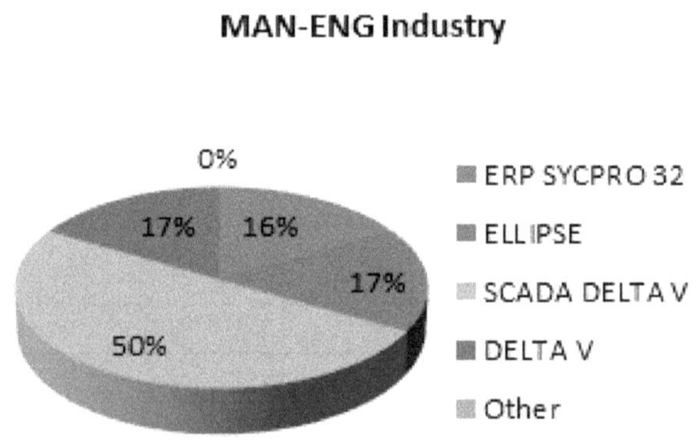

Figure 6. Software packages used by the academic industry

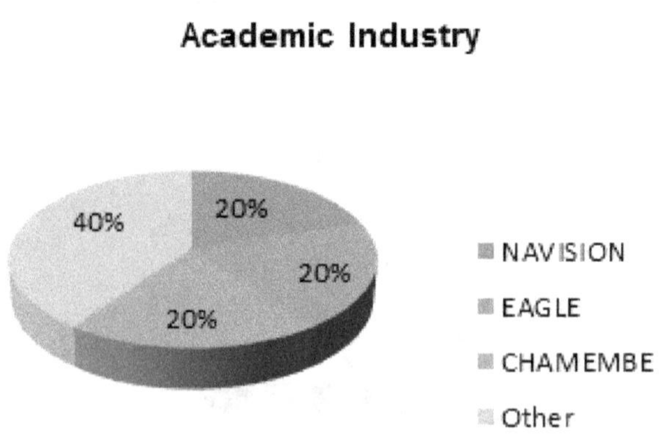

The power generation uses the SAP system. There is 100% usage of SAP because there is one power generation company in the case study country, with various branches in different parts of the country. All the branches are linked through one main sever for the organization. This again proves that SAP software is the mostly used as in the mining industry.

Age Analysis of Installed Software Packages

The analysis was made to determine the importance attached by companies to investment on new software. Using latest software technology enables companies to be competitive. The research findings show that generally most organizations across different industry categories have installations that are within the range of 0-5 years. This shows the endeavour by the industry to keep pace with fast changing technology. Figure 7 shows the age analysis of the software installations in the mining industry. Seven companies had their installations within the last five years, whilst only two installed within the last ten years.

In the manufacturing chemicals industry as shown in Figure 8, three companies installed their software within the last five years, whilst two installed between the last ten years and only one installed in the last twenty years. In the manufacturing

Figure 7. Installation age of software used in the mining industry

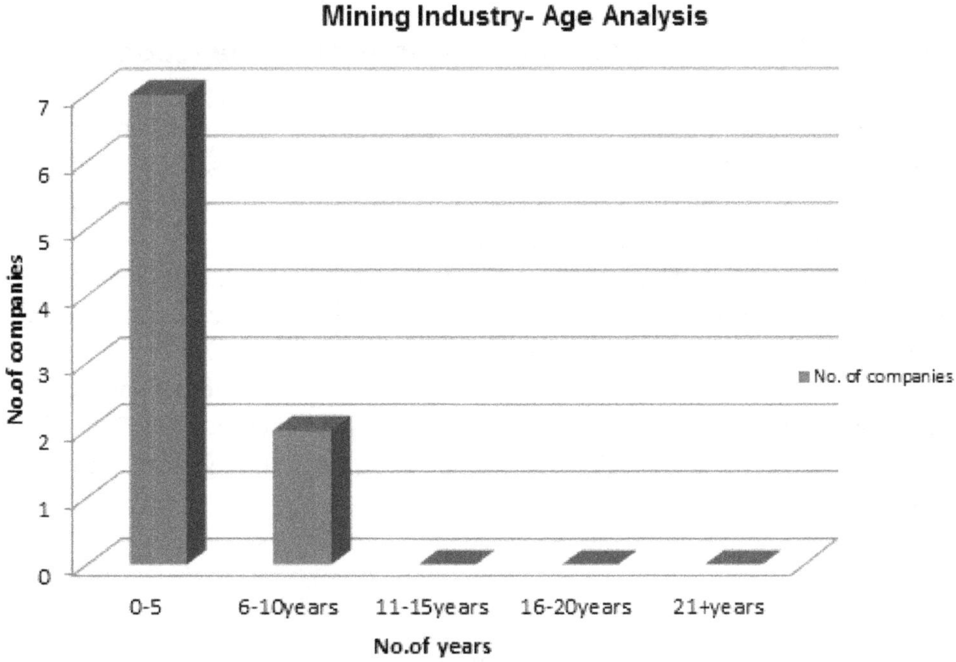

Figure 8. Installation age of software used in the manufacturing chemicals industry

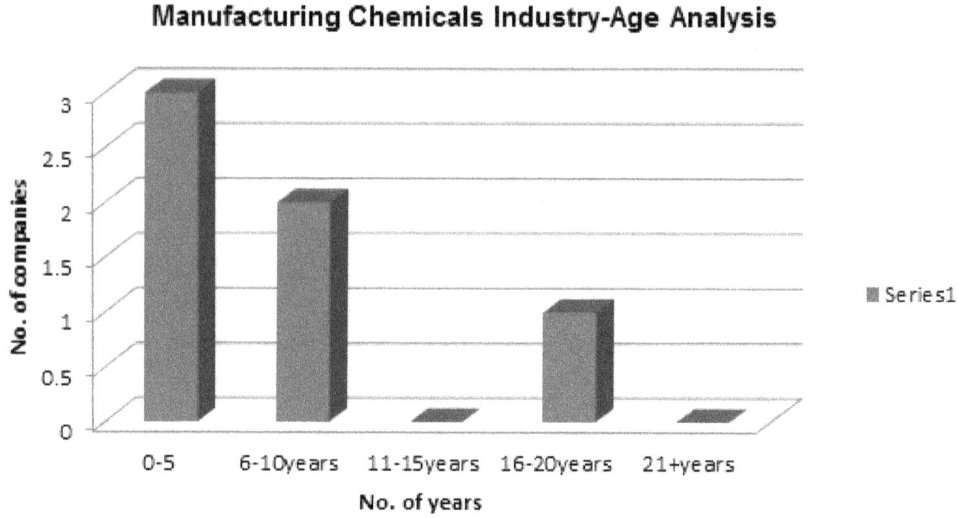

engineering industry two companies installed their software between the last ten years, and then four companies had installations more than ten years back as shown in Figure 9.

In the academic industry as shown in Figure 10, three institutions installed between the last five years, the other two between the last ten years.

Network Platform Used to Support Software Installations

Most companies in the case study country are installing multi-user applications on different sites to cater for various branches for the companies. Only small-scale companies are installing single user software because of the installation costs. Figure 11 shows the types of installations in various industries.

Many companies in the case study country have not yet embraced the technique of cloud computing as shown in Figure 12. 73% of the companies own their server and only 27% of the companies are hosted on the cloud.

Despite most of the companies having their own servers, internet connectivity has enabled employees to log on to their LAN using the internet. Only 20% of the companies surveyed are unable to access their network through the internet as shown in Figure 13.

Figure 9. Installation age of software used in the manufacturing engineering industry

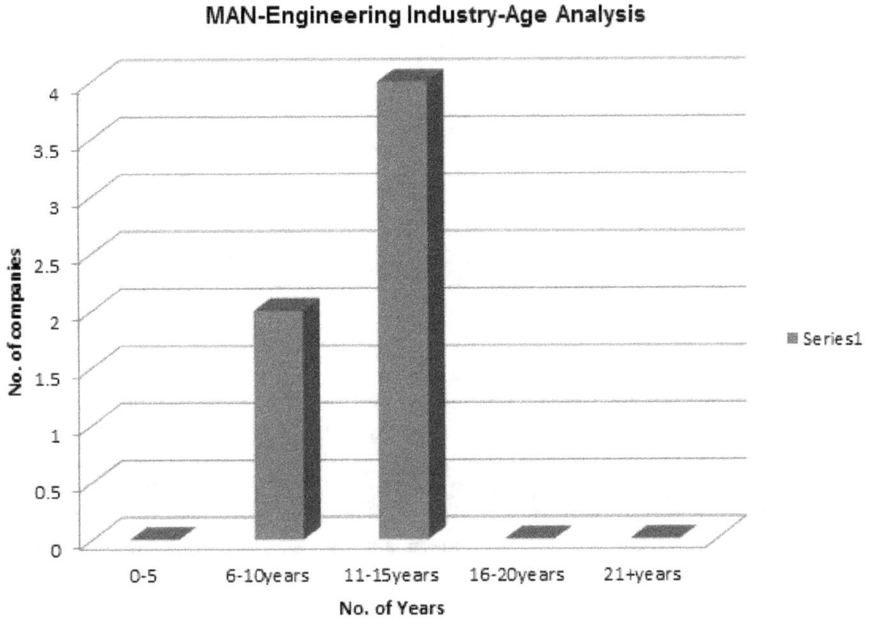

Figure 10. Installation age of software used in the academic industry

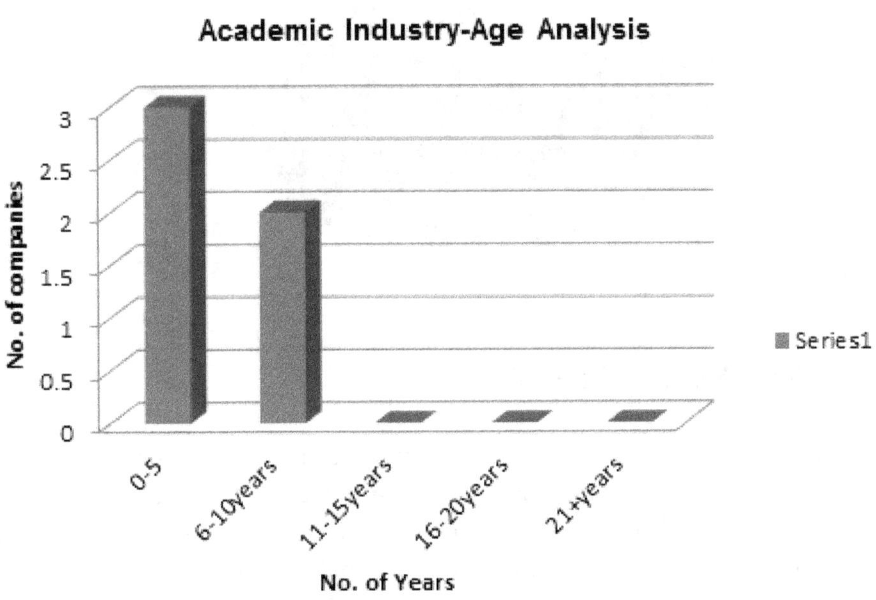

Figure 11. Software users in various companies

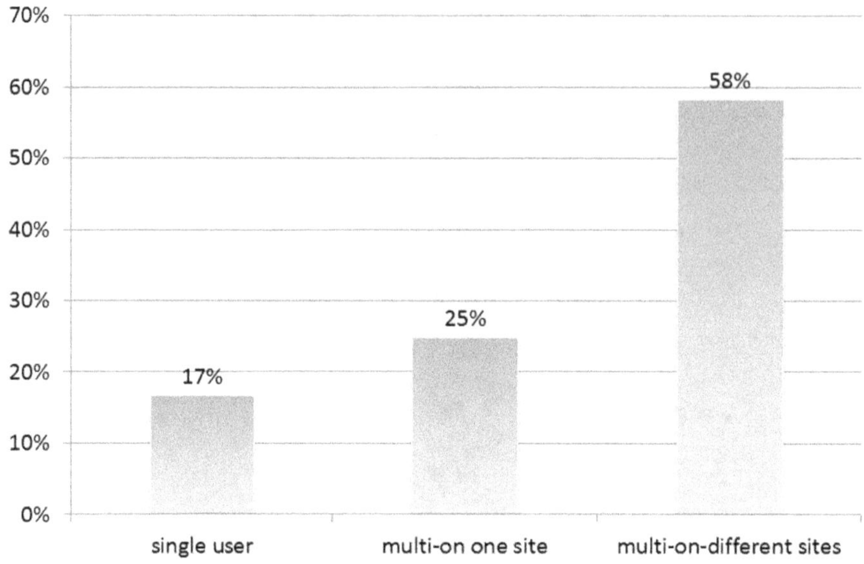

Figure 12. Database hosting

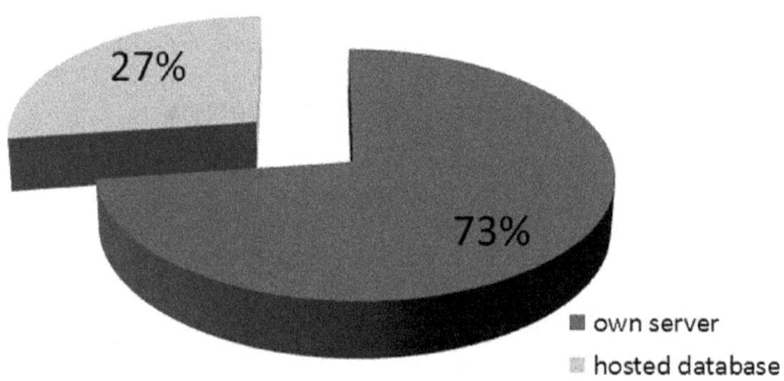

BENEFITS AND CHALLENGES

The respondents stated that having the ERP software improved customer satisfaction, improved process control, reduction of wastes, improved inventory control and labour reduction. Surveyed companies cited the following as benefits of installing these software packages; reduced level of inventory, reduced manpower requirements, better handling of customer orders and reduced cycle time according to the software users in companies in the case study country. The statistics of the responses are shown in Figure 14.

Figure 13. Database accessing in companies

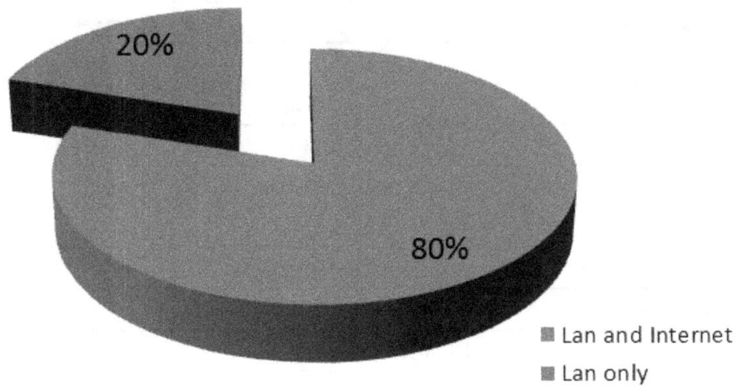

Figure 14. Benefits of installing software packages

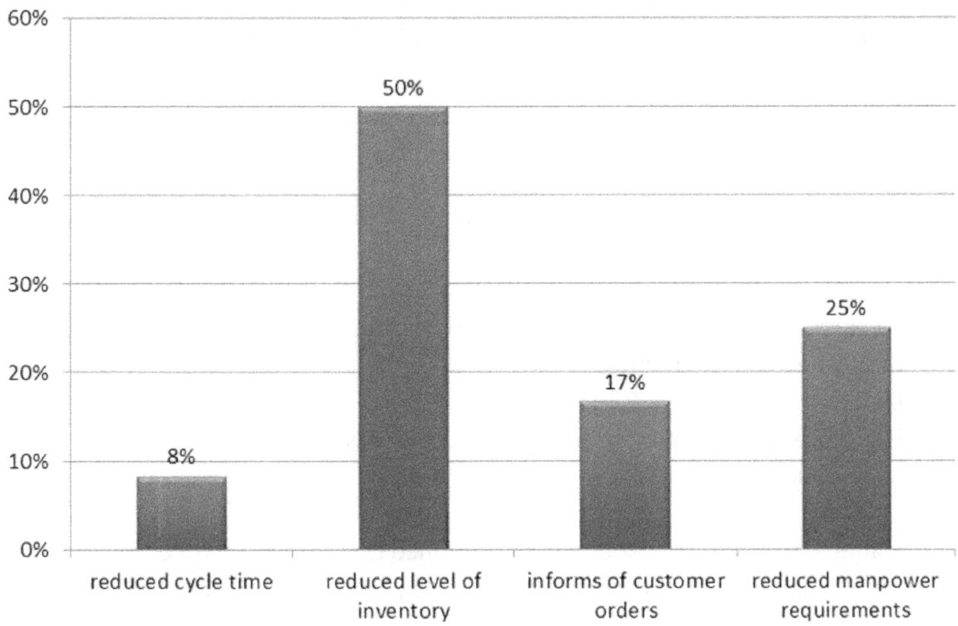

Whilst licensing, upgrade cost and loss of employment were among the challenges faced by the respondents in the different sectors of the different industry. The respondents also noted that to log on to the system was very slow especially when using a wireless internet connection. The other challenge which was noted by respondents is that the systems need to be financially maintained. Table 2 summarises the findings.

Table 2. Challenges faced in using ERP software

IT Department Challenges	End User Challenges
Low computer literacy of users	Network fluctuations
Users' resistance to change	Low system response speed
Hardware upgrade	Reports do not give the correct on the ground situation
Infrastructure robustness	Strong knowledge base required
Maintaining link-up connection	User congestion
Delays in performing tasks	Non user-friendly system
Business Process Re-engineering	Too much centralisation
Infrastructure compatibility	Budget constraints
Inventory grouping	Limited user licences
Budget constraints	Limited storage capability
Identification of assets into the system	Little communication between management and end-users
Poor network systems	Poor integration with other existing systems e.g. ONKEY
Power outages	Hierarchical decision making deems process slow
Lack of backup information	Low computer literacy
Customisation of the Dispatch function	Bar coding

The following are the challenges consultants face in implementing ERPs in companies:

1. Lack of commitment and resistance to change from users.
2. Lack of adequate funding.
3. Unfair competition.
4. High turnover of top management in companies.
5. Lack of adequate support resources.

In Table 3, fifteen (15) respondents mentioned that they purchased the software package as it suited well with their processes, ten said they purchased the software

Table 3. Selection criteria to purchase the software package

Selection Criteria	Respondents Rate	Percentage Response
Cost	10	36
Suitability to processes	15	53
Familiarity with personnel	3	11
Total	**28**	**100**

because of its cost and only three showed that they purchased the software as it was familiar with its personnel. All the respondents stated that training was offered when the MIDs was installed in their company.

SOLUTIONS AND RECOMMENDATIONS

The following recommendations were derived from research and analysis of findings:

1. Top Management Support

Top management support remains a highly rated factor fundamental for the success of any project in general and MIDs projects in particular

2. Process Management

There should be informed decisions regarding the selection of the software vendor and the consultant for the achievement of implementation vision and goals. Companies should choose a vendor with knowledge of the local trading environment.

3. Technology Management

Proper budgets and allowance for additional indirect costs should be made for the acquisition of adequate infrastructure for the efficiency of the MIDs system.

4. Change Management

There should be continuous involvement of end users throughout all project stages and culture change activities such as training should be carried out.

5. Benefits Evaluation

Companies should carry out project evaluation and introduce continuous improvement activities such as software upgrades and end user training.

CONCLUSION

The chapter indicated that there are top five software packages used by companies in the developing countries. Syspro, Sage, ELLIPSE, SAP, and Navision command the greatest market share. The research also concluded that less than half of the population were satisfied with their software. The age analysis of installed software packages revealed a great need to upgrade the technology in various companies. This will improve the competitiveness of the companies in the developing country, both in service and product delivery.

REFERENCES

Akkermans, H., & van Helden, K. (2002). Vicious and virtuous cycles in ERP implementation: A case study of interrelations between critical success factors. *European Journal of Information Systems*, *11*(1), 35–46. doi:10.1057/palgrave.ejis.3000418

Doan, T., & Kennedy, M. L. (2009). Innovation, creativity, and meaning: Leading in the information age. *Journal of Business & Finance Librarianship*, *14*(4), 348–358. doi:10.1080/08963560802424011

Gargeya, V. B., & Brady, C. (2005). Success and Failure Factor of Adopting SAP in ERP System Implementation. *Business Process Management Journal*, *11*(5), 501–516. doi:10.1108/14637150510619858

Gifford, C. (2013). *The MOM Chronicles: ISA-95 Best Practices Book 3.0*. International Society of Automation.

Gosain, S. (2004). Enterprise Information Systems as Objects and Carriers of Institutional Forces: The New Iron Cage. *Journal of the Association for Information Systems*, *5*(4), 151–182.

Gwangwava, N., Nyanga, L., Moyongewenyu, B., & Mpofu, K. (2013) A framework for a Technology Transfer Decision Support System (TTDSS). *SAIIE25 Proceedings*.

IEC 62264-3. (2007). *Enterprise-Control System Integration – Part 3: Activity Models of Manufacturing Operations Management*. IEC.

Komninos, I. (2002). *Product life cycle management, Urban and regional Innovation research unit, faculty of engineering*. Aristotle University of Thessaloniki.

Monk, E. F., & Wagner, B. J. (2006). *Concepts in Enterprise Resource Planning* (2nd ed.). Thomson.

Nah, F., Lau, J., & Kuang, J. (2001). Critical factors for successful implementation of enterprise systems. *Business Process Management Journal*, *7*(3), 285–297. doi:10.1108/14637150110392782

Npd-Solutions. (2011). Retrieved from http://www.npd-solutions.com/pdm.html

Otieno, J. (2008). ERP implementation challenges: A Kenyan case study. *Lecture Notes in Business Information Systems*, 399-409.

Otieno, J., & Abeysinghe, G. (2006). A framework for ERP implementation in Kenya: An empirical study. *Proceedings of Information Resource Management Association (IRMA) International Conference*.

Saenz, B., Artiba, A., & Pellerin, R. (2009). Manufacturing Execution System-A Literature Review. *Production Planning and Control*, *20*(6), 525–539. doi:10.1080/09537280902938613

ADDITIONAL READING

Al-Fawaz, K., Al-Salti, Z., & Eldabi, T. (2008). Critical success factors in ERP implementation. *RE:view*.

Harris, B., & Williams, B. (2005). *System Dynamics Methodology Using Systems Concepts*. Kellogg Foundation.

KEY TERMS AND DEFINITIONS

Enterprise Resource Planning (ERP): Automation and integration of a company's core business to help them focus on effectiveness and simplified success. ERP software applications can be used to manage product planning, parts purchasing, inventories, interacting with suppliers, providing customer service, and tracking orders. ERP can also include application modules for the finance and human resources aspects of a business.

Machine-to-Machine (M2M): A term used to describe any technology that enables networked devices to exchange information and perform actions without the manual assistance of humans. M2M is considered an integral part of the internet of things (IoT) and brings several benefits to industry and business in general as it has a wide range of applications such as industrial automation, logistics, smart grid, smart cities, health, defence, etc. mostly for monitoring but also for control purposes.

Manufacturing Execution Systems (MES): A control system for managing and monitoring work-in-process on a factory floor. An MES keeps track of all manufacturing information in real time, receiving up-to-the-minute data from robots, machine monitors and employees.

Manufacturing Information Systems: A management information system designed specifically for use in a manufacturing environment. The role of manufacturing information systems is to support manufacturing operations by providing relevant and timely information for decision making at different levels of the company hierarchy. It also automates and secures the sequencing of manufacturing and business processes.

Manufacturing Operations Management (MOM): The subsequent system to manufacturing execution system (MES) software which expands focus from a single facility to the entire supply network and monitors a variety of aspects of the manufacturing process, including production capacity analysis, work-in-process (WIP), inventory turns, and standard lead times.

Product Data Management (PDM)/Product Life-Cycle Management (PLM): PDM is a category of computer software used to control data related to products. PDM creates and manages relations between sets of data that define a product, and store those relationships in a database. It is an important tool in product lifecycle management. PLM is a strategic business approach that applies a consistent set of business solutions that support the collaborative creation, management, dissemination, and use of product definition information throughout the lifecycle of the product.

STEP/STEP-NC: Standard for the exchange of product model data, a comprehensive ISO standard (ISO 10303) that describes how to represent and exchange digital product information. STEP is a means by which graphical information is shared among unlike computer systems around the world. It is designed so that virtually all essential information about a product, not just CAD files, can be passed back and forth among users.

Supply Chain Management (SCM): An integrated approach to planning, implementing, and controlling the flow of information, materials, and services from raw material and component suppliers through the manufacturing of the finished product for ultimate distribution to the end customer. It includes the systematic integration of processes for demand planning, customer relationship collaboration, order fulfillment/delivery, product/service launch, manufacturing/operations planning and control, supplier relationship collaboration, life cycle support, and reverse logistics and their associated risks.

Chapter 5
A Comparative Assessment of Enterprise Content Management Maturity Models

Shadrack Katuu
International Atomic Energy Agency, Austria & University of South Africa, South Africa

ABSTRACT

There is increasing interest in maturity models dating back to the 1980s. A maturity model is a management tool designed to help organizations implement effective processes in a given management discipline. This chapter provides comparative assessment of three enterprise content management (ECM) maturity models that have been developed and implemented in different parts of the world. While the maturity model concept has existed for several decades, ECM maturity models are a more recent phenomenon. This study explores the theoretical foundations and functionality drawing lessons that could contribute to future developments. It outlines a history of maturity models and describes three ECM models developed in different parts of the world and compares the different levels and dimensions of maturity levels within the models. The chapter demonstrates that the different maturity models have varied naming conventions, different levels of complexity, and vary in the approach to maturity assessment.

INTRODUCTION

Many organizations are under intense pressure from shareholders as well as clients to improve their processes in order to have better products and/or services. The larger organizations—for instance, in the manufacturing sector—have sought to improve their processes through business reengineering efforts that include the

DOI: 10.4018/978-1-5225-3628-4.ch005

adoption of enterprise-wide systems such as enterprise resource planning (ERP) software applications. Through the use of ERPs, large organizations have been able to integrate vastly different business activities, including finance, inventory management, marketing and sales, materials management, product planning, production planning, and service delivery, as well as shipping and payment.

Enterprise-wide systems such as ERPs generate vast amounts of data gathered from forms and managed through workflows and result in large volumes of digital information assets. In order to effectively manage this digital content effectively as well as meet legal or compliance requirements, organizations have also implemented enterprise content management (ECM) systems. ECM systems offer a number of functional services, including document management, collaboration, and records management and ensure an institution's content is managed throughout its life cycle.

Organizations that have ECM applications have an expectation that the systems will result in both hard benefits, such as improved process efficiencies, and soft benefits, such as improved recordkeeping. However, assessing the success of implementation is often based on subjective rather than objective criteria. Maturity models are an option in providing an objective assessment of ECM implementation. A maturity model is a "structured collection of elements that describe characteristics of effective processes" (Murray & Ward, 2007, p. 5).

This chapter uses a literature review as its research design in order to provide a background to maturity models in general and then critically examines three ECM maturity models. The motive is to examine how they have been developed, how they are structured, and what value they offer to contemporary institutions.

BACKGROUND INFORMATION

This section provides background information on both ECM and maturity model concepts.

ECM

Organizations have a variety of business systems to help them manage their records. Depending on the institution, the digital records connected to these business systems could either be managed in network drives or by using specialized business applications such as ECM software applications or, although not desirable, left unmanaged. The term ECM has been used for more than a decade by professional service institutions such as Gartner and Forrester (Shegda, Chin, Logan, & Lundy, 2004) as well as in research projects such as InterPARES (Rogers, Malmas, & Enns, 2011). The Association for Information and Image Management (AIIM) defines

ECM as constituting "strategies, methods and tools used to capture, manage, store, preserve and deliver content and documents related to organisational processes" (AIIM, 2010).

Although the ECM concept is quite broad, in the majority of cases it has been used in the narrow sense of software applications used to manage information within organizations (Franks, 2016b; Koehler-Kruener, Chin, & Hobert, 2015). ECM software applications offer functional business services that include document management, document collaboration, records management and archiving, business process management (BPM) and workflow (including case management), electronic forms, and file synchronization, as well as imaging (Real Story Group, 2016, p. 2). This information should be managed throughout its life cycle from the time it is generated or received, through its use and storage and eventual destruction or preservation for historical purposes (Brown, Katuu, Sebina, & Seles, 2009; Goh, Katuu, Miller, Nordland, & Sebina, 2009; Katuu, 2002b).

Many institutions face numerous challenges in the complex processes required to implement ECM software successfully (Nordheim & Päivärinta, 2006; Päivärinta & Munkvold, 2005; Usman, Muzaffar, & Rauf, 2009). There are different approaches to providing ECM implementation guidance, such as business process oriented (Vom Brocke, Simons, & Cleven, 2011) or providing a framework for consolidating systems and databases into an ECM framework (Smits & O'Callaghan, 2014). These approaches demonstrate the need for a "detailed deployment and validation plan" that is critical to achieving "timely implementation of ECM" (Hullavarad, O'Hare, & Roy, 2015, p. 263). In addition, ECM implementation is costly and, therefore, organizations should expect a return on their investment, but "determining ECM success is difficult because many of the ECM benefits are intangible and difficult to measure" (Vom Brocke, Herbst, & Urbach, 2014).

As a result there is a lack of a comprehensive means to assess whether ECM implementation or deployment has achieved the goals. Ideally, organizations should be able to rely on standards and best practice guidelines to provide this desired return on investment. However, a recent assessment of the available standards and best practice guidelines reveals that there is a gap once an ECM application has been rolled out within an organization (Katuu, 2012a). Therefore, organizations are not optimally benefiting from ECM implementation (Salamntu & Seymour, 2014, p. 299), partly because there is not enough guidance to practitioners regarding ECM implementation (Herbst, Simons, vom Brocke, & Derungs, 2014, p. 109). For this reason, maturity models offer a structured framework for identifying and assessing capability through an objective assessment mechanism (Durga, 2015; Katuu, 2016a; Pelz-Sharpe et al., 2010).

Maturity Models

A maturity model is a management tool designed to help organizations implement effective processes in a given management discipline (Simon, Schoeman, & Sohal, 2010). Maturity models have existed as a concept for more than three decades (Cameron, 2011, p. 21). Some scholars contend that the maturity models concept started in the late 1970s and early 1980s under the auspices of the Total Quality Management (TQM) philosophy (Baskarada, 2009; Pfeifer, 2002). TQM is an integrative philosophy of management for continuously improving the quality of products and processes (Ahire, 1997).

The maturity model concept became most prominent through computer software engineering in the 1980s and 1990s (Baskarada, 2009; Liu, 2002). The concept has since spread to several subject domains, including:

- E-government (Maseh & Katuu, 2017; Valdes et al., 2011)
- E-learning (Marshall, 2005)
- Financial management (McRoberts & Sloan, 1998)
- Human resources management (Curtis, Hefley, & Miller, 1995)
- Health (Gillies, 2000; Katuu, 2017)
- Information systems (Guédria, Naudet, & Chen, 2015; Saleh & Alshawi, 2005)
- Project management (Kerzner, 2011; Sukhoo, Barnard, Eloff, & Van der Poll, 2007)

Maturity models are developed on the basis that organizations do not move from zero capability to optimum capability instantaneously, but rather progress along a journey of maturity (Murray & Ward, 2007, p. 5). The number of levels for each model may vary from three to seven, as illustrated in Table 1.

More than 20 subject domains utilize the maturity model concept (Wendler, 2012, p. 1328). This chapter examines ECM maturity models, basing the analysis on two research questions directed through a literature review research design.

RESEARCH DESIGN AND RESEARCH QUESTIONS

According to Hickson (2008, p. 3), research is the "systematic investigation of a specific question in order to establish new facts and draw new conclusions." It involves the interpretation and revision of current knowledge and the discovery of new knowledge. There are two basic missions in research: either deriving new knowledge through theoretical research or dealing with practical problems or issues

Table 1. Diversity of maturity levels in maturity models

Maturity Levels	
3 levels	• Capability Immaturity Model (Finkelstein, 1992)
4 levels	• Capability Maturity Model (Kana & Karni, 2004) • Gartner Security Process Maturity Model (Van Mien, 2001) • Health Normative Standards Framework Maturity Model (Department of Health [South Africa] & Council for Scientific and Industrial Research [South Africa], 2014)
5 levels	• Capability Maturity Model for Decision Making (Kana & Karni, 2004) • COBIT Maturity Model (Walker, McBride, Basson, & Oakley, 2012) • eHealth Maturity Model (Vital Wave Consulting, 2009) • Gartner IT Asset Management Maturity Model (Adams, 2005) • People Capability Maturity Model (Curtis, et al., 1995)
6 levels	• Gartner Business Continuity Planning Maturity Model (Mingay, 2002) • Gartner Knowledge Management Maturity Model (Harris, 2006) • General Practice Information Maturity Model (Gillies, 2000) • General Practitioner Information System Measurement Model (Saleh & Alshawi, 2005)
7 levels	• Gartner Infrastructure Maturity Model (Bittman, 2004)

through applied research (Powell & Connaway, 2004, p. 2). This chapter's research mission is a theoretical investigation of ECM and maturity models.

Wendler (2012) argued that there are four ways of categorizing studies on maturity models:

- **Maturity Model Development:** An article whose main objective is developing or constructing a new maturity model.
- **Maturity Model Application:** An article that principally applies maturity models in a specific context.
- **Maturity Model Validation:** An article whose purpose is to empirically or conceptually validate one or more maturity models.
- **Meta-Article:** An article that assesses models through literature reviews, process models, or other theoretical considerations.

After assessing more than 200 articles on maturity models, Wendler (2012, p. 1324) discovered that the distribution within the four categories was 46%, 35%, 14%, and 6%, respectively. The study demonstrated the scarcity of meta-articles dealing with theoretical issues that constitute maturity models. Therefore, this chapter's theoretical contribution is as a meta-study on ECM maturity models. It explores the three models developed in different parts of the world, analyzing their conceptual framework, and draws lessons on naming conventions, maturity dimensions, and assessment criteria.

Any research process, such as a meta-study, involves identifying a question or questions "collating and integrating current knowledge on the topic; designing a method to collect information" to develop the research question; and "finally developing new conclusions from the evidence" (Hickson, 2008, p. 3). This meta-study has two distinct research questions namely:

1. What are the different ECM maturity models that exist globally?
2. What are the comparative strengths and weaknesses of these ECM maturity models?

The research process was undertaken through a review of literature that addresses the two research questions. Table 2 provides a summary of the research design.

DISCUSSION ON THE VARIETY OF ECM MATURITY MODELS

As noted in Table 2, this section addresses the first research question. The review of literature revealed at least six maturity models that are tangentially related to ECM because they cover a significant aspect or module of ECM such as document management and digital asset management. The maturity models are outlined briefly as follows:

- Records Management Capacity Assessment System that was developed by the International Records Management Trust in the early 2000s as an objective means of assessing the strengths and weaknesses and associated risks of records management systems using criteria based on records and information management international standards (Demb, 2008; Griffin, 2004).
- Records and Information Risk Management Maturity Model that was published in 2004 in order to assist organizations in managing records and

Table 2. Research design

	Research Design	
Research mission	Theoretical	
Research method	Literature review	
Research questions	(1) What are the different ECM maturity models that exist globally?	Discussion of data analysis in Section 4.0
	(2) What are the comparative strengths and weaknesses of these ECM maturity models?	Discussion of data analysis in Section 5.0

information risk holistically and systematically through its five maturity levels (Lemieux, 2004, pp. 38-44).

- McLaren Document Management Maturity Model, which is one of the earliest maturity models developed covering document management. It has nine maturity levels that "do not describe the potential offered by ECM technology," but rather "characterize the level of understanding of the people who work with documents" within an institution (Goodwin, 2006, p. 4). This maturity model does not have assessment criteria.
- ARMA Generally Accepted Recordkeeping Principles that was developed for information governance with five maturity levels related to eight principles (ARMA International, 2013). ARMA developed and published the model to "foster general awareness of information governance standards and principles and to assist organisations in developing information management systems that comply with them"(ARMA International, 2014, p. 3). The model has assessment criteria for the different maturity levels.
- Digital Asset Management Maturity Model was published in 2012 in order to help enterprises striving to align their business and technology efforts for improved digital asset management (Durga, 2012; Lipsey & Regli, 2012). The most current version uses 15 maturity dimensions organized in four categories to define the digital asset management ecosystem (Real Story Group & DAM Foundation, 2014). The model has assessment criteria for the different maturity dimensions and levels.
- Digital Preservation Capability Maturity Model was published in 2014 and was developed to help practitioners "identify at a high level the capabilities of their organisation relative to optimal digital preservation capabilities" in order to chart the evolution from "disorganized and undisciplined management of electronic records" to "increasingly mature stages of digital preservation capability" (C. Dollar & Ashley, 2014, p. 1). DPCMM has been used as a self-assessment tool by the United States Council of States Archives and has been incorporated in a tool development project by the Section of International Organisations of the International Council on Archives (C. M. Dollar & Ashley, 2015).

However, these models do not cover all of the functional business services that constitute ECM applications, nor do they address issues of ECM implementation. The three maturity models found to have covered ECM application implementation were developed over the last decade and a half. The first one was developed by a South African institution, the second developed in the United Kingdom, and the third developed by a conglomeration of institutions based in the United States of America.

The ECM-Knowledge (ECMk) Maturity Model

While working to enhance its own ECM implementation program between 2004 and 2007, one South African institution developed a basic maturity framework it called the ECM-knowledge (ECMk) Maturity Model. The institution requested anonymity but shared details on the development and features of the maturity model for purposes of this discussion.

The rationale for developing the model was the realization that implementation maturity was not dependent on the accumulation of years, but rather the accumulation of experience, that is, chronological years offered little evidence of capability if not accompanied by acumen. For ease of understanding, the design of ECMkMM mirrored the organic stages of human development. The aim of the model was to assess levels of organizational knowledge and concomitant reduced dependence on the service provider supporting the ECM application implementation. The relationship between an organization and its service provider was analogous in the model to the dependent interaction between a human child and its guardian.

The model has five levels of maturity that draw heavily from the human development subject domain (Louw, Van Ede, & Louw, 1998). They are:

Level 1: Newborn
Level 2: Toddler
Level 3: Adolescent
Level 4: Adult
Level 5: Senior citizen

Table 3 shows the different levels of maturity and the related knowledge criteria that characterize each level.

ECMkMM was used within a very limited context within the South African institution that developed it. The model helped the institution asses the maturity of its interactions with ECM service providers.

The Content Maturity Model (CMM)

The Content Maturity Model (CMM) was developed in the United Kingdom with the objective of laying out a baseline of ECM understanding throughout an organization (Cameron, 2011, p. 40). The model has five levels of maturity:

Level 1: Individual
Level 2: Team

Table 3. The ECM-knowledge Maturity Model

Level	Knowledge Characteristics
Newborn	• This first phase of maturity in the model is the point at which the ECM application has initially been installed. • As with human newborns, organizations will have unremarkable vision with regard to their ECM implementation program, only being able to focus on short-term gains, if at all (Rodriguez, 2009, p. 80). This would mean that decision-making capabilities are mostly done with high reliance on the service provider. • Again as with human newborns, organizations tend to require a lot of emotional reassurance from service providers because initial crises cause anxiety, akin to the cuddling and soft stroking that newborns get (Rodriguez, 2009, p. 80).
Toddler	• The second phase is characterized by increasing rates of ECM implementation problems and waning of any initial enthusiasm. • As with toddlers who cry as a form of basic instinctive communication, organizations often lack sophistication in communicating implementation problems (Miller, 1999, p. 242). With human infants, crying may express a variety of feelings, including hunger, discomfort, overstimulation, boredom, wanting something, or loneliness. With institutions, communication is unsophisticated in expressing exactly what the problems are because the causes are often not understood and symptoms not well expressed. • During this phase human infants seek proximity to a specified attachment figure in situations of alarm or distress for the purpose of survival (Rodriguez, 2009, p. 83). The forming of attachments is considered to be the foundation of the infant/child's capacity to form and conduct relationships throughout life. The attachment is usually to adults who are sensitive and responsive in social interactions with infants. In a similar fashion, organizations at this level tend to seek proximity to service providers as they face their first situations of distress and will remain attached to those service providers who are sensitive and responsive to their interactions, demonstrating continued dependence.
Adolescent	• This is a transitional phase in human development characterized by a number of physical, emotional, cognitive, and attitudinal changes (Rowe & Miller, 2011, p. 35). • For organizations, physical changes may be the result of software upgrades, configuration adjustments, and changes in the technological environments on which ECM applications reside. • The emotional changes in organizations have to do with starting to move away from the total dependence of the service provider. Cognitive changes in this regard include being able to, among other things, develop strategic and tactical plans for the ECM program with little or no assistance from the service provider. • Attitudinal changes in the organization may result from the increasing influence of the peer group rather than the service provider in providing advice and guidance. In human development, adolescents start to view their peer group as more important and influential than their guardians (Rowe & Miller, 2011, p. 197). This is because peer groups offer their members the opportunity to develop various social skills and will provide either positive or negative influences (Rodriguez, 2009, p. 98). For organizations, this attitudinal shift from service providers to peer group may be the result of such things as drastic changes in the physical environment where the service provider's knowledge is no longer relevant and the realization that other peer institutions provide much better learning opportunities, having undergone similar challenges.
Adult	• In human development, this stage is when one is regarded as independent, self-sufficient, and responsible (Arnett, 2004, p. 209). For organizations, the same could be true. • At this point a human adult has the capacity to commit to concrete affiliations and partnerships. In addition, they have the ability to face the fear of losing their ego in situations that call for self-abandonment. For organizations this would mean that they can address their problems honestly and take responsibility for decisions made. Additionally, they would be able to provide advice and guidance to other institutions that are undergoing the earlier levels of development. • In humans, as the years advance, strength and flexibility decrease. A parallel may be found in organizations where flexibility to adapt to changes wanes with time and responsiveness to new challenges may be much slower.
Senior citizen	• In humans, this is a period of general physical decline. The individual has slower reaction times and fading agility (Rodriguez, 2009, pp. 108-109). For organizations the same may be true with reduced cognitive sharpness and difficulty in recalling lessons learned from the past in order to address current realities.

Level 3: Enterprise
Level 4: Optimize for acquisition
Level 5: Innovate for growth (Cameron, 2011, p. 22)

The maturity levels are mapped against three dimensions: people, processes, and systems.

- People relates to human aspects that include “skills, training, relationship to risk and reward and performance management” (Cameron, 2011, p. 23).
- Processes relate to “standards, guidelines, policies and the workflow of everyday processes as information passes through the business” (Cameron, 2011, p. 25).
- Systems relate to the “hardware, software and applications which create, store and manage information” (Cameron, 2011, p. 25).

Table 4 illustrates the components of the CMM (Cameron, 2011, p. 25).

According to Cameron (2011, p. 23), most organizations are positioned between stages 2 and 3 because “ECM tools and techniques are emerging in these organizations but struggling to make the single department solution universal.” Martinez (2013, pp. 19-20) echoed these sentiments based on results from a composite assessment of a number of institutions that showed that the People dimension was at the Individual level, the Process dimension was at the Enterprise level, and the Systems dimension was at the Team level.

The ECM Maturity Model (ECM3)

The ECM Maturity Model (ECM3) was developed as an open-source product by individuals from a number of consulting firms primarily in the United States of America. The first edition was published in March 2009 and a second edition in March 2010 (Pelz-Sharpe, et al., 2010).

As with all maturity models, the ECM3 has graded levels of capabilities ranging from rudimentary information collection and basic control through increasingly

Table 4. The Content Maturity Model

	Individual	Team	Enterprise	Optimize	Innovative
People					
Process					
Systems					

sophisticated levels of management and integration, finally resulting in a mature state of continuous experimentation and improvement (ECM Maturity Model, 2012). There are five levels of maturity:

Level 1: Unmanaged
Level 2: Incipient
Level 3: Formative
Level 4: Operational
Level 5: Proactive (Pelz-Sharpe, et al., 2010, p. 6)

ECM3 is structured on the premise that the way to measure the maturity of ECM implementation would be through 13 dimensions of maturity across three categories: human, information, and systems (Pelz-Sharpe, et al., 2010, p. 8). The human category relates to people and all their attributes that enhance maturity, such as the expertise of different types of professionals within an organization, as well as their interactions to ensure strategic alignment with institutional goals for success. The information category relates to attributes affecting the content in application, including the ability to manage that content. The systems category relates to the application's technical abilities.

In addition to these three categories, there are 13 maturity dimensions, as well as five levels of maturity. This very exhaustive set of evaluation criteria results in a sophisticated means of assessing maturity. Figure 1 provides a graphical representation of the maturity dimensions across three categories (Pelz-Sharpe, et al., 2010).

Figure 1 shows that the human category has four maturity dimensions: business expertise, IT expertise, process, and alignment. Business expertise relates to employee and executive education and understanding of core ECM precepts. IT expertise entails the ability to properly take advantage of incumbent and new systems. The process dimension is the extent to which the enterprise has analyzed its content-oriented business processes. Alignment relates to the extent of effective business/IT collaboration, understanding, and synchronization (Pelz-Sharpe, et al., 2010, p. 9).

The information category has five dimensions: content/meta-data, depth, governance, reuse, and findability. The content or meta-data dimension is the extent to which the enterprise has analyzed its content and meta-data. Depth relates to the completeness of content life cycle management. Governance is the extent of policies and procedures addressing information management. Reuse is the extent to which content within ECM applications can be reused for purposes other than what they were initially created for. Findability is the ability to find the right content at the right time (Pelz-Sharpe, et al., 2010, p. 9).

Figure 1. The ECM3 maturity dimensions

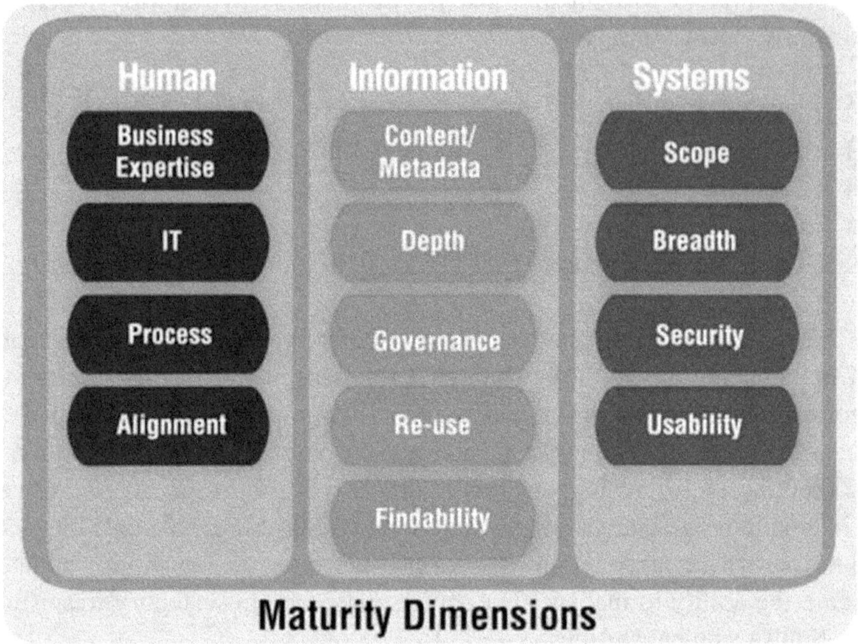

The final category is systems and has four dimensions: scope, breadth, security, and usability. Scope is the relevant range of ECM functional capabilities adopted, for example, document management, business process management, digital asset management, etc. Breadth relates to the evolution from departmental to enterprise-wide management applications where necessary. Security is the extent to which actual content access reflects enterprise entitlements, and usability is the ECM application's fitness to purpose (Pelz-Sharpe, et al., 2010, pp. 9-10).

DISCUSSION COMPARING DIFFERENT ECM MATURITY MODELS

As noted in Table 2, this section addresses the second research question. Maturity models are developed in order to assist institutions in assessing their development paths in order to reach the ideal level in a particular domain (Wendler, 2012). The three ECM models outlined in the preceding section were developed within different institutional contexts in order to achieve varied objectives. Each of them has particular strengths and weaknesses. The following sections briefly describe each of the models and provide a composite analysis of the characteristics among all of them.

The ECM-knowledge (ECMk) Maturity Model

The ECMkMM model was developed for an organization that wanted to look specifically at the relationship between the institution and its ECM service provider. In this narrow perspective the model may have fulfilled the institution's objective. Although this narrow perspective provides the opportunity for in-depth validation of one objective, it hinders the chance for generalizability over a wide range of perspectives. For example, the analogy of human development could not be sustained when looking at other aspects of human life such as courting and marriage and the bearing of children. Therefore, this analogy was limited to making parallels between a child and a guardian. Additionally, although there were strong parallels between human and organizational knowledge characteristics in the first three stages of development, these parallels waned significantly when in the adult and senior citizen levels.

The Content Maturity Model (CMM)

The CMM was developed to provide a baseline for understanding ECM in an organization (Cameron, 2011, p. 40). The model provides a more sophisticated analysis mechanism than ECMkMM because it not only has maturity levels, but also provides three assessment dimensions: people, processes, and systems. Although these dimensions provide conceptual differentiation, it would be difficult for an organization to assess itself without any specified assessment criteria. This potentially limits its implementation and may suggest one reason for the small number of real-life case studies (Martinez, 2013; Muppidi, 2016; Wicaksono, Shihab, & Sandhyaduhita, 2015).

The ECM Maturity Model (ECM3)

The ECM3 model was developed by a consortium of institutions and has the most sophisticated structure of the three models. In addition to the levels of maturity, there are three categories, for a total of 13 dimensions, making it a very detailed model. Such sophistication ensures exhaustive assessment from multiple perspectives. This may suggest the reason for a growing number of real-life case studies in published sources that have applied the model within individual countries (Katuu, 2016c; Santos & Pinto, 2015) or alternatively applied through assessment tools or in global surveys (Durga, 2015, 2016).

Consolidated Assessment of Models

In examining the three maturity models related to ECM implementation, at least four different patterns emerge. Table 5 demonstrates that all three models have the same number of levels, have a varied number of dimensions, and only one model has assessment criteria.

Table 5 also shows that all three models have five levels of maturity. However, the models have different naming conventions for each of the levels, as illustrated in Table 6.

Third, as Table 5 has illustrated, there is a progression in complexity of the maturity dimensions from one model to the next. Figure 2 is a graphical illustration of the single dimension for ECMk, the 3 dimensions for CMM, and the 13 dimensions for ECM3.

Fourth, Table 5 reveals that ECM3 has the most sophisticated assessment criteria. The fundamental difference between the ECMkMM and CMM, on the one hand, and the ECM3, on the other, is the degree of detail. ECMkMM has one perspective to the assessment process: using knowledge. The CMM, on the other hand, uses three maturity dimensions. In contrast to both these models, the ECM3 allows an assessment of maturity levels using 13 maturity dimensions. Additionally, the latter provides specific assessment criteria within each of the maturity dimensions at each of the maturity levels.

Table 5. Summary information on ECM maturity models

Maturity Model	Number of Maturity Levels	Number of Maturity Dimensions	Assessment Criteria
ECMkMM	5	1	Simple
CMM	5	3	Average
ECM3	5	13	Sophisticated

Table 6. Naming conventions in ECM maturity models

	ECMkMM	CMM	ECM3
1	Newborn	Individual	Unmanaged
2	Toddler	Team	Incipient
3	Adolescent	Enterprise	Formative
4	Adult	Optimize	Operational
5	Senior citizen	Innovative	Proactive

Figure 2. Progressive sophistication in ECM maturity models

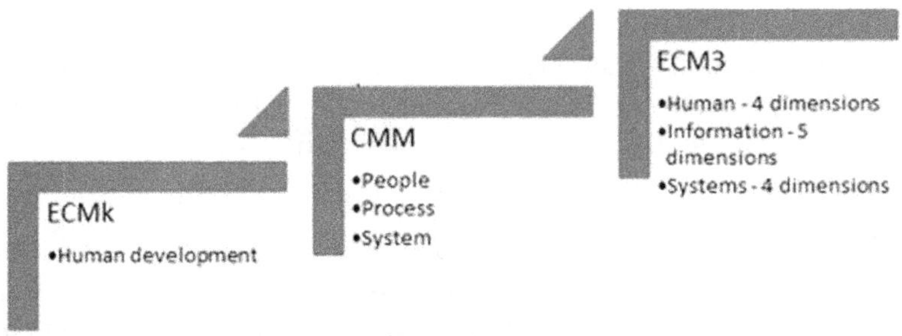

Although the level of detail is commendable, ECM3 does not prescribe any method for computing or presenting the results of the assessment. What it does offer is a graphical representation of possible assessment results for each dimension and level as shown in Figure 3 (Pelz-Sharpe, et al., 2010, p. 14).

Figure 3 shows that the most mature dimensions at "Enterprise ABC" are IT Expertise, Scope, Breadth, and Security, whereas the least mature dimensions are Findability and Usability. Therefore, based on a counting scheme, the most mature dimensions would have a score of three and the least mature dimensions would

Figure 3. Possible assessment results for ECM3 dimensions and levels

← "Enterprise ABC" ECM Maturity Levels

	Level: / Dimension:	1) Unmanaged	2) Incipient	3) Formative	4) Operational	5) Proactive
HUMAN	IT Expertise	No experience managing formal repository and workflow systems	Struggling 1.0 implementations of some systems	More advanced version 2.0+ implementations of systems, with focus on business-critical content	Managing repository and workflow systems is a core IT skill	Pro-active experimentation and learning about emerging content technologies
	Business Expertise	Ignorance about value and role of ECM	Growing sense of awareness about lack of management services	Communication plans include updates to key stakeholders about ECM business value	Executive sponsorship of ECM as a practice; process and content analysis are core skills	Content management designated a core employee skill and part of their HR reviews
	Process	Few or no standardized procedures around content	Basic process analysis leads to some ad-hoc workflows	Initial modeling of inter-departmental processes to prep for automation	Automated processes span systems and departments	Robust exception-handling and experimentation within framework
	Alignment	Key business drivers are not well understood by IT strategists, resulting in ECM gaps in IT portfolio	Gaps still exist between technology and core business processes; IT-metrics not evaluated by business outcomes	IT and Business both understand their information management roles and their respective strategies are no longer developed in a vacuum	Execution of IT & Business strategies become more cohesive, but still follow push-pull model	Strategy development between IT and the Business is done in collaborative and concurrent manner with frequent reviews using proper metrics
INFORMATION	Content/metadata	No formal inventory; no formal classification	Departmental inventories and initial content tagging	Enterprise inventory underway; controlled vocabularies (CVs) initiated	All new repositories and content types registered; global taxonomies created	Pervasive ROT elimination; Folksonomy development; Ongoing metadata reviews
	Depth	No lifecycle management	Most content archived haphazardly; some departmental RM efforts	Development of formal electronic retention, RM, and disposition schemes	Implementation of electronic and paper-based RM across the enterprise	All content types go through formal lifecycles.
	Governance	No policies and procedures	Scattered policies; few or no formal procedures	Development of information governance structure and codification of procedures	Policies and procedures widely disseminated; Enterprise ownership in place	Active review and adaptation; Voice of Customer key to feedback process
	Re-use	Content routinely duplicated	Content still routinely duplicated	Initial content analysis and structuring	Documents repurposed across systems and channels	Content components re-used across systems and channels
	Findability	Employees spend excessive time searching using various internal search engines	Search indexes tuned and basic metadata applied	Rationalization of search technology; analysis of search logs and further tuning, leveraging CV terms	Development of specific enterprise and/or federated search applications	Search and classification become a central service, with business-driven variants
SYSTEMS	Scope	No understanding of core content types	Some basic DM implementations with ad hoc workflow	Identification of core content types, locales; pilot projects for DAM, BPM, etc.	Business-critical information systems prioritized	Broad availability of diverse management systems
	Breadth	No systems	Scattered departmental efforts	Initial attempts to combine or integrate systems across departments	Successful departmental initiatives have been scaled enterprise-wide	Encourage and adopt innovations from departmental levels
	Security	No security regime in place	Dependent on individual systems	Formal projects initiated to address gaps & redundancies due to multiple solutions	Standardized policies and procedures exist and are system enabled	Security is treated as a centralized shared service
	Usability	Lack of systems make end user usability considerations moot	Employee adoption rates measured, but dissatisfaction unanalyzed	Some initiatives use Scenario Analysis and User Persona techniques to guide design	User-centered design underpins all system designs, with formal collection of user feedback	Usability is a guiding principle in all system activity

Measurement / Monitoring and Feedback Processes

have a score of zero. For this sample assessment, the graphical representation in Figure 3 allows for a general set of conclusions to be drawn. When one compares the three general categories, the greatest weakness is in the information category (with only one dimension at Level 2 maturity) and the greatest strength in the systems category (with three dimensions at Level 3 maturity). For Enterprise ABC, the two dimensions requiring the most urgent attention are Findability and Usability because they don't even register Level 1 maturity. This kind of granular assessment, as well as the ability to represent the results graphically through the use of the red line, provides the opportunity for communicating sophisticated assessment results very easily. It also allows the sharing of information with an audience that may not fully understand the complexities, yet provide the gist of the assessment. This is particularly useful when communicating with decision makers, such as an organization's senior management.

In 2010 ECM3 was used by representatives of six South African institutions that had convened at an ad hoc meeting where they discussed their varied experiences in implementing ECM software applications (Katuu, 2016a). These representatives constituted mainly records and information management professionals as well as a few information technology professionals who were all responsible for the introduction and/or administration of ECM systems in their own institutions. The representatives were from two state-owned entities, a charitable institution, and national-level and provincial-level government ministries. In order to maintain their requested anonymity, results of the assessment from the six institutions have been numbered 1 to 6. Figure 4 illustrates the ECM3 maturity dimension's result for each of the six institutions and the average score for each dimension (Katuu, 2016a, p. 226).

As Figure 4 shows, most results fluctuate between two and three levels of maturity, depending on the dimension examined.

CONCLUDING REMARKS

The maturity model concept has been discussed for several decades. Maturity models provide graded levels of capabilities ranging from rudimentary aspects to increasingly sophisticated levels of management and integration, finally resulting in a mature state of continuous experimentation and improvement. Maturity models span different disciplines, including e-government, e-learning, the health sector, project management, and most recently, the ECM domain.

This study relied on a review of literature to address the two research questions that directed the investigation. The research process was informed by a brief background on both ECM applications and maturity models. The research questions resulted in

Figure 4. Detailed results of all South African institutions' ECM3 maturity dimensions as well as an average result

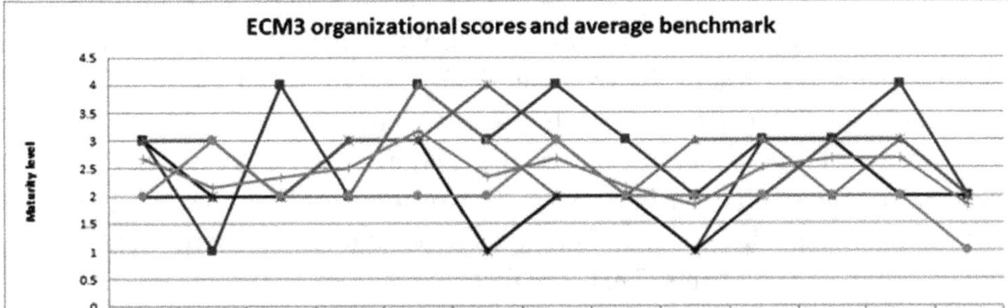

	IT expertise	Business expertise	Process	Alignment	Content/metadata	Depth	Governance	Re-use	Findability	Scope	Breadth	Security	Usability
Org 1	2	2	2	3	3	1	2	2	1	3	3	2	2
Org 2	3	1	4	2	4	3	4	3	2	3	3	4	2
Org 3	3	2	2	2	4	3	2	2	3	3	2	3	2
Org 4	3	2	2	3	3	1	2	2	1	2	3	2	2
Org 5	3	3	2	3	3	4	3	2	2	2	3	3	2
Org 6	2	3	2	2	2	2	3	2	2	2	2	2	1
Average	2.67	2.17	2.33	2.50	3.17	2.33	2.67	2.17	1.83	2.50	2.67	2.67	1.83

a critical analysis of three ECM maturity models: ECMkMM, CMM, and ECM3. The discussions lead to three concluding remarks.

First, the number of maturity levels in all three ECM models are similar even though the naming conventions could be different. Second, the three models differ significantly in the number of maturity dimensions. ECMkMM has one maturity dimension that mirrors the stages of human development. The CMM has three dimensions: people, process, and systems. ECM3 provides the most sophisticated assessment with 13 dimensions that are divided into three categories of maturity: human, information, and systems. Third, only ECM3 has sophisticated assessment criteria that stem from the granular nature of its maturity dimensions.

This chapter described how ECMkMM was developed within an organization for its own purposes and although it offered utility for a given time period, it doesn't have efficacy in the long term. Although CMM is a little more generalizable than ECMkMM, it lacks widespread use. On the other hand, ECM3 has been used within national as well as global contexts over the course of the last decade.

ECM3 is architecturally the most sophisticated of the models, but this does not mean it is perfect. There are two issues of concern with ECM3. Recent studies in the utility of ECM3 show that it does not adequately cover all organizational aspects, and the most glaring gaps relate to organizational culture and change management (Katuu, 2013, p. 4). In addition, regardless of how objectively one wants to conduct an ECM3 assessment, the collection of data relies heavily on self-reported assessments within the organization (Katuu, 2016a, pp. 232-233). In order to increase objectivity, one would have to request documentary or other demonstrable evidence whenever collecting data (Demb, 2008, pp. 135-136). A number of information-related subjects such as the management of classified information and facilitation of access to information legislation may need to be incorporated specifically within ECM maturity models or generally in information management maturity models (Desouza & Vanapalli, 2005; Kastenhofer & Katuu, 2016; Katuu, 2002a; McGraw & Fick, 2011).

In conclusion ECM maturity models have not been used extensively and therefore are still in their formative years (Durga, 2015; Franks, 2016a; Katuu, 2012b). This meta-research study offers insight on how to assess current ECM models, identifies the conceptual gaps that exist, and provides opportunities on improvements to be made in future developments. This may include incorporating standards and good practice guidelines as the basis for assessment criteria in the models (Katuu, 2016b).

ACKNOWLEDGMENT

The author would like to thank Laura Millar, David Wallace, and John McDonald for the very helpful suggestions during the development of this chapter. The author wishes to thank the editor and anonymous reviewers of this book for their constructive comments to improve the chapter. The views expressed herein are those of the author and should not be attributed to his current employer, the International Atomic Energy, or any of his previous employers.

REFERENCES

Adams, P. (2005). *Update to the IT asset management maturity model.* Retrieved 11 February 2017, from https://www.gartner.com/doc/484622/update-it-asset-management-maturity

Ahire, S. L. (1997). Management Science: Total Quality Management interfaces: An integrative framework. *Interfaces*, *27*(6), 91–105. doi:10.1287/inte.27.6.91

AIIM. (2010). *What is Enterprise Content Management?* Retrieved 11 February 2017, from http://www.aiim.org/What-is-ECM-Enterprise-Content-Management.aspx

ARMA International. (2013). *General Accepted Recordkeeping Principles®: Information Governance Maturity Model.* Retrieved 3 March 2016, from http://www.arma.org/docs/bookstore/theprinciplesmaturitymodel.pdf?sfvrsn=2

ARMA International. (2014, February 20). *General Accepted Recordkeeping Principles®*. Retrieved 3 March 2016, from http://www.arma.org/docs/sharepoint-roadshow/the-principles_executive-summaries_final.doc

Arnett, J. J. (2004). *Emerging adulthood: The winding road from the late teens through the twenties*. Oxford, UK: Oxford University Press.

Baskarada, S. (2009). *Information Quality Management Capability Maturity Model*. Wiesbaden, Germany: Vieweg and Teubner. doi:10.1007/978-3-8348-9634-6

Bittman, T. J. (2004). *Gartner introduces the infrastructure maturity model*. Retrieved 26 March 2017, from https://www.gartner.com/doc/461820/gartner-introduces-infrastructure-maturity-model

Brown, A., Katuu, S., Sebina, P., & Seles, A. (2009). Module 4: Preserving electronic records. *Training in Electronic Records Management* Retrieved 26 March 2017, from http://www.irmt.org/documents/educ_training/term%20modules/IRMT%20TERM%20Module%204.pdf

Cameron, S. A. (2011). *Enterprise content management - A business and technical guide*. Swindon, UK: The Chartered Institute for IT.

Curtis, B., Hefley, W. E., & Miller, S. (1995). *Overview of the people capability maturity model*. Retrieved 26 March 2017, from http://www.sei.cmu.edu/reports/95mm001.pdf

Demb, S. R. (2008). A case study of the use of the Records Management Capacity Assessment System (RMCAS) Tool across the London Museum Hub. *Records Management Journal, 18*(2), 130–139. doi:10.1108/09565690810882986

Department of Health [South Africa] & Council for Scientific and Industrial Research [South Africa]. (2014). *National Health Normative Standards Framework for Interoperability in eHealth in South Africa*. Retrieved 26 March 2017, from http://www.samed.org.za/Filemanager/userfiles/hnsf-complete-version.pdf

Desouza, K. C., & Vanapalli, G. K. (2005). Securing knowledge in organizations: Lessons from the defense and intelligence sectors. *International Journal of Information Management, 25*(1), 85–98. doi:10.1016/j.ijinfomgt.2004.10.007

Dollar, C., & Ashley, L. J. (2014). *Assessing digital preservation capability using a maturity model process improvement approach*. Retrieved 3 March 2016, from http://static1.squarespace.com/static/52ebbb45e4b06f07f8bb62bd/t/53559340e4b058b6b2212d98/1398117184845/DPCMM+White+Paper_Revised+April+2014.pdf

Dollar, C. M., & Ashley, L. J. (2015). Ensuring Long-term availability, usability and trustworthiness of digital information. *The Sedona Conference Journal, 16*, 178–199.

Durga, A. (2012). A maturity model-based approach for DAM implementations. *Journal of Digital Media Management, 1*(2), 183–194.

Durga, A. (2015). *State of ECM technology: Customer survey results*. Retrieved 26 March 2017, from https://www.youtube.com/watch?v=obn6qFFqeN4

Durga, A. (2016). *Benchmarking with Excel and paper? There's a better way...* Retrieved 26 March 2017, from https://www.realstorygroup.com/Blog/2990-Benchmarking-with-excel-and-paper-Theres-a-better-way

ECM Maturity Model. (2012). *About*. Retrieved 26 March 2017, from https://web.archive.org/web/20130825194705/http://ecm3.org/about/

Finkelstein, A. (1992). *A software process immaturity model*. Retrieved 26 March 2017, from http://www0.cs.ucl.ac.uk/staff/A.Finkelstein/papers/immaturity.pdf

Franks, P. (2016a). *Enterprise content management and digital preservation online education: A case study*. Paper presented at the European Conference on Knowledge Management, Kidmore End, UK.

Franks, P. (2016b). Integrated ECM solutions: Where records managers, knowledge workers converge. *Information & Management, 50*(4), 18–22.

Gillies, A. (2000). Information support for general practice in the new NHS. *Health Libraries Review, 17*(2), 91–96. doi:10.1046/j.1365-2532.2000.00229.x

Goh, E., Katuu, S., Miller, A., Nordland, L. P., & Sebina, P. (2009). Module 3: Managing the Creation, use and disposal of electronic records. *Training in Electronic Records Management*. Retrieved 26 March 2017, from http://www.irmt.org/documents/educ_training/term%20modules/IRMT%20TERM%20Module%203.pdf

Goodwin, M. (2006). *The enterprise document management maturity model whitepaper*. London: McLaren Software.

Griffin, A. (2004). Records Management Capacity Assessment System (RMCAS). *Archival Science, 4*(1-2), 71–97. doi:10.1007/s10502-005-6991-9

Guédria, W., Naudet, Y., & Chen, D. (2015). Maturity model for enterprise interoperability. *Enterprise Information Systems, 9*(1), 1–28. doi:10.1080/17517575.2013.805246

Harris, K. (2006). *A knowledge management maturity model explains where you're going and how to get there*. Retrieved 26 March 2017, from https://www.gartner.com/doc/489448/knowledge-management-enables-highperformance-workplace

Herbst, A., Simons, A., vom Brocke, J., & Derungs, R. (2014). Critical success factors in enterprise content management: Toward a Framework for readiness assessment. In J. vom Brocke & A. Simons (Eds.), *Enterprise content management in information systems research* (pp. 109–124). Berlin: Springer. doi:10.1007/978-3-642-39715-8_7

Hickson, M. (2008). *Research handbook for health care professionals*. Chichester, UK: Wiley-Blackwell.

Hullavarad, S., O'Hare, R., & Roy, A. K. (2015). Enterprise content management solutions—Roadmap strategy and implementation challenges. *International Journal of Information Management*, *35*(2), 260–265. doi:10.1016/j.ijinfomgt.2014.12.008

Kana, M., & Karni, R. (2004). A capability maturity model for knowledge-based decision-making. *Information, Knowledge, Systems Management*, *4*(4), 225–252.

Kastenhofer, J., & Katuu, S. (2016). Declassification: A clouded environment. *Archives and Records: The Journal of the Archives and Records Association*, *37*(2), 1–27. doi:10.1080/23257962.2016.1194814

Katuu, S. (2002a). Are we information providers or the information police? The uneasy marriage between access and security. In G. E. Gorman (Ed.), *International yearbook of library and information management 2002/2003* (pp. 361–378). London, UK: Facet Publishing.

Katuu, S. (2002b). Sustainable management of electronic records: The challenge for developing countries. *World Forum Proceedings of the International Research Foundation for Development*. Retrieved 26 March 2017, from https://www.researchgate.net/publication/315381902_Sustainable_management_of_electronic_records_-_the_challenge_for_the_public_sector_in_developing_countries

Katuu, S. (2012a). Enterprise content management – An overview of phases, standards and best practices guidelines. *Bilgi Dünyası*, *13*(2), 457–476.

Katuu, S. (2012b). *Enterprise content management and digital curation applications – Maturity model connections*. Paper presented at the Memory of the World in the Digital Age: Digitization and Preservation, Vancouver. Retrieved from http://www.unesco.org/new/fileadmin/MULTIMEDIA/HQ/CI/CI/pdf/mow/VC_Katuu_28_D_1130.pdf

Katuu, S. (2013). The utility of maturity models – The ECM maturity model within a South African context. *iPRES 2013 Workshop: Digital Preservation Capabilities - How to Assess and Improve Capabilities in Digital Preservation*. Retrieved 3 March 2016, from http://benchmarkdigitalpreservation.files.wordpress.com/2013/09/caiw2013katuu.pdf

Katuu, S. (2016a). Assessing the functionality of the enterprise content management maturity model. *Records Management Journal*, *26*(2), 218–238. doi:10.1108/RMJ-08-2015-0030

Katuu, S. (2016b). Managing digital records in a global environment – A review of the landscape of international standards and best practice guidelines. *The Electronic Library*, *34*(5), 869–894. doi:10.1108/EL-04-2015-0064

Katuu, S. (2016c). Transforming South Africa's health sector - The eHealth strategy, the implementation of electronic document and records management systems (EDRMS) and the utility of maturity models. *Journal of Science and Technology Policy Management*, *7*(3), 330–345. doi:10.1108/JSTPM-02-2016-0001

Katuu, S. (2017). Health information systems, eHealth strategy and the management of health records: The quest to transform South Africa's public health sector. In K. H. Moahi, K. J. Bwalya, & P. M. Sebina (Eds.), *Health information systems and the advancement of medical practice in developing countries* (pp. 237–261). Hershey, PA: IGI Global. doi:10.4018/978-1-5225-2262-1.ch014

Kerzner, H. R. (2011). *Using the project management maturity model: Strategic planning for project management*. Hoboken, NJ: Wiley.

Koehler-Kruener, H., Chin, K., & Hobert, K. A. (2015). *Magic quadrant for enterprise content management.* Retrieved 26 March 2017, from http://www.project-consult.de/files/Gartner_MQ_ECM_Enterprise_Content_Management_2015.pdf

Lemieux, V. (2004). *Managing risks for records and information*. Lenexa, KS: ARMA International.

Lipsey, D., & Regli, T. (2012). The DAM maturity model: An interview with David Lipsey and Theresa Regli. *Journal of Digital Media Management*, *1*(2), 202–205.

Liu, R. Y. L. (2002). *Capability maturity model integration: Origins and applications* (Master's thesis). California State University, Long Beach, CA.

Louw, D. A., Van Ede, D. M., & Louw, A. E. (Eds.). (1998). *Human development* (2nd ed.). Cape Town, South Africa: Kagiso Tertiary.

Marshall, S. (2005). Determination of New Zealand tertiary institution e-learning capability: An application of an e-learning maturity model. *Journal of Open, Flexible and Distance Learning*, *9*(1), 58–63.

Martinez, N. (2013). *Localisation workflows: The impact of process well-handledness on automation* (Master's thesis). University of Limerick, Limerick, Ireland.

Maseh, E., & Katuu, S. (2017). The Kenyan judiciary's open government initiative: Prospects and challenges. *Journal of Science and Technology Policy Management*, *8*(1), 76–94. doi:10.1108/JSTPM-04-2016-0007

McGraw, G., & Fick, N. (2011). Separating threat from the hype: What Washington needs to know about cyber security. In K. Lord & T. Sharp (Eds.), American's cyber future: Security and prosperity in the information age volumes I and II (pp. 41-53). Washington, DC: Center for a New American Security.

McRoberts, H. A., & Sloan, B. C. (1998). Financial management capability model. *International Journal on Government Auditing*, *25*(3), 8–11.

Miller, K. (1999). *Simple steps - Developmental Activites for infants, toddlers, and two-year-olds*. Beltsville, MD: Gryphon House, Inc.

Mingay, S. (2002). *Outlining the Gartner BCP maturity model.* Retrieved 26 March 2017, from https://www.gartner.com/doc/367647/outlining-gartner-bcp-maturity-model

Muppidi, H. (2016). *Email archive retention using enterprise content management* (Master's thesis). St Cloud State University, St. Cloud, MN. Retrieved from http://repository.stcloudstate.edu/cgi/viewcontent.cgi?article=1052&context=mme_etds

Murray, A., & Ward, M. (2007). *Improving project performance using the PRINCE2 Maturity Model (P2MM)*. Norwich, UK: The Stationary Office.

Nordheim, S., & Päivärinta, T. (2006). Implementing enterprise content management: From evolution through strategy to contradictions out-of-the-box. *European Journal of Information Systems*, *15*(6), 648–662. doi:10.1057/palgrave.ejis.3000647

Päivärinta, T., & Munkvold, B. E. (2005). Enterprise content management: An integrated perspective on information management. *Proceedings of the 38th Annual Hawaii International Conference on System Sciences (HICSS)*. doi:10.1109/HICSS.2005.244

Pelz-Sharpe, A., Durga, A., Smigiel, D., Hartman, E., Byrne, T., & Gingras, J. (2010). *ECM 3 - ECM Maturity Model*. Retrieved 26 March 2017, from http://ecmmaturity.files.wordpress.com/2009/02/ecm3-v2_0.pdf

Pfeifer, T. (Ed.). (2002). *Total Quality Management: Strategies, methods, techniques*. Munich, Germany: Cal Hanser Verlag. doi:10.3139/9783446224025

Powell, R. R., & Connaway, L. S. (2004). *Basic research methods for librarians*. Westport, CT: Libraries Unlimited.

Real Story Group. (2016). *Enterprise content management and cloud file sharing: Comprehensive product evaluations*. Olney, MD: Real Story Group.

Real Story Group, & DAM Foundation. (2014). *DAM maturity model*. Retrieved 3 March 2016, from http://dammaturitymodel.org/

Rodriguez, T. J. (2009). *Understanding human behavior - A psychology worktex*. Manila, Philippines: Rex Publishing.

Rogers, C., Malmas, S., & Enns, L. (2011). *CS14 - City of Surrey: (Surrey) policies, guidelines and procedures for a drive migration project as part of an enterprise content management program*. Retrieved 26 March 2017, from http://interpares.org/ip3/ip3_cs14_report.cfm

Rowe, W.-A., & Miller, C. (2011). *My skills, my money, my brighter future in Rwanda - An assessment of economic strengthening interventions for adolescent girls*. Baltimore, MD: Catholic Relief Services.

Salamntu, L. T. P., & Seymour, L. F. (2014). *A review of organisational benefits through the use of enterprise content management (ECM) System in public sector organisations*. Paper presented at the Third International Conference on Informatics Engineering and Information Science (ICIEIS), Lodz, Poland.

Saleh, Y., & Alshawi, M. (2005). An alternative model for measuring the succes of IS projects: The GPIS model. *Journal of Enterprise Information Management, 18*(1), 47–63. doi:10.1108/17410390510571484

Santos, Â. S. S., & Pinto, M. M. (2015). *O repositório digital seguro: Diagnóstico e especificação*. Paper presented at the 12 Congreso Nacional BAD, Évora, Portugal. Retrieved from http://bad.pt/publicacoes/index.php/congressosbad/article/viewFile/1329/pdf_100

Shegda, K. M., Chin, K., Logan, D., & Lundy, J. (2004). *Building the magic quadrant for ECM 2004*. Retrieved 26 March 2017, from https://www.gartner.com/doc/461108/building-magic-quadrant-ecm-

Simon, A., Schoeman, P., & Sohal, A. S. (2010). Prioritised best practices in a ratified consulting services maturity model for ERP consulting. *Journal of Enterprise Information Management, 23*(1), 100–124. doi:10.1108/17410391011008923

Smits, M., & O'Callaghan, R. (2014). Strategy development for enterprise content management. In J. vom Brocke & A. Simons (Eds.), *Enterprise content management in information systems research* (pp. 91–107). Berlin: Springer. doi:10.1007/978-3-642-39715-8_6

Sukhoo, A., Barnard, A., Eloff, M. M., & Van der Poll, J. A. (2007). An evolutionary software project management maturity model for Mauritius. *Interdisciplinary Journal of Information, Knowledge, and Management, 2*, 99–118. doi:10.28945/103

Usman, M., Muzaffar, A. W., & Rauf, A. (2009). *Enterprise content management (ECM): Needs, challenges and recommendations.* Paper presented at the 2nd IEEE International Conference on Computer Science and Information Technology (ICCSIT), Beijing, China. doi:10.1109/ICCSIT.2009.5234541

Valdes, G., Solar, M., Astudillo, H., Iribarren, M., Concha, G., & Visconti, M. (2011). Conception, development and implementation of an e-government maturity model in public agencies. *Government Information Quarterly, 28*(2), 176–187. doi:10.1016/j.giq.2010.04.007

Van Mien, A. D. (2001). *The Gartner security process maturity model.* Retrieved 26 March 2017, from https://www.gartner.com/doc/326006/gartner-security-process-maturity-model

Vital Wave Consulting. (2009). *Health information systems in developing countries: A landscape analysis.* Retrieved 26 March 2017, from http://vitalwave.com/wp-content/uploads/2015/09/Gates-Foundation-HIS-Analysis-2009.pdf

Vom Brocke, J., Herbst, A., & Urbach, N. (2014). *How to measure ECM success.* Retrieved 26 March 2017, from http://www.bptrends.com/bpt/wp-content/uploads/06-03-2014-COL-ClassNotes-How-to-Measure-ECM-Success-vom-Broke-et-al-final-1.pdf

Vom Brocke, J., Simons, A., & Cleven, A. (2011). Towards a business process-oriented approach to enterprise content management: The ECM-blueprinting framework. *Information Systems and e-Business Management, 9*(4), 475–496. doi:10.1007/s10257-009-0124-6

Walker, A., McBride, T., Basson, G., & Oakley, R. (2012). ISO/IEC 15504 measurement applied to COBIT process maturity. *Benchmarking: An International Journal, 19*(2), 159–176. doi:10.1108/14635771211224518

Wendler, R. (2012). The maturity of maturity model research: A systematic mapping study. *Information and Software Technology, 54*(12), 1317–1339. doi:10.1016/j.infsof.2012.07.007

Wicaksono, S., Shihab, M. R., & Sandhyaduhita, P. (2015). *Formulating implementation strategy for enterprise content management system using soft system methodology: A case of a marine logistics company in Indonesia.* Paper presented at the 2015 International Conference on Advanced Computer Science and Information Systems (ICACSIS), Depok, Indonesia. doi:10.1109/ICACSIS.2015.7415178

ADDITIONAL READING

Bititci, U. S., Garengo, P., Ates, A., & Nudurupati, S. S. (2015). Value of maturity models in performance measurement. *International Journal of Production Research*, *53*(10), 3062–3085. doi:10.1080/00207543.2014.970709

Donovan, P. O., Sullivan, D. T. J. O., Donovan, P. O., Bruton, K., & Sullivan, D. T. J. O. (2016). IAMM: A Maturity Model for Measuring Industrial Analytics Capabilities in Large-scale Manufacturing Facilities. *Internation Journal of Prognostics and Health Management*, *7*(32), 1–11.

Fath-Allah, A., Cheikhi, L., Al-Qutaish, R. E., & Idri, A. (2014). E-government maturity models: A comparative study. *International Journal of Software Engineering and Its Applications*, *5*(3), 71.

KEY TERMS AND DEFINITIONS

Business Process Management: An integrated system for managing business performance by managing end-to-end business process.

Chapter 6
Appraisal of Transactional Data Through Visualisation for SMEs

Wajid Khan
Bournemouth University, UK

Siti Aishah Mohd Selamat
Bournemouth University, UK

Manoharan Ramachandran
Bournemouth University, UK

ABSTRACT

E-commerce has proven to play a pivotal role in the economy growth. One of the key e-commerce functions is the collection of the vast amount of useful transactional data to help businesses in understanding their consumers' behaviour. With the rapid and large volume of data collected, it is posing a great challenge for businesses to analyse the data on a day-to-day basis. The key issue is not in the generation or collection of data; it is in the manipulation of the collected data to churn out new and insightful information. Information visualisation is an effective tool in converting data into interactive interfaces to unearth hidden trends. It provides a platform to explore the data in a more rapid and intuitive approach. There are several existing techniques to analyse multidimensional data. This chapter seeks to introduce a comprehensive and robust visualisation model and framework for adoption. The visualisation model consists of four major layers, which include acquisition and data analysis, data representation, user and computer interaction, and result storage.

DOI: 10.4018/978-1-5225-3628-4.ch006

INTRODUCTION

The cloud based systems are getting popularity amongst individual, institutes and small and medium size businesses across the globe - exploiting the Web 2.0 true potential. Web 2.0 where the focus is on the content which is mainly produced or created by its users (Graham & Caverlee, 2008).

Similarly, data mashup systems are getting the same reception from various small and medium size businesses on the World Wide Web. The industry leading companies such Google, Yahoo, Microsoft are providing free cloud based services, such emails, mapping tools and cloud based storage (Yu, Benatallah, Casati, & Daniel, 2008). A mashup application can be characterised as a lightweight (simpler and faster) and a tactical (competitive activities) presentation layer that uses the web platform – Web-Oriented Architecture (WOA) (Ketter, Banjanin, Guikers, & Kayser, 2009) - in order to integrate multiple sourced applications into one web-based offering.

Enterprise 2.0 contributes greatly to generate, coordinate, collaborate and communicate content, applications, and data in business and enterprise environment (McAfee, 2009). Enterprises such as small and medium size businesses require tools which could address its core values (success and failure prediction), but also assist in decision making and providing extensive research based solutions from the existing data streams which will lead to effective resource utilisation and not much has been done to answer these burning questions.

However, information visualisation and data analysis is a field which holds the key value in taking any business to the next level (expected and unexpected outcomes readiness). A business could do well and produce some quality figures in sales, but after all, it's about the techniques, how sustainable the practice is, and what further could be learned.

Information visualisation is converting data into interactive interfaces in order to easily understand problems, hidden patterns, scope and explore data with a more rapid and intuitive approach - usually abstract data are transformed into visual images (Tufte, 1997) to see the big picture at a glance. In addressing data problems, a number of techniques, (Pillat & Freitas, 2006), (Eick, 2000), (Stolte, Tang, & Hanrahan, 2002) have been invented to explore data for trends, patterns and to analyse multidimensional data. The famous among these techniques are pixel-oriented visualisation (Keim, Hao, Ladisch, Hsu, & Dayal, 2001), prisma (Godinho et al., 2007), pixel bar charts (Keim, Hao, Dayal, & Hsu, 2002) snap-together (North & Shneiderman, 2000) - but the stand out theoretical model which was introduced by Fry (Fry, 2004) is a complete model for analysing data, however the procedure is very complex and almost impossible without help from specialists on a regular basis to process and analyse data to generate business reports.

Information visualisation is a relatively new practice and comprises of creating visual artifacts aimed at amplifying understanding and focusing on the development and empirical analysis in order to present information in an easy to understand, visual form. By presenting information in a visual format decision makers are able to forge a strong connection between the information and the various outcomes of their decisions. This method is perfect for handling multi-faceted and complex data and business transactions that would otherwise be difficult to understand. Research into multiple coordinated visualisation methods such as pixel charts, bar charts and value cell visualisation have contributed vastly to the understanding of in depth data through various aspects that can be seen today.

As with every business, business analytical tools have a specific focus, and in this case, the focus is held on the data used in the process. Simple visualisation models Figure 1 are the most basic visualisations of data that are often included in other software packages, and these serve the very basic data needs of a business. The more complex visualisation models are mostly the result of Ph.D. research and higher and have been carefully designed to prove the in depth analysis the businesses crave, but in reality are very difficult to practice in the modern real time enterprise environment.

The strides forward in this field have been substantial, there is still a lot of work to be done. It is for this reason that this research focuses on an easy to use and integrate

Figure 1. Compound Visualisation Model

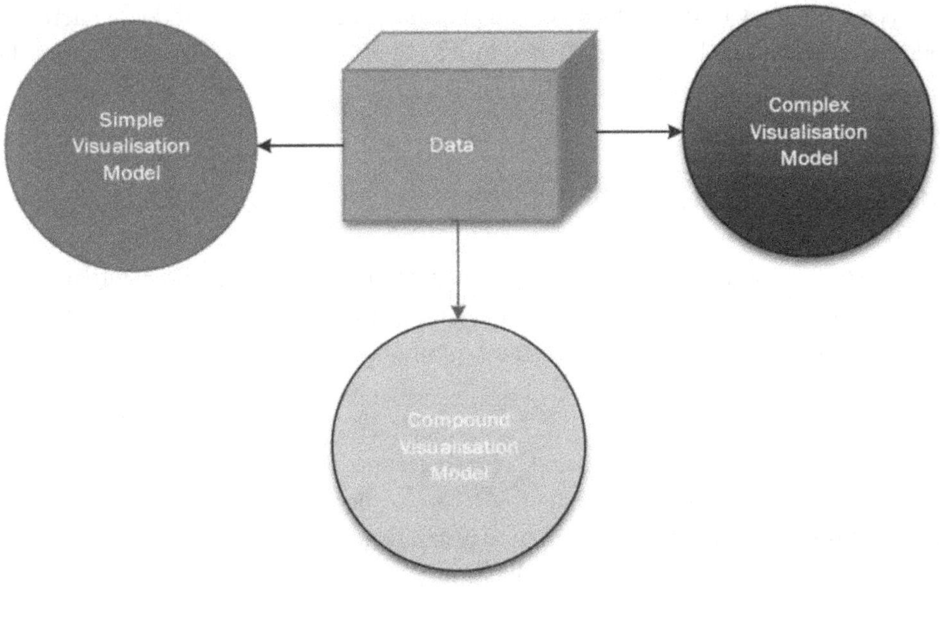

visualisation model, which focuses on a non-aggregated, multi-attribute, multiple dimensional and multiple coordinated visualisation – which is no small task. This fresh and innovative approach in the field of information visualisation, and one that will make it possible for key decision makers to not only identify the important information within the data, but also allow them to uncover unexpected new data patterns already within the data, but inaccessible without this new visualisation model.

Over the past decade, the world has been the epicentre of an extraordinary information explosion - one that has been fuelled by the increased use of the internet coupled with the increased numbers of connected devices worldwide. This rate of data growth has accelerated much faster than at any other point throughout history. It is getting difficult day by day to analyse business and consumer behaviour with such huge volume of data generated by enterprises. There is a greater need for analytical tools to help decision makers to understand data properly - there is no better option than information visualisation to understand and explore large data. The understanding of data will lead to the identification of trends, effective resource utilisation, educated decision making and understanding business and its core values (success and failure anticipation).

PROBLEM STATEMENT

Small and medium size businesses (SMEs) have the resources and ability to store and create a huge amount of transactional data, data and the purpose for which it was collected could be quite easily be disassociated. In such cases, patterns and trends are not easy to be explored because of the nature and size of data. Therefore there is a greater need for an analytical tool to identify the expected and predict the unexpected from big data sources.

AIMS AND OBJECTIVES

The aim of this research was to develop a business analytical tool, which will analyse highly complex data through information visualisation. The research objectives were to represent data with simple graphs and complex business scenarios are achieved through:

- Multiple Attribute Information Visualisation
- Multiple Coordinated Information Visualisation
- Transactional Tagging Information Visualisation
- Linked Data Information Visualisation

RELATED WORK

Web 2.0

The concept of web 2.0 is one that is relatively new on the scene, and adds to the new generation of web applications for business use, and was introduced for the first time during a conference between O'Reilly and Media Live (Tim, 2005). The whole concept of web 2.0 has become a benchmark practice in the industry as the most dynamic and interactive form of web application, allowing users to access content from the website, contribute to it at the same time and even keep up with the new content on the website without having to visit the site constantly. Not only that, but it also allows developers to create exciting new web applications quickly and easily by utilizing the data available on the Internet as a resource.

The concept of web 2.0 revolves primarily around the re-use of data using the open standards based technology within a scalable, flexible and open source environment. Web 2.0 addresses many of the current issues in the data accumulation and analysis space, including information overload and the singular dimensional problems you can sometimes get when multifaceted data is displayed in a single pattern. By using web 2.0, businesses are able to take advantage of the modern dimensions and techniques of data visualisation, ensure a user-friendly experience without sacrificing commercial accessibility and reusability of data. In order to be able to deliver this easy to use environment, Web 2.0 technologies have been positioned on uncomplicated programming platforms, dramatically speeding up the time to market of new, inventive web applications with new uses and approaches and constantly increasing the variety of new enterprises coming to market.

Data Mash-Up

The term 'mash-up' has been around for many years and usually refers to the musical editing of vocals or instruments from different songs combined into a single track. The term is now being used more and more in the world of data to represent the re-use of various data sourced and web applications together to serve a single function (Yu et al., 2008). The idea behind this is simple – there are thousands of applications on the web that all specialise in different things, and by bringing a collection of them together and combining them for the user to be able to create a programme that meets their exact requirements.

The main characteristics of a successful mash-up application are the use of a lightweight and tactical presentation layer that utilizes a web platform while using Web-Oriented Architecture (WOA) in order to allow a multi-sourced application into a single web based offering. The basic architecture of a mash-up application can be seen below in Figure 2.

The data used in any mash-up application is usually found and retrieved from several different sources, usually from local databases or datasets or larger external data repositories, however data can also be harvested from websites through crawling bots, or even through a more service oriented approach (SOA0 through the use of API's and content brokers.

Mash-Up Architecture

In modern times mash-up systems are fairly common, and can be defined by architecture and placed into two category types:

Browser Based Mash-Up

A browser based mash-up application is usually designed to compile information on the browser side of the process, collecting data and resources to be immediately released and transferred straight to the client (Ngu, Carlson, Sheng, & Paik, 2010). One of the main advantages of this kind of architecture is its lightweight nature, allowing it to function on reduced resources and maintenance that is cost effective and productive. However, it is not without its drawbacks, in this case, the fact that all data is compiled on the browser side means that the user will run into problems when it comes to re-using data or other elements, as it is very difficult to re-use processed data for any other purpose. Not only that, but browser based mash-up

Figure 2. Basic Mash-up Architecture

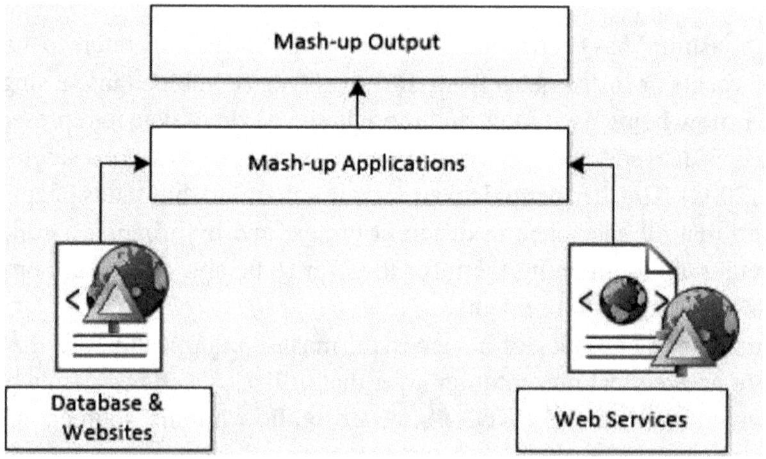

applications are not without their security risks, - as most of the processing takes place on the client side it is considered the less secure type of mash-up development.

Server Based Mash-Up

On the other side of the coin is server based mash-up application development. This is the most popular form of mash-up application architecture as it has the most benefits and advantages for the least security risk. This type of mash-up application compiles, formats and processes the data found from various sources in a single, remote location and only transmits the finished data from the mash-up application to the client. By using this method you are ensuring you can re-use the data collected for multiple purposes while ensuring high security and limited unauthorized access to the architecture.

Mash-Up Types

While the architecture of a mash-up application can only be categorised into 2 types, the application itself can fit into several categories. The determining factors for these categories can vary from the purpose of the application to the method of development, but the beating heart of all these types, and of any mash-up application, is data.

Data Mash-Up

The purpose of a data mash-up application is to combine similar or relevant types of data into a singular source to represent it in a visual format. Every data mash-up application that focuses on 'similar data types' as its key data input falls into this category. An example of this could be a news mash-up application – these applications collect new stories from various different news sites and papers, aggregate them and present them in a singular package.

Consumer Mash-Up

A consumer mash-up application is one that combines different data types from different data sources in order to present it in an integrated way that is easy to view and understand. There are several types of application that have been developed to serve the consumer market – the most well-known of which is 'housingmaps.com', which combines Google maps and Craigslist housing data into one website and one integrated view, allowing you to see housing listings and their locations on a map in one viewer.

Enterprise Mash-Up

Enterprise mash-up applications differ from standard and consumer data mash-up applications. This is primarily because enterprise mash-ups not only collect resources and data from external web services but also combine them with their own resources to establish a strong, collaborative relationship between businesses and developers whilst they work with the application.

Web Mining

Web mining is a technical term relating to the application of various data mining techniques to discover patterns and trends from the internet, and from a variety of enterprise datasets that are available to mash-up applications. There are typically three different forms of data mining that cover different aspects of the internet - these are web usage mining, web content mining and web structure mining. While there are many different definitions of web mining, the definition found in the Gartner Group is the most comprehensive (Mulvenna, Anand, & Büchner, 2000). They define data mining as 'the process of discovering meaningful new correlations, patterns, and trends by sifting through large amounts of data stored in repositories and by using pattern.'

Data mining has many different applications, most of which are based on the industry they are being utilized for. Within higher education, data mining is the process of uncovering hidden trends and patterns (that leads them towards predictive modelling) and using a combination of explicit knowledge bases, sophisticated analytical skills, and academic domain knowledge. This is also known as the process of producing new observations from pre-existing observations. This has also been described as 'the process of automatically extracting useful information and relationships from immense quantities of data'. However, if you were to take the purist approach, data mining does not involve looking for specific data - it simply finds patterns that are already present (Cooley, Mobasher, & Srivastava, 1997). This means that data mining cannot be used to unearth information in response to a question or hypothesis.

The continued increase in growth of online information combined with the unstructured nature of web data requires the development of powerful and efficient web data mining tools. Instead, web mining tools are designed to dig out useful facts and knowledge from web pages, hyperlinks, page content and usage logs. It can be used in a number of ways to enable the streamlining of the business process. For example, the web mining of an e-bank service can enable the employees of a bank to support e-business, helping them to understand various marketing dynamics, new promotions or suggestions on the internet. Because of this there is now more of a tendency amongst banking companies and individuals to collect banks of information through web data mining and to use that information for their own interests to gain business intelligence, and this helps them make enlightened business decisions.

Web mining didn't come about overnight. Web mining techniques are the result of an incredibly long process combining years of research with product development. They saw their beginnings when business data was first being stored on computers and on the internet, and their evolution continued with drastic improvements in data access, and now the developments in real-time technologies that allow internet users to navigate through data quickly. Data that is gathered from surveys, manual input or from independently networked locations to define the data collection. Semantic Web has been developed to address current web problems, and it does this by methodically structuring the content of the internet and adding semantics before extracting the maximum benefits from the power of the internet. Sir Tim Berners-Lee defined the semantic web as 'an extension of the current web in which information is given well defined meaning, better enabling computers and people to work in co-operation.' It is, in a word, a vision. The idea of having all of the data on the web systematically defined and linked together in a way that can be used and understood by machines. And not just for display purposes, but for automation, integration and the reuse of data across a whole liturgy of applications. Web mining also plays a pivotal role in achieving this, as it lets users quickly and easily find the information they need without digging for it themselves.

By definition, web mining is the discovery and analysis of useful information from the internet - and is mainly used in bulk to access this data from a large number of websites. It can also be regarded as enabling the continued use of the internet - from which can draw automatically, standardizing and explaining data. But web mining doesn't come in just one form - there are various types, and these can be used to discover and enable automatic knowledge mining of user access patterns from various different web servers. This is known as web usage mining - and is subtly different from web data mining, in the sense that it is used to extract meaningful user patterns from web server access logs using data mining techniques.

Web based documents are split into three categories based on their structure: un-structured, semi-structured and fully-structured. There have been studies into these categories, specifically on un-structured and fully-structured data, to look into their methods of extracting semantics for ontology learning. In the research that focuses on fully-structured web documents benefit from a standardized syntax such as XML. However most of the web documents in existence today are in a semi-structured format, but only a few references are made to research that specializes in this format in extracting semantics for ontology learning. In most cases, the plain text has been extracted from the semi-structured pages in pre-processing, therefore neglecting embedded information within the semi-structured format. Some other researchers focus on extracting semantics from more template is driven web pages, and so these methodologies are limited both in their usage and their applicability.

Web Usage Mining

The term 'web usage mining' was originally coined by Cooley in 1997, and is the application of various techniques used in data mining to discover and analyse the usage patterns of web data. It analyses user interactions with web servers (including web logs or other database transactions from a group of related sites). However, it has sparked privacy concerns and is currently the topic of much passionate debate. This is because the process includes mining data from the web server access logs, proxy server logs, browser logs, user profiles, registration data, user sessions or transactions, cookies, user queries, bookmark data, mouse clicks and scrolls and any other data as a result of interactions. Its aim is to discover the general patterns in web access logs, but in order to discover these patterns it is necessary to perform a few processes, including pre-processing, pattern discovery and pattern analysis. Web servers automatically record and store data about all of their user interactions whenever requests for resources are received. By analysing these web server logs, all kinds of different websites can better understand user behaviour and web structure, which in turn improves the overall design of this immeasurable collection of resources (Srivastava, Cooley, Deshpande, & Tan, 2000). By mining web usage pattern data, we can help in the progression of internet based studies- including those based on how internet browsers are used, and how the user's interaction with the browser interface changed. These patterns can be put to use in gathering business intelligence, which will, in turn improve customer attraction, retention and other aspects of customer behaviour.

Web Scraping

Web scraping is defined as the technique of automatic web data extraction, used in order to extract data from the core HTML of the website by parsing (the scouring and analysis) the web pages. It does this by using programs specially coded for manipulation of web pages. Examples of this are the converting the web page to another format such as XML, or by embedding the browsers. It can be seen as fairly close to web indexing (this indexes web content using a soft-bot mainly adopted mainly by search engines) but instead, web scraping focuses on the transformation of completely unstructured web content into formally structured data, which can be stored and analysed in a central database (Vargiu & Urru, 2012). It can be used for a multitude of things, including online price comparison, weather data monitoring, web research, web content mash-up and web data integration. Not only that, but it can provide various levels of scrape like human copy and paste, text gripping (based on the UNIX command or regular expression matching facilities of programming languages like Pearl), HTTP programming (HTTP requests to the remote web server), embedding web browsers, HTML parsers, and web scraping software tools.

It works by allowing prolog (a language used in artificial intelligence, which has the capability to interact with web services and clients) to keep the data and extract the required information (with the help of Prolog Server Pages and some inference rules). PSP accepts the arguments generated by HTML and generates the responses from the web server so that it can interact with Prolog, therefore generating an output which is passed to the HTML. To complete the process of web scraping what you need is a web server, prolog compiler and a web browser.

Web Structure Mining

Web structure mining is a field of research entirely focused on identifying the preferable documents by using the analysis of the web link structure. The idea is simple: a hyperlink from document A to document B implies that the author of document A thinks that document B contains worthwhile and useful information. Web structure mining seeks out and exploits the additional information that can be found within the structure of the hypertext. Because of this one of the most important areas of application in this area is the identification of the relevance of these linked pages - which appear equally important when analysed, especially when you look at their content in isolation (Kosala & Blockeel, 2000). To illustrate: a hyperlink induced topic search analyses the hyperlink's topology by discovering authoritative information sources for broader search topics. To find this information, it goes to authority pages, which are defined in relation to hubs as their counterparts. Hubs are otherwise known as pages that link to multiple related information authorities. For example, Google owes its massive success to its page ranking algorithm, which states that the relevance of any page increases with the number of outbound links (hyperlinks to it from other pages) - more so if those other pages are relevant.

Web Content Mining

Web content mining is an automatic process of the internet which extracts patterns from web contents, data and documents (such as HTML files, images or emails). This already goes beyond simple keyword or key phrase extraction – instead, it can take full advantage of the semi-structured nature of the web page text and can use it to detect co-occurrences of terms within the texts (Mobasher, Dai, Luo, Sun, & Zhu, 2000). For example, we can also discover trends over time, which could indicate a surge or decline of interest in certain topics (such as the programming language

Java). Yet another area this can be applied in is event detection - using web content mining in the identification of stories in multiple continuous news streams that correspond to new or previously unidentified events.

Enterprise 2.0

The concept of Enterprise 2.0 was first introduced over 5 years ago by McAfee as a way of showcasing how effective and useful web 2,0 technologies could be within the business landscape. In particular, he wished to bring to light the way work practices and knowledge efforts could be made more visible to decision makers within business (Tufte, 1997). When it was first introduced the concept was fascinating to a great many people, but the concept didn't take off with enterprises at that time. Instead, enterprises demanded to see real evidence and case studies of enterprise 2.0 and web 2.0 technologies within the workplace before they would commit to using it. These case studies needed to be analysed from multiple angles and viewpoints including security, consistency and scalability capability. Not only that, but they also required best practice policies for its usage and specific guidelines on the introduction and support of web 2.0 applications in their environment.

But how was the new technology of Enterprise 2.0 studied and discussed in other areas? While there haven't been many in depth qualitative analysis on Enterprise 2.0 within the work context and its impact, there have been multiple studies on the direction, detail, quality and format of social networking systems (particularly Facebook and Twitter). There are even qualitative studies focused on their application and real use within the business. However in recent years, some managerial journals have published new results of the applied Enterprise 2.0 technology, and some books have even been published with a specific emphasis on providing explanatory guidelines for management in the enterprise. While these sources were a considerable jump forward, they still did not exhaustively and qualitatively describe how and in which areas this new technology is more efficient for enterprises - or the additional value it can have to cooperative work. This is something that the research literature unfortunately still lacks - an in depth analysis and explanation of cooperation processes in very complex development environments. This is the sort of thing that needs investigating, by basing the analysis on central concepts such as awareness, trust, openness, sharing through data mash-up tools and information visualization.

Information Visualisation

Information visualisation is a widely recognised sub section of science and mathematics, processing statistics, data and graphics into a visual form, and has been officially recognised as its own entity within the fields since the mid to late

80's. While the depth of the work is consistent with a fairly young field of study, it is compensated by a background formed by years of research, statistics, and graphical design elements. In simple terms – when you visualise information you are converting complex data and statistics into a visual interactive interface, allowing it to be easily understood at a glance. This process also reveals hidden patterns and codes within the data that might not have been visible before, increasing the productivity and usefulness of the data.

A successful and effective data visualisation method is defined according to the following user tasks (Carr, 1999):

- To analyse or examine entire datasets at once.
- The ability to see non-aggregated elements or subsets within the data.
- Customisations and filtering in order to analyse data from various different angles.
- Additional information available to each action performed by the user.
- The ability to relate data items to other data items with similar characteristics.
- The ability to undo an unwanted action and visually show the steps performed up until that current point.

In order to address some of the underlying problems of effective data visualization a number of different techniques have been developed to explore data and mine for new trends and patterns. For example, Tableua's visual spread sheet and SpotFire's visual data exploration interfaces utilize qualified tree maps and screen filling curves, 2 techniques that are now utilised by many business managers for data analysis. There are generally 3 methods of visualization that have proven to be effective and popular among business owners. These are:

- **Pixel Oriented Visualization**: This particular method of visualizing data is gaining in popularity due to the fact that large amounts of data can be displayed in a single pixel (Keim, 2000). Pixel based techniques have also been used to build fully interactive decision trees for decision makers, allowing them to see the predicted outcomes of multiple different decisions. Pixel visualizations succeed where other approaches fail due to their diverse nature, allowing them to use multiple views instead of restricted outputs.
- **Pixel Bar Charts**: Pixel bar charts, along with value cell bar charts are most successful when used to visualize non-aggregated, multiple viewpoint data, but when it comes to multiple coordinated views it leaves a lot to be desired (Keim et al., 2001).
- **Snap Together**: This approach encompasses a general-purpose system which supports the use of multiple data set types, which means coordination

is incredibly limited. DEVise is a common tool within snap together, and represents multi-dimensional datasets by separating them into several bi-dimensional visual representations (North & Shneiderman, 2000). Users are then able to zoom or focus in on data and analyses complex data at a lower, deeper level.

- **Multi-Coordinated Visualization:** This form of visualization provides the user with a clear understanding of the data, and helps users make massive strides with data validation and cross examination to find hidden patterns and links within the datasets (Pillat & Freitas, 2006). However, in order to see the data source within data, even multi-coordinated practices fall short of the mark. To combat this, we suggest link data visualization allowing a link to be forged between various data aspects.

RESEARCH METHODOLOGY – THE DATA VISUALISATION APPROACH

The proposed information visualization model is illustrated and presented through design flow charts. In this section, it will cover a brief introduction to the existing information visualization model, together with the proposed model. The proposed model will be further discussed by covering the individual steps entails of the model – which consist of, the acquisition and data analysis, represent++, interact++ and history step.

The process of understanding the complex world of data all starts with setting goals and answering questions about the data. This process can be broken down into the following steps:

1. **Acquire:** The process of obtaining data from local and remote sources through either web based services of other, more direct forms of dataset access.
2. **Parse:** Parsing organizes the data collected and gives it structure within your personal environment.
3. **Filter:** Passing the data through pre-set filters to remove any data that is not of use for that specific analysis.
4. **Mine:** Once the data has been filtered the user can then apply various data analysis tools to start discerning patterns in the data.
5. **Represent:** The selection and determination of simple representation mediums, discerning what shape or form the finished visualization will take.
6. **Refine:** Refinement makes improvements to the basic representation, making it clearer and easier to engage with visually.

7. **Interact:** The final stage of visualization, the addition of various methods for manipulating the data once it has been visualized, making the controlling features visible to the user.

These seven stages as shown in Figure 3 highlight a comprehensive picture of data visualisation, and with the addition of several subtle sub steps a user is able to further refine their visualisation experience and link various sets of data together in new and innovative ways.

Proposed Methodology

The concept of this new research all revolves around a single concept – information visualisation through mash-up technologies. By using these technologies users are able to understand vast quantities of complex transactional data and use the information generated to make informed and well researched decisions for the future of their business, while keeping the application lightweight.

The proposed methods of data visualisation as depicted in Figure 4 will follow the same basic steps as outlined above, but with a few minor alterations, the process can be dramatically improved. Within the enterprise 2.0 environment, the reports in comparison function play a key role in the process. In order to accommodate this extra function, the new model will add 'History' to the seven basic steps. This history layer will be stored in XML, which can then be shared with internal and external applications as required.

Figure 3. Seven Stages of Information Visualisation

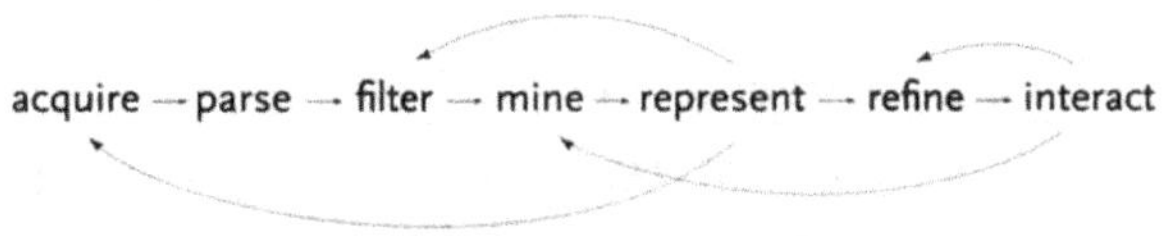

Figure 4. Proposed 4 Stage/Layer Visualisation Model

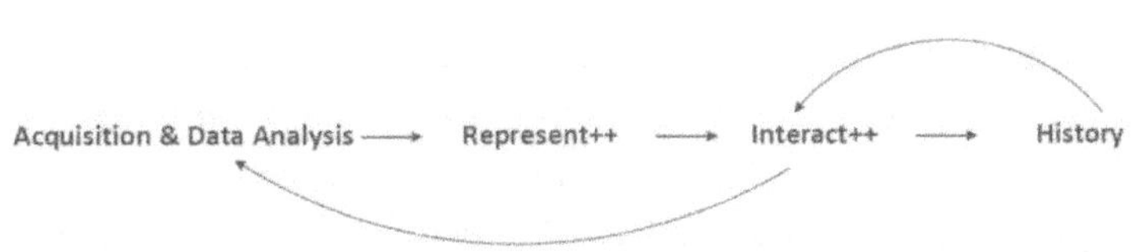

Acquisition and Data Analysis

The structure of data to be analysed is the main focal point for any visualisation practice and this is the section that needs the most attention. In the proposed system, computers will establish whether the request for information visualisation was received from an enterprise 2.0 or non-enterprise environment, and decide how to proceed from there.

If the request has come from an enterprise environment then the proposed framework will already be in place, and the systems will already have direct access to datasets that are in the correct file formats for visualisation. By processing the data through a known environment the first 4 steps of the process (acquire, parse, filter and mine) can all be achieved through the use of API's and web based services., and this is a key feature of working within a known environment.

If the data request has come from a non-enterprise environment then the process becomes slightly more complicated. This means the datasets are not easily accessible through the use of API's and other web based tools, but will instead need assistance from third party tools such as Weka. Weka is a collection of machines learning algorithms for data mining tasks, and helps the system with the pre-processing, classifications, regressions, clustering, association rules and even some of the visualisation elements. A similar open source tool is RapidMiner, which can be accessed via any data mash-up application. By using these open source tools the framework can acquire, parse, filter and mine datasets from any unfamiliar sources before being moved on to visualisation.

Represent++

In the proposed framework, represent layer of the visualisation process will remain unchanged, as this is a key step in the effective understanding of complex transactional data. This layer can be extended into further elements, deepening the users understanding of the data. By using the innovative approach of multi-coordinated visualisation users are also able to update data visualisations automatically after applying changes, allowing the data to be viewed from different angles and perspectives.

Another important feature of the represent later in web 2.0 is tagging, which allows the user to assign keywords to particular elements in a non-hierarchical fashion. In advanced versions users are able to introduce transactional tagging, which will help decision makers to gain real time insights into their business transactions. Tags such as units sold, time, space and monetary value can be placed on any chosen attribute, either by the system itself of manually by the end user. The visualisation will then use the tags to identify the different data elements on the chart or graph, and this will allow further cross linking between sectors and data.

Interact++

In order to allow the user full interactive reign within the data visualisation customisable widgets have been introduced to enhance the interactive capability of the original model. To make this work a separate layer of user friendly mash-up interfaces have been designed, each of which is highly customisable to allow freedom and flexibility to the user. This allows different users to access different data visualisations depending on their needs – for example, a business manager might need to see a complete company overview, whilst a staff member would only be interested in seeing their own teams' performance.

History

Finally, the addition of the history layer is a new contribution to the theoretical visualisation model shown in Figure 4, where sensitive information will be stored in XML where it can be re-used for analysis and comparison whenever needed.

FRAMEWORK DESIGN

Acquisition and data analysis layer process and prepare raw data or refined data for the represent++ layer. It has been named represent++ because there are add-on features as compare to the original model, all these features are explained in the previous chapter, however the improvements are based on providing variety of graphical representation along with multiple coordinated visualisation, user friendly interfaces and interaction at the interact++ layer while history section stores and record generated reports of reference and comparison purposes.

Acquisition and data analysis is the first step where data is exposed to the system for parsing, filtering and mining, thus removing unwanted data and transforming data into a structural design; in other words converting data into information. The information is then passed onto the represent++ layer where system visualise information in various formats and forms requested by the user from the interact++ layer as demonstrated in Figure 5.

History layer is responsible for recording the generated graphs for re-use and com- parison purposes in future. The system always flows from left to right as shown in above figure, however, if a user request, to re-visualise data then the system sends

Figure 5. System Flow

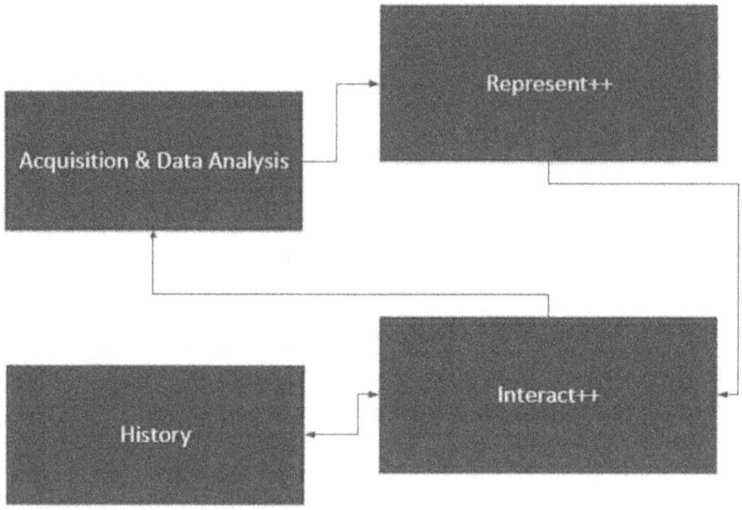

an instruction to data analysis layer to filter data for the represent++ layer. The user then views the information on interact++ layer and the cycle continues as per user demand and requirements.

This research framework is a dedicated tool specially created for individuals and for data sets which are not related to any business application. In other words, data is directly imported into the data analysis model for processing, filtering, mining and then pushed into a database for re-use at the representation layer of our visualisation model through bespoke web intelligent.

As shown in the Figure 6, the data first imported by the users through interface layer and it goes into the data analysis model - where its refined, unwanted information is removed and normalised and clean data is passed onto the MySQL database where it could retrieved by the system functions at a later stage for visualisation.

A sample import code has been shown in Figure 7, however the tool is created in a PHP framework, which comes with thousands of lines of code, and it is not possible to demonstrate the full system code. Instead, system features are demonstrated.

PERFORMANCE ANALYSIS

The novelty of this framework is that the focus is on the 'represent++' layer, with ancillary layers such as interact, refine and history included in the list. The represent++ layer is such an important part of the model, as a data analysis tool,

Figure 6. Data Analysis Process

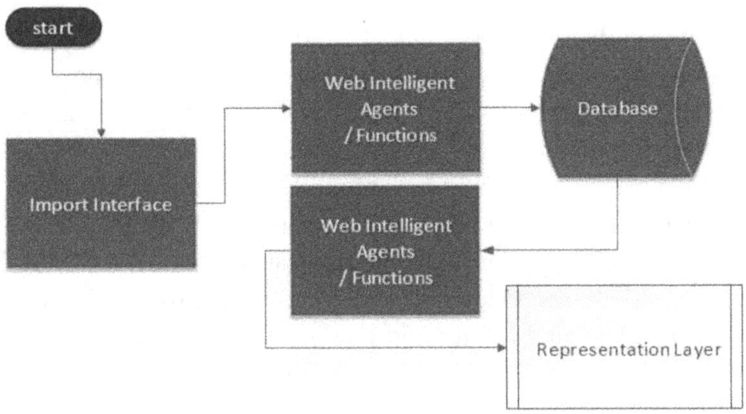

Figure 7. Visualixer Import Function Highlights

```
.....
        public function upload()
        {
            $this->load->model('import_model');
            $this->load->helper('file');
            $user_data = $this->session->userdata('id');

            //$path = './upload/'.$user_data;
            $path = './upload/temp';
            if(!is_dir($path)) //create the folder if it's not already exists
            {
              mkdir($path,0777,TRUE);
            }
            $this->load->helper('form');
            $config['upload_path'] = $path;
            $config['allowed_types'] = '*';
            $config['max_size']        = '';
            ......
```

various visualised representation of data has been produced to make it easy for the users to understand complex scenarios through visual outputs. Performance analysis will also showcase the processing of unstructured datasets using a third party application to support the process.

Acquisition and Data Analysis Layer

The acquisition and data analysis elements of the information visualisation model are discussed in this section. A transaction can mean many things depending on the context, but in this particular instance a transaction is defined as a sequence of information exchanges and related work (such as database updating) that is treated as a unit for the purposes of satisfying a request (Williamson, 1979).

Transactional data can cover finances, logistics or work related data, involve anything from a purchase order and shipping status to employee work hours and insurance costs and claims. Once it has been processed, transactional data is then grouped with its associated master and reference data to create the transactional records. A relevant date and time are added to this, along with the relevant reference data needed for each particular transaction record.

Most studies conducted resulting in research and working papers use synthetic data sets in order to produce their control group (Susaki, 2012). This is normally done using random number generators to help create more data on the fly, and is usually adequate for research conditions into new data analysis techniques. In the first part of this experiment, these large data sets were made available directly to the visualisation framework for purely business and data analysis purposes. The data sets which have been obtained for the first part of the experiment are consist of several million records, collected from various sections of the existing enterprise systems.

The data consists of various different factors and user behaviors, including:

- Geographical Location
- IP Address
- Transaction amount
- Transactional type
- Shopping Times
- Sales Transactions
- Staff Behavior
- Conversion Figures

Along with resources such as drivers and vehicle information for transport companies, all provided in huge masses of raw data sets for our analytical purposes. The process of this information visualization model involves various different data analysis techniques, followed by representation (or visualizing the data), couples with the interactive layers in the visualization model.

As can be seen in Figure 8, the acquire, parse, filter and mining steps of the process are all handled via web intelligent and API agents, and this means the system can directly access the data it needs to begin the visualization process. For

Figure 8. Acquisition and Data Analysis Layer

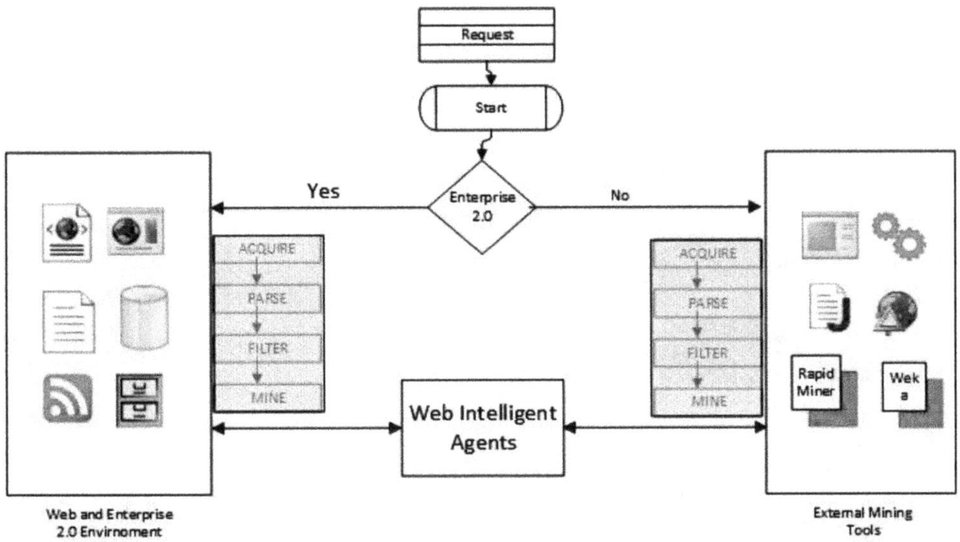

this example a dataset of roughly half a million rows of data with over five hundred attributes was processed. The first 4 stages of the visualization process (acquire, parse, filter & mine) were all done in the space of a few seconds using the robust framework put in place. This process is intelligent enough that it can be customized by end users to create their own set of roles based on what they need to see, and using a system primarily focused on visualization but with the capacity to perform all other aspects of the process is vital.

Data could be imported to the system through import function, once the data is imported, the acquisition and analysis model refines and filter information and then present choices to the users on the visual representation of data as shown in the Figure 9 and Figure 10.

Represent++

All visualisation outputs are processed & presented to the end user at this stage. The user can adopt various styles of visualisation, including bar charts, area charts and pie charts. The framework is flexible enough that it allows the user to filter the information they want to see with a few simple clicks.

The visualisation framework used in this example has processed a dataset representing staff sales. The extracted datasets used for the columns can be seen in the data processing stage above. Because of its flexibility the system can represent the extracted data in various different formats – in this case various different staff

Figure 9. Data Import option at the Acquisition and Analysis Layer

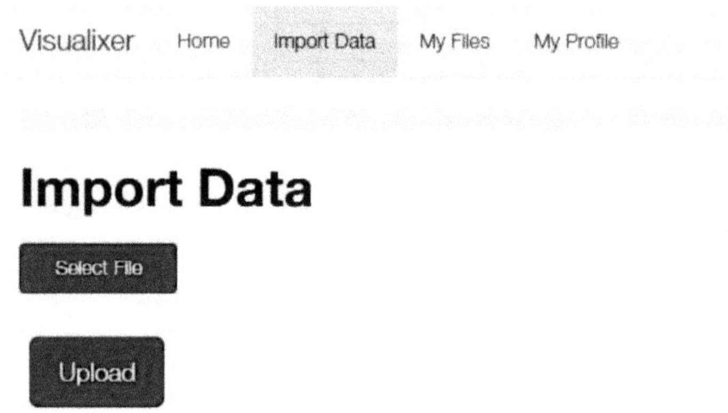

Figure 10. Represent++ layer options

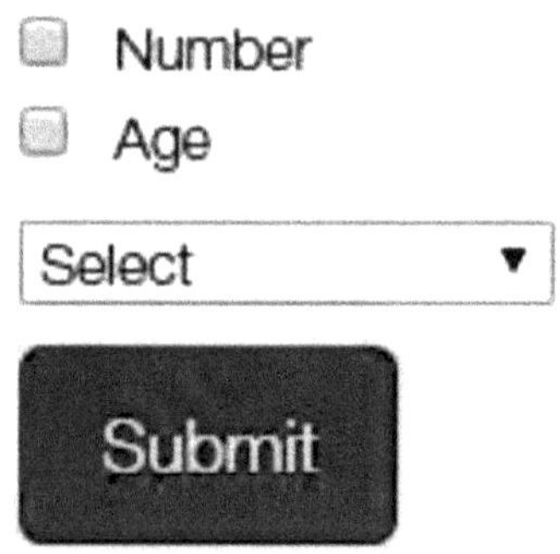

members are shown in a pie chart in Figure 11, which depicts which staff member is selling the most stock.

Information is visualised at the represent layer in this particular framework, and this allows a business manager of key decision maker to quickly identify the key information they need and act on it – allowing them to improve company sales by offering the correct coaching and training. Being able to share this data with said staff members can also serve as motivation, and the process can be applied to multiple uses including home finances or bank transfers.

Multiple Coordinated Visualisation

Multiple coordinated visualisation, where same information is visualised in two different styles at same time. The benefits of such visualisation give more in-depth and different prospect to the decision makers. Figure 12 and Figure 13 show multiple

Figure 11. Pie Chart Showing Staff Sales Information

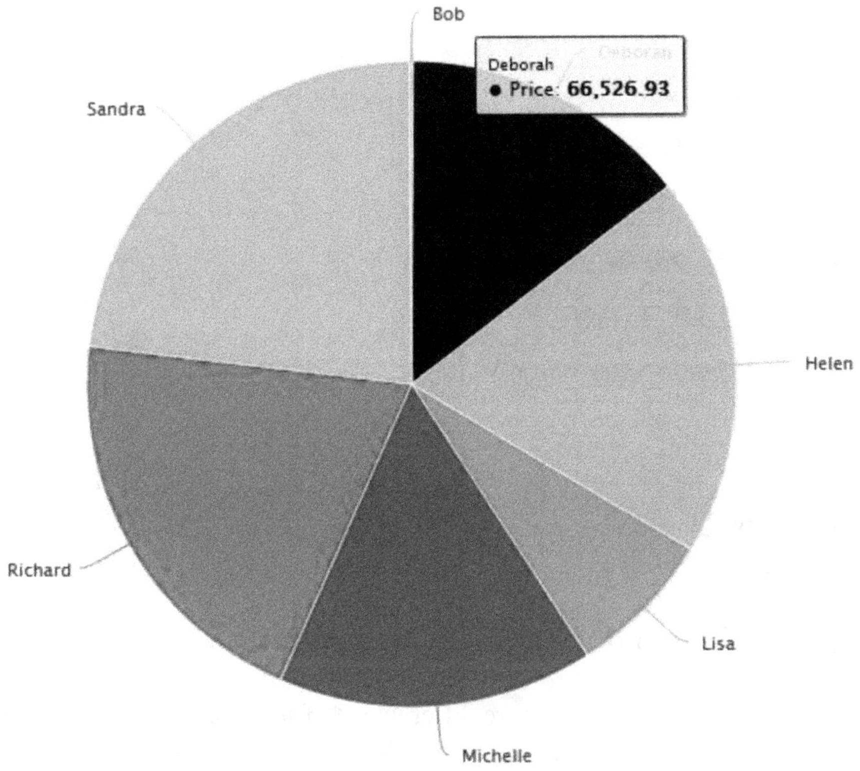

Figure 12. Multiple Coordinated Information Visualisation

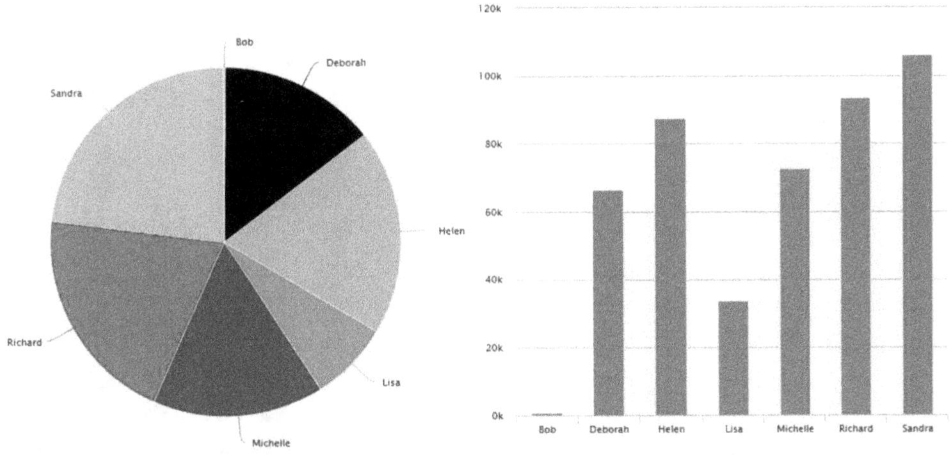

Figure 13. Multiple Coordinated Visualistaion Through Bar/Line Charts

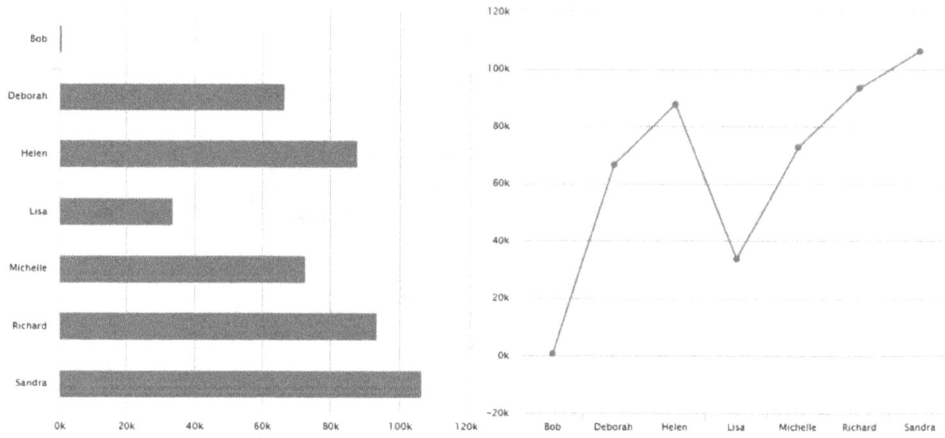

coordinated visualisaiton generated by the framework while processing the same set of information.

Multiple Attribute Information Visualisation

Multiple attribute information where different set of information is visualised in a multiple coordinated view. This type of visualisation helps the decision make to see data connections in various business and data attributes. Figure 14 shows staff and customers, visualised from the collected and processed sample of data. This type

Figure 14. Multiple Attribute Information Visualisation

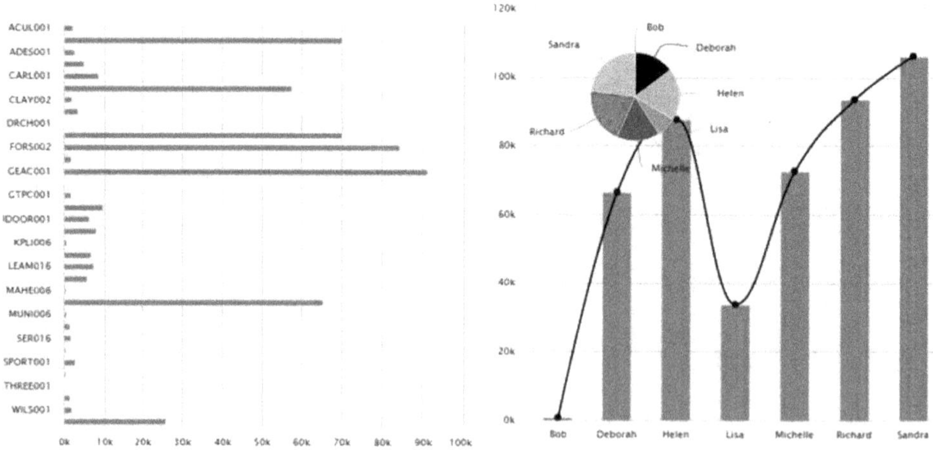

of visualisation indicate glimpse to decision makers to see where their business is coming from and which staff members are contributing.

Transactional Tagging Information Visualisation

Transactional tagging information visualisation is a new type of visualisation, where transactional in the data are tagged by the web intelligent agents based on medium and source where it's been generated or originated. It also consider various links within the data such which staff member had converted or initiated the transactional and which business category does it belongs. All information stored in a raw format and analysed by the system in a visual form to give more precise definition to decision makers in order to predict and pay attention to various resource and business revenues.

Figure 15 demonstrates staff members and type of business activities in a visualised form, the business mediums are tagged and information is shown in bubble charts.

The relationship between staff members and business activities is highlighted in the same colour. For example, staff member John has undertaken weddings and chauffeur activities in the sales forecast and Karen is good with city transfers and cricket events transfers. This type of information visualisation helps management in effective resource utilisation, jobs could be assigned to resources who feel confident in achieving the business goals.

Figure 15. Transactional Tagging

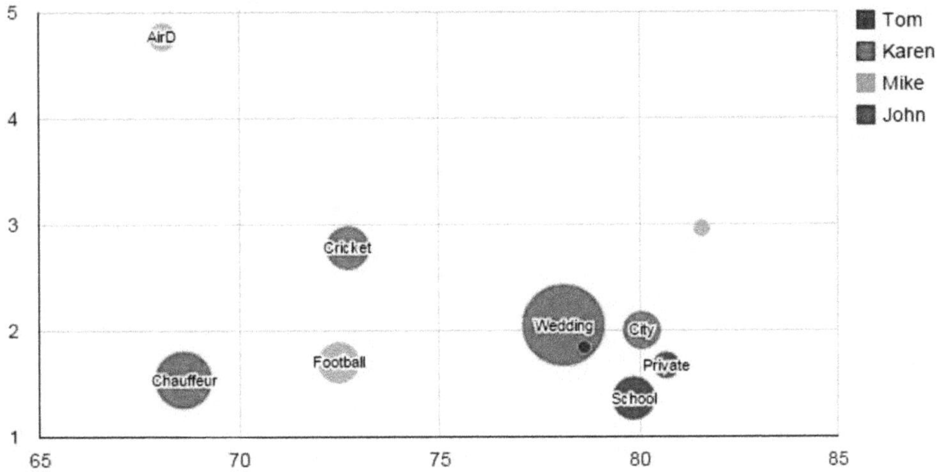

Linked Data Information Visualisation

Linked data visualisation explore linked information in transaction and then visualise it further, this approach helps the users to dig into the transaction detail to analyse the data in more depth and find out the exact sources, all this done with visualise representation of data. The linked data visualisation works with transactional tagging visualisation and by clicking on any of the bubbles in transactional visualisation. It shows a pop-up boxes which further analyse the information.

Figure 16 shows transactional tagged information through bubble chart, when the users click on any of the items on the chart it shows another box which has additional information for the users as shown in Figure 17.

The pop-up as shown in Figure 17, shows bar chart and with additional information giving client IDs, which shows the all these clients have booked wedding transfer with the business. This kind of visualisation gives further and extensive analysis of business transactions.

The following Figures shows various type of information visualisation generated through the proposed framework (Figures 18-20).

The advanced stages of this visualisation model allow business managers and end users to see the same set of data in various different forms, allowing in depth analysis from multiple angles which in turn can reveal interesting patterns and trends in the data which would have previously remained hidden. This has been demonstrated here with a pie chart representing sales staff information, which we can view in a number of different methods.

Figure 16. Transactional Tagging Visualisation for Linked Data

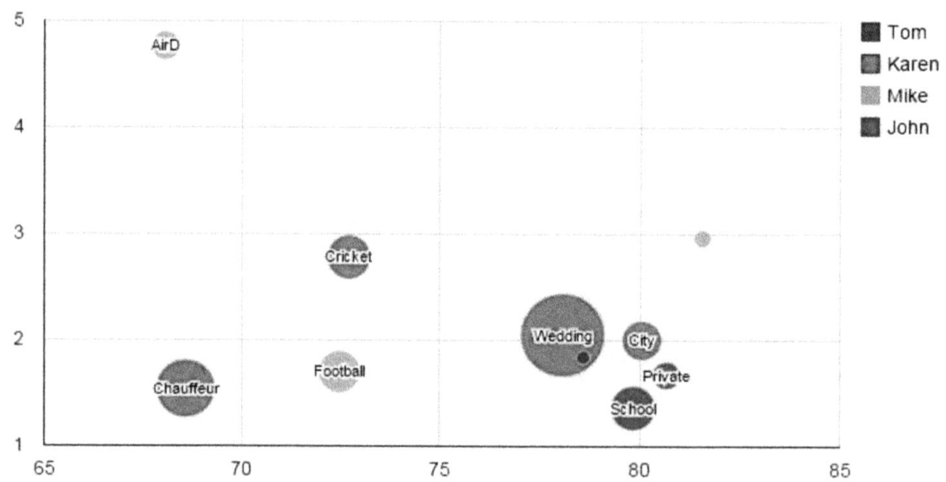

Figure 17. Linked Data Information Visualisation

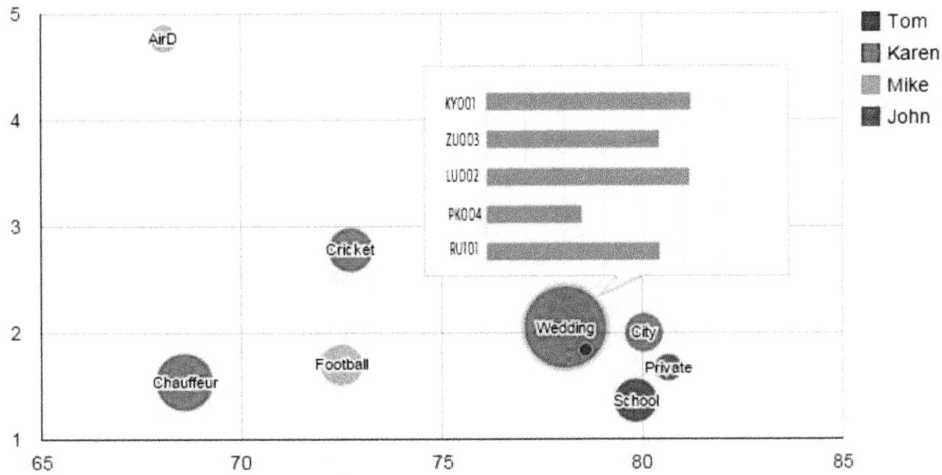

Figure 18. Highlighting transaction a particular date

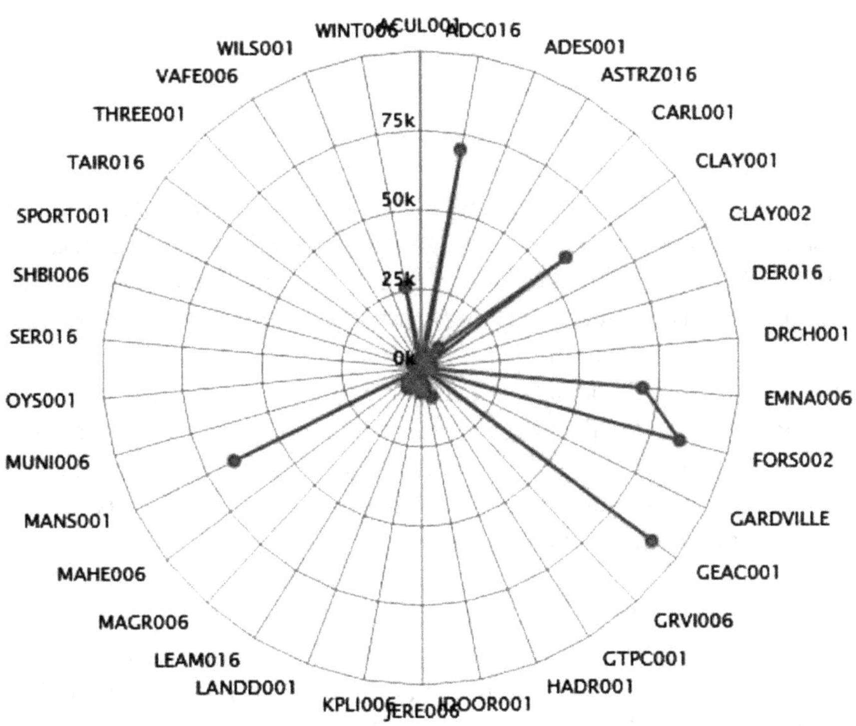

Figure 19. Vehicle Transactions in Sand Signika Chart

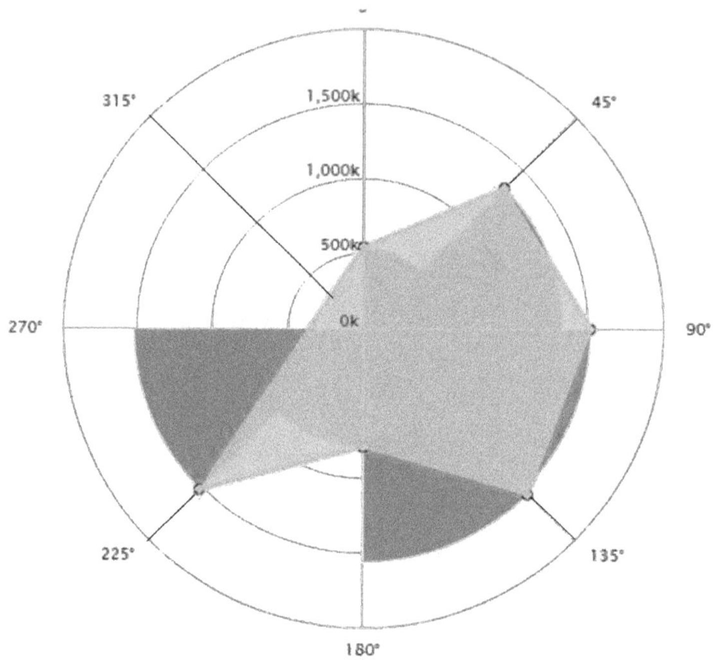

Interactivity

The fundamental aspect is accepting input from a human in the form of user. Interactive computer systems are functions and programs which allow users to enter data or instruction to the application to various interfaces. The applications respond to user requests through various interfaces where system accepts instructions or command as what the users want to see or expect from the system.

Computer program or devices that (in response to a user's action or request) presents choices (paths) depending on where in the program the user initiated the action. The following different choices, the user can accordingly control or change the action of the device or outcome of a game or program. Interactivity is key element in our computational information framework.

In this framework, the user has the unique ability to tweak the input or change the data questions, and the system, through to its intelligent design is able to accept the changes easily (no matter how challenging) and alter the outcomes, producing output based on the users instructions. All of these various outputs are in the form of interactive charts and graphs of data, all of which is quickly processed by the framework without any problems. With the addition of interactive graphs into the mix, application is able to give additional information which is not always available

Figure 20. Transaction Dates Through Pie Charts

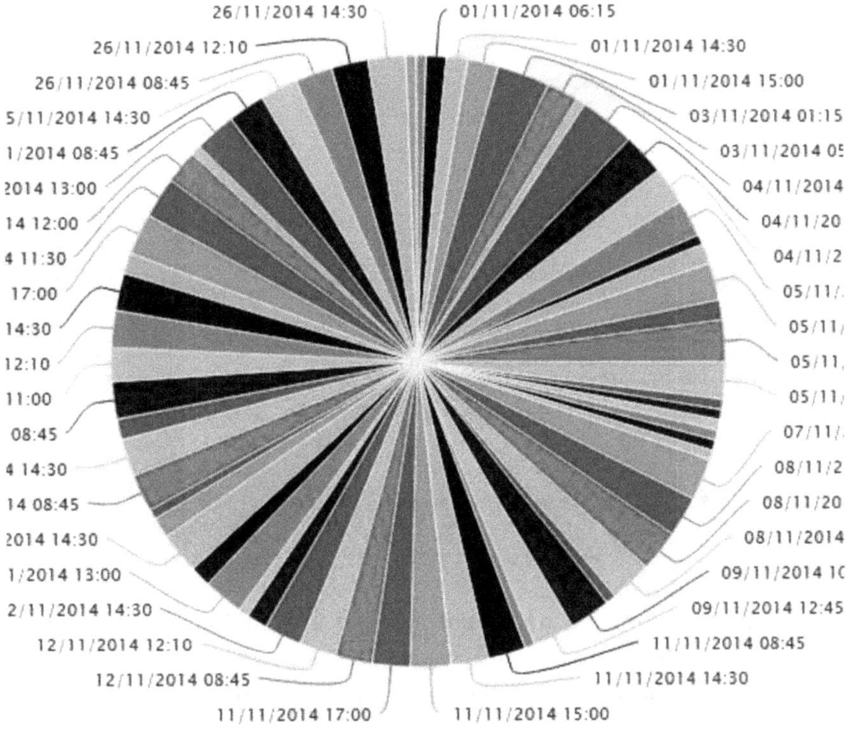

or visible on the static graphs produced, but is none the less connected or associated with an event. With interactive graphs, such events are just a mouse click (or hover) away - and these graphs will generate more data and information for the analysis purposes. As shown in the Figure 21.

The refine and interact layers have been developed with user friendly interfaces, allowing the end user or business manager to quickly request and access data in various visual forms. The application intelligent agents involved in the process have been designed to convert complex queries into a simple task within the framework. The framework used is highly optimised, and can not only give real time output but also deliver the requested information in a matter of seconds.

History

In the final stage of the framework, the system will generate reports and store them within the system in XML format. By utilizing this unique storage system data is made available or comparison and even re-purpose as can be seen in the above

Figure 21. Interactive Layer of Visualisation Model

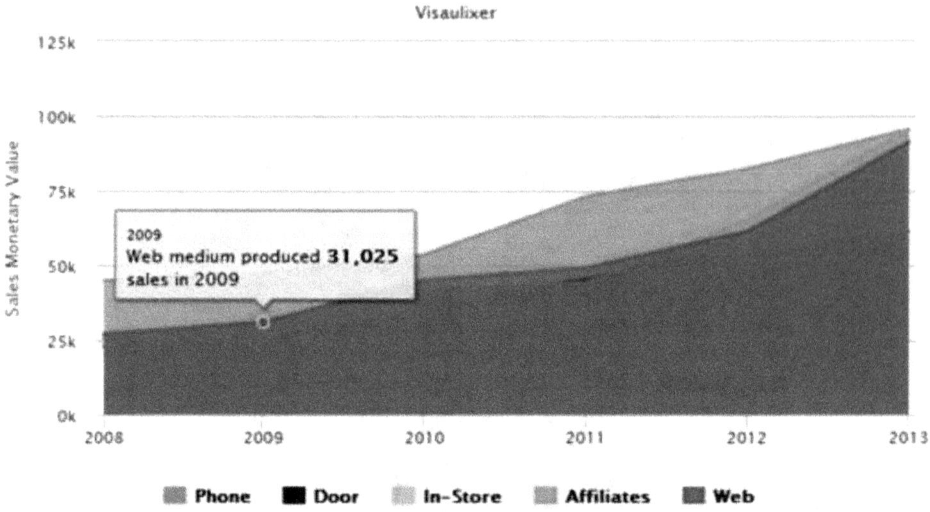

example. The application also allows users to export visual information into more popular file formats (such as JPG, GIF and PDF's). Figures stored by the system remain accessible by the system at all times, and can be called upon for the interact and refine layers of the system at any time.

CONCLUSION AND CONTRIBUTIONS

This research has contributed in various facets of data manipulation problems and information visualization in Web 2.0 through data mashup technologies. The contributions of this study are highlighted below:

- **Information Visualization Model:** The proposed model is a novel and comprehensive four-layer model addressing data related problems and its solution through information visualization. The model application and results are tested and validated with help of various experiments. The transactional data explored various aspects of the business in performance analysis. The model has also explored various privacy or data protection related aspects as different level of usage with variant privileges. The contribution of this research revolves around the four layers of the visualization model. The first layer, acquisition and data analysis demonstrate how complex data could be refined, filtered and mined for data representation purposes. The system models and flow charts demonstrate how data could be retrieved by the users in an enterprise environment and passed on to the data representation layer. There are various important contributions achieved on the data representation layer: (i) the introduction of existing mash-up system for data representation; (ii) the variety of data expression which helped in exploring data from various angles.

- **Multiple Attribute Information Visualization:** The multiple attributed data visualization achieved through data mash-up system with resource friendly technologies in web 2.0 environment. The end users are given more options to explore data in more detail for in-depth analysis. The multiple attribute visualization techniques helps users to relate data with other elements in the data sets. For instance, sales staff performance measured against product sales. This comparison helps explore which staff member is best in selling what products? This gives a good understanding to the management for effective resource utilisation. The users are empowered to inspect data with multiple attribute investigation leading to the exploration of hidden facts and trends.
- **Transactional Tagging Information Visualization:** Transactional tagging is a very fresh and novel contribution of this research and specially in Enterprise 2.0 environments. Transactional tagging visualization - where each data element is tagged by the data analysis model and then visualized at the representation layer; assisting users to distinguish visualized data elements through tagging. Various elements are tagged which are then visualized at the representation layer showing vital relation attributes. For example, visualizing staff members' sales of various sports tours. The analyzer can quickly identify what tags have an influence on staff or which staff have an influence on tags. This shows, which tags are popular in certain areas (cities and towns) in the country, where the business could focus and taking it to the next level.
- **Multiple Coordinated Information Visualization:** The multiple coordinated visualization is another very unique aspect of the research. The purpose of this type of visualization is to enable users to see same retrieved information from the data set or database but with different styles at the same time. For instance, the output could be shown in a pie chart, but the same information next to the graph could be visualized in an area chart. The user can analyses data from a picture which is easy on the eye or more meaningful for the data elements. This technique has been used as a fresh approach on transactional data sets facilitating small and medium size businesses in exploring complex data sets.

- **Linked Data Visualization:** Linked data visualization is an exclusive and novel contribution of this study. In the database or any other form other than the visualized form of the data, linking elements of data should not be a big problem as we often normalize relational database in making relations from one table to another so that the database or data elements make more sense. The linked data feature highlighted the need for such tools to understand data more precisely for more trends and analysis exploration. Linked data visualization works with multiple coordinated and multiple attribute visualization, where similar or related data elements visualized and presented highlighting the relations between different visualized aspects. The linking of data without visualization could be done with various approaches but linking relation elements of data in visualized form is a fresh and novel approach.

Finally, the information visualization model proposed and validated in this research is a complete set of facility for data related problems and questions. The technique will assist in analyzing any type and size of data, helping small and medium size businesses and organizations to explore data further for trends and progressions.

REFERENCES

Almeida, L. H. (2009). WebPrisma: An Interactive Web-Based Tool for Exploratory Visualization Using Multiple Coordinated Views. *Information Visualisation, 2009 13th International Conference*. doi:10.1109/IV.2009.17

Carr, D. (1999). *Guidelines for designing information visualization applications.* Paper presented at the ECUE'99.

Cooley, R., Mobasher, B., & Srivastava, J. (1997). *Web mining: Information and pattern discovery on the world wide web.* Paper presented at the Tools with Artificial Intelligence. doi:10.1109/TAI.1997.632303

Eick, S. G. (2000). Visualizing multi-dimensional data. *Computer Graphics, 34*(1), 61–67. doi:10.1145/563788.604454

Fry, B. J. (2004). *Computational information design*. Massachusetts Institute of Technology.

Godinho, P. I. A., Meiguins, B. S., Meiguins, A. S. G., do Carmo, R. M. C., de Brito Garcia, M., Almeida, L. H., & Lourenco, R. (2007). *PRISMA-A multidimensional information visualization tool using multiple coordinated views.* Paper presented at the Information Visualization, 2007. IV'07. 11th International Conference. doi:10.1109/IV.2007.90

Graham, R., & Caverlee, J. (2008). *Exploring feedback models in interactive tagging*. Paper presented at the Web Intelligence and Intelligent Agent Technology, 2008. WI-IAT'08. IEEE/WIC/ACM International Conference on. doi:10.1109/WIIAT.2008.419

Keim, D. A. (2000). Designing pixel-oriented visualization techniques: Theory and applications. *IEEE Transactions on Visualization and Computer Graphics*, *6*(1), 59–78. doi:10.1109/2945.841121

Keim, D. A., Hao, M. C., Dayal, U., & Hsu, M. (2002). Pixel bar charts: A visualization technique for very large multi-attribute data sets. *Information Visualization*, *1*(1), 20–34. doi:10.1057/palgrave.ivs.9500003

Keim, D. A., Hao, M. C., Ladisch, J., Hsu, M., & Dayal, U. (2001). *Pixel bar charts: A new technique for visualizing large multi-attribute data sets without aggregation*. Paper presented at the IEEE Symposium on Information Visualization, 2001. INFOVIS 2001. doi:10.1109/INFVIS.2001.963288

Ketter, W., Banjanin, M., Guikers, R., & Kayser, A. (2009). *Introducing an agile method for enterprise mash-up component development*. Paper presented at the Commerce and Enterprise Computing, 2009. CEC'09. IEEE Conference on. doi:10.1109/CEC.2009.40

Kosala, R., & Blockeel, H. (2000). Web mining research: A survey. *ACM Sigkdd Explorations Newsletter*, *2*(1), 1–15. doi:10.1145/360402.360406

Mobasher, B., Dai, H., Luo, T., Sun, Y., & Zhu, J. (2000). *Integrating web usage and content mining for more effective personalization*. Paper presented at the International Conference on Electronic Commerce and Web Technologies. doi:10.1007/3-540-44463-7_15

Mulvenna, M. D., Anand, S. S., & Büchner, A. G. (2000). Personalization on the Net using Web mining: Introduction. *Communications of the ACM*, *43*(8), 122–125. doi:10.1145/345124.345165

Ngu, A. H., Carlson, M. P., Sheng, Q. Z., & Paik, H.-y. (2010). Semantic-based mashup of composite applications. *IEEE Transactions on Services Computing*, *3*(1), 2–15. doi:10.1109/TSC.2010.8

North, C., & Shneiderman, B. (2000). Snap-together visualization: a user interface for coordinating visualizations via relational schemata. *Proceedings of the working conference on Advanced visual interfaces*. doi:10.1145/345513.345282

Pillat, R. M., & Freitas, C. M. (2006). Coordinating views in the InfoVis toolkit. *Proceedings of the working conference on Advanced visual interfaces.* doi:10.1145/1133265.1133368

Srivastava, J., Cooley, R., Deshpande, M., & Tan, P.-N. (2000). Web usage mining: Discovery and applications of usage patterns from web data. *Acm Sigkdd Explorations Newsletter*, *1*(2), 12–23. doi:10.1145/846183.846188

Stolte, C., Tang, D., & Hanrahan, P. (2002). Polaris: A system for query, analysis, and visualization of multidimensional relational databases. *IEEE Transactions on Visualization and Computer Graphics*, *8*(1), 52–65. doi:10.1109/2945.981851

Tim, O. (2005). *What is web 2.0? design patterns and business models for the next generation of software*. Academic Press.

Tufte, E. R. (1997). *Visual explanations: images and quantities, evidence and narrative* (Vol. 36). CT: Graphics Press Cheshire.

Vargiu, E., & Urru, M. (2012). Exploiting web scraping in a collaborative filtering-based approach to web advertising. *Artificial Intelligence Review*, *2*(1), 44.

Williamson, O. E. (1979). Transaction-cost economics: The governance of contractual relations. *The Journal of Law & Economics*, *22*(2), 233–261. doi:10.1086/466942

Yu, J., Benatallah, B., Casati, F., & Daniel, F. (2008). Understanding mashup development. *IEEE Internet Computing*, *12*(5), 44–52. doi:10.1109/MIC.2008.114

KEY TERMS AND DEFINITIONS

Data Mash-Up: Bringing a collection of data from various sources and combining them for the end-user.

Enterprise 2.0: The combine implementation of social software and collaborative technologies.

Enterprise Mash-Up: A collection of data from external web services with the combination of the enterprise internal data resources.

Information Visualization: The process of processing statistics, data, and graphics into a visual form.

Web 2.0: The second stage of the online technology development.

Web Mining: The application of various data mining techniques to discover patterns and trends from the internet, and from a variety of enterprise datasets that are available to mash-up applications.

Web Scraping: The technique of automatic web data extraction, used in order to extract data from the core HTML of the website by parsing the web pages.

Chapter 7
An Analysis of Customers' Acceptance of Internet Banking:
An Integration of E-Trust and Service Quality to the TAM – The Case of Turkey

Yakup Akgül
Alanya Alaaddin Keykubat University, Turkey

ABSTRACT

The aim of this chapter is to provide insights into the causal factor of people's internet banking acceptance. It endeavors to address a research need for extending the technology acceptance model (TAM) by adding contextual variables and its implication on e-customer satisfaction and e-customer loyalty. As trust and service quality has never failed to be a significant predictor in e-commerce research, this study proposes to integrate trust and service quality into the TAM to get a better understanding of e-banking adoption. A research model reflecting the effect of e-trust and internet banking service quality dimensions on TAM constructs is proposed.

INTRODUCTION

Innovations and technological changes have come with great benefits to traditional banking. The development of internet usage and the huge funding initiatives in internet banking have drawn the attention of researchers towards internet banking adoption. In order to get financial stability in firm's steer and greater competitive

DOI: 10.4018/978-1-5225-3628-4.ch007

advantages in a rigid competitive environment, recently, businesses have switched their focus on making information technology an essential part of their functions and processes (Oliveira et al., 2014). In this way, internet banking plays an important role in reducing operating and fixed costs (Chen et al., 2012; Fonseca, 2014; Zhao et al., 2010) and helps the bank in building better relationship with their customers (DeYoung et al., 2007; Ribbink et al., 2004; Rod and Ashill, 2010; Rod et al., 2009). Important amounts have been spent on building internet banking systems, reports show that potential customers may not use them, despite their availability (Reid and Levy, 2008; Wang et al., 2003).

In today's commerce, internet banking has gained significant importance and the growth of the field is accelerating (Lin, 2011). Internet banking provides the customers to carry out a range of banking activities electronically at any time and place with low handling cost (Angelakopoulos and Mihiotis, 2011; Narayanasamy et al., 2011; Nor and Pearson, 2008; Weir et al., 2006; Yoon and Steege, 2013). Internet banking is merely the usage of smart phones, tablets and personal digital assistants (PDAs) to communicate with banking system through wireless application protocol (WAP). With the aid of mobile banking, bank customers can carry out banking facilities such as information inquiry, account managing, bill payment and money transfers etc. (Luarn and Lin, 2005). Despite the growing amount of internet banking users, the customer adoption of internet banking has not yet reached the expected levels. One of the common concerns that have been emphasized about the adoption of internet banking is poor service quality and customer dissatisfaction (Calisir and Gumussoy, 2008; Li-hua, 2012; Zhao et al., 2010). In fact, the main barrier for customer intention to use internet banking services is relate to the user habits. Additionally, customers are struggling to change their habits, behaviors, the way they interact with its internet banking services offer (Alsajjan and Dennis, 2010; Chemingui, 2013; Chen and Teng, 2013; Hanafizadeh et al., 2014; Khalil et al., 2010; Nor and Pearson, 2008). The potential customers' readiness to adopt the technology is also a significant factor (Al-Ajam and Md Nor, 2015; Al-alak, 2014; Chiou and Shen, 2012; Giordani et al., 2014; Harrison et al., 2014; Rezaei et al., 2014). Interestingly, Karjaluoto et al. (2002) found lack of consumer skills and experience with computers and technologies, which influence the consumer attitude toward internet banking. In addition, interpersonal, social network, security concerns (Hasim and Salman, 2010), habit and personality traits (McNeish, 2015) are the main criteria which influence customers to use internet banking. Given the open nature of the internet and the lack of sufficient regulations concerning e-commerce activities, the importance of trust and trust-related concepts such as transaction security and risk is likely to emerge as the biggest concern among the actual and potential internet banking users (Agarwal et al., 2009; Alalwan et al., 2015; Al-smadi, 2012; Candra, 2013; Ezzi, 2014; Maduku, 2013; Nasri, 2011; Ozdemir et al., 2008; Suh and Han, 2002).

Trust was shown to have a striking influence on user willingness to engage in online exchanges of money (Reid and Levy, 2008; Wang et al., 2003). Trust is not merely a short-term issue, but also the most significant long-term barrier for realizing the potentials of e-commerce (Gefen, 2000). As trust has never failed to be a significant predictor in e-commerce research, Agarwal et al. (2009) proposed to include trust in behavioral models to better explain internet banking adoption. Few studies have investigated business e-trust in banking context and its impact on internet banking adoption, and even fewer focused on emergent countries (Alalwan et al., 2015; Al-smadi, 2012; Ezzi, 2014). This research aims to address this knowledge gap by identifying significant specific variables that affect the adoption of internet banking services by users in Turkey. Distinguishing such variables will advance the probability of raising the adoption rate of these services, by deepening the knowledge about the variables which facilitate or interrupt business adoption process. For this purpose, this study develops and examines a theoretical model integrating e-trust to the TAM to provide a more comprehensive model of user's internet banking adoption. In Turkey, the development of internet banking has a great potential and has grown significantly along with the use of smart phones and tablets; and the penetration rate of internet is half of the country's population. As per June 2016, the number of internet banking subscribers is over the 50 million (AA, http://aa.com.tr/tr/ekonomi/internet-bankaciligi-bireysel-musteri-sayisi-50-milyonu-asti/631514 03.05.2017)

In addition to; this study attempts to investigate the structural relationship between internet banking service quality, e-customer satisfaction, and e-loyalty based on distinct constructs. In this study, the construct and model of e-satisfaction developed by Herington and Weaven (2009) and Ribbink et al. (2004) and e-customer loyalty by Ramseook-Munhurrun and Naidoo (2011) and Zeithaml et al. (1996) are applied to the context of internet banking industry. e-customer satisfaction is measured as a multi-items measure to specify customer satisfaction and will be defined to transaction-specific judgments from internet banking perspectives (Bodet, 2008; Bressolles et al., 2014; Cronin and Taylor, 1992; Herington and Weaven, 2009). Thus, customer loyalty is defined as consumer's intention to revisit the website of internet banking in the future (Amin et al., 2013; Ramseook-Munhurrun and Naidoo, 2011; Zeithaml et al., 1996). This research study makes a significant contribution to literature of bank marketing and academicians by explaining how internet banking service quality dimensions are used as a predictor of e-customer satisfaction and consequently increase e-customer loyalty In this study, four key dimensions of internet banking service quality (personal need, site organization, user friendliness and efficiency of websites) and e-trust (benevolence, credibility, integrity and orientation to resolve problems) have been found as the main factors influencing e-customer satisfaction and loyalty. Therefore, a better understanding of these relationships will help the internet banking industry in developing the marketing strategies, maintaining long-term the relationship with their customers, and achieving competitive advantages in the rigid global market. Consequently, customer satisfaction and loyalty have been identified as an important factor in building and maintaining the relationship with their customer in order to reduce the perceived risk of using internet banking (Aldas-Manzano et al., 2011; Chen, 2013; Chen et al., 2012; Dahlstrom et al., 2014).

A more recent concern of IS related research are varied to the field of internet banking arising as the latest development of IS area (Al-Jabri and Sohail, 2012; Gu et al., 2009; Shih et al., 2010; Yu, 2012; Zhou et al., 2010). Many studies have examined the consumers' resistant factors for adopting internet banking (Agarwal et al., 2009; Candra, 2013; Eriksson et al., 2005; Ezzi, 2014; Gerrard and Cunningham, 2003; Jaruwachirathanakul and Fink, 2005; Maduku, 2013; Sayar and Wolfe, 2007). Among these, relatively few studies have been conducted in the Turkish context. Therefore, it would be interesting to identify specific factors that determine acceptance of internet banking by Turkish users. There is a prospering body of academic research being focused on investigating the determinants of computer technology acceptance and use. The technology acceptance model (TAM) (Davis, 1989), among the distinct models that have been proposed, which appears to be the most widely accepted among information systems researchers. According to the TAM, adoption behaviour is determined by the intention to use a particular system, which in turn is determined by the perceived usefulness and the perceived ease of use of the system. Despite the information systems researchers have examined and replicated the TAM, and agreed to its validity in predicting the individual's acceptance of various corporate information technologies, the TAM's fundamental constructs do not fully reflect the specific influences of technological and usage-context factors that may convert user acceptance (Agarwal et al., 2009; Candra, 2013; Chen and Barnes, 2007; Eriksson et al., 2005; Ezzi, 2014; Guriting and Ndubisi, 2006; Howcroft et al., 2002; Laforet and Li, 2005; Maduku, 2013; McKechnie et al., 2006; Ozdemir et al., 2008; Pikkarainen et al., 2004; Reid and Levy, 2008; Wang et al., 2003). As Davis (1989) noted, we have a need for future technology acceptance research to address how other variables affect usefulness, ease of use, attitude and user acceptance. However, factors affecting the acceptance of a new information technology are likely to vary with the nature of the technology, target users and context (Wang et al., 2003). Agarwal et al. (2009) noted that there is a pressing need to validate integrated technology acceptance models that include universal variables, such as trust, across cultures. Ezzi (2014) as well as Candra (2013) and Reid and Levy (2008) highlighted the complexity nature of trust and the necessity to explore this concept more to get a better understanding of internet banking adoption.

HYPOTHESES DEVELOPMENT

Technology Acceptance Model

The theoretical background for this book chapter extract from the TAM. Davis (1989) proposed to interpret and predict the acceptance of information systems by end-users, the TAM has been validated different information technology applications over time. The TAM is an extension of the theory of reasoned action (TRA), Fishbein and Ajzen (1975) developed to describe the psychological determinants of behavior. The TRA is based on the attitude–behaviour paradigm from cognitive psychology. According to the TRA, an individual's performance of a specific behaviour is determined by his or her behavioral intentions, which are a function of individual attitudes and subjective norms (Fishbein and Ajzen, 1975).

Based on the TRA, the TAM assumes that two personal beliefs about usefulness and ease of use of the technology influence attitude, which in turn leads to behavioral intention and then generates behavior to use the system (Venkatesh and Davis, 1996; Davis et al., 1992). Perceived usefulness is defined as the degree to which individuals believe that using a particular system would enhance their job performance (Davis, 1989). Perceived ease of use relates to the degree to which individuals believe that using a particular system would require no particular effort (Davis, 1989). Perceived ease of use affects attitude either directly and indirectly through its effect on perceived usefulness: [. . .] even if potential users believe that a given application is useful, they may at the same time believe that the systems are too hard to use and that performance benefits of usage are outweighed by the effort of using the application (Davis, 1989).

Davis (1989) debates that people adopt an application primarily because of the functions it performs and secondarily because of the ease or difficulty associated with making the system perform these functions. A significant body of research provides empirical support to show that perceived usefulness and perceived ease of use play a critical role in predicting and determining the adoption and the use of new information technology, including the adoption of internet banking (Alalwan et al., 2015; Ezzi, 2014; Charfeddine and Nasri, 2013; Nasri, 2011; Norazah and Norbayah, 2009; Norazah et al., 2008; Venkatesh et al., 2003; Moon and Kim, 2001;

Venkatesh and Davis, 2000; Lin and Lu, 2000; Karahanna et al., 1999; Teo et al., 1999; Igbaria et al., 1996; Adams et al., 1992).

Within the TAM, perceived usefulness is the vital factor having a direct effect on behavioural intention, determining a large ratio of attitude and mediating the effects of perceived ease of use on behavioural intention (Chen and Barnes, 2007; Davis, 1989; Davis et al., 1989; Eriksson et al., 2005; Fusilier and Durlabhji, 2005; Gefen et al., 2003; Guriting and Ndubisi, 2006; Hu et al., 1999; Jaruwachirathanakul and Fink, 2005; Lin and Lu, 2000; Venkatesh and Davis, 2000; Venkatesh, 2000; Venkatesh and Davis, 1996). Recent information system acceptance studies reveal that the influence of perceived ease of use on attitude and intentions remains speculative and suggest that perceived ease of use operates mainly through perceived usefulness (Ezzi, 2014; Ha and Stoel, 2008; Gefen and Straub, 2000). Previous empirical research suggests that perceived usefulness and perceived ease of use might not be the only beliefs that mediate the impact of the external environment on attitude and intentions (Alsajjan and Dennis, 2010). To improve this absence, it is recommended to incorporate into the TAM, other underlying behavioral constructs and external variables to fit the user's context, and therefore to demonstrate better predictive power regarding user's acceptance of e-commerce (Alsajjan and Dennis, 2010; Lin et al., 2007; McFarland and Hamilton, 2004). The TAM undertakes that behavior – "the manifest, observable response in a given situation" (Ajzen, 2006) – is volitional. Behavioral intention demands a person's readiness to perform the given behavior (Ajzen, 2006), which makes it the primary predictor of the actual behavior. In the TAM, intention is a perform of attitude and perceived usefulness. Attitude is "the degree of evaluative affect that an individual associates with using the target system" (Davis, 1993). Also, attitude shows people's predisposition to respond either favorably or unfavorably to a particular behavior (Triandis, 1979). The TAM suggests that attitude is based on the prominent beliefs (perceived usefulness and perceived ease of use), which a person has about the consequences of a given behavior and his or her evaluation of those consequences. Polatoglu and Ekin (2001) indicated that customer attitude is composed of one's beliefs about the object and perceived importance of that attribute in making the decision to adopt. Some works regard the relation between attitude and behavioral intention proposed in the TAM as problematical, because of partial mediation of attitude in the relationship between beliefs and intentions, whereas it is assumed that beliefs are better predictors of intentions (Alsajjan and Dennis, 2010; Venkatesh et al., 2003). However, an extensive body of literature on the adaptation of electronic commerce forms, including internet banking, reports the significant relationship between attitude and behavioral intention (Ajzen and Fishbein, 1980; Bajaj and Nidumolu, 1998;Bobbitt and Dabholkar, 2001; Cao and Mokhtarian, 2005; Celik, 2008; Cheng et al., 2006; Dick and Basu, 1994; Eriksson et al., 2005; Hausman and Siekpe, 2008; Hernandez and Mazzon, 2007; Jaruwachirathanakul and

Fink, 2005; O'Cass and Frenech, 2003; Shih, 2004; Suh and Han, 2002; Venkatesh and Davis, 2000). Although the results of many empirical studies have verified the validity of the TAM, the original TAM constructs explain less than 45 percent of intention and usage variance in an e-commerce-related context, including internet banking (Celik, 2008; Chen and Wells, 2002; Suh and Han, 2002). As considered by Yang and Yoo (2004), attitude resides in the mind, precedes and produces behavior and thus can be used to predict behavior. Thus, attitude seems to be a well-established predictor of behavior. Moreover, Alsajjan and Dennis (2010) remarked that the impact of relative attitude should become more important in predicting consumer behavior because internet banking increases banking customers' choice set (Alsajjan and Dennis, 2010). To examine the TAM more, we put forward the following five basis hypotheses:

H1: There is a positive effect of perceived ease of use on attitude.
H2: There is a positive effect of perceived ease of use on perceived usefulness.
H3: There is a positive effect of perceived usefulness on attitude.
H4: There is a positive effect of perceived usefulness on the intention.
H5: There is a positive effect of attitude on the intention.

E-Trust

In line with the previously notable studies, this research reinforces the idea that the pure application of the TAM is not adequate to explain the adoption of internet banking (Alalwan et al., 2015; Candra, 2013; Charfeddine and Nasri, 2013; Cheong and Park, 2005; Ezzi, 2014; Luarn and Lin, 2005; Ozdemir et al., 2008; Pikkarainen et al., 2004). Indeed, due to the open internet technology infrastructure, the distance separating partners and the absence of human interactions, the importance of trust and trust-related concepts is increasing for internet banking (Charfeddine and Nasri, 2013; Ozdemir et al., 2008). Trust is at the heart of all kinds of relationships, it is an important catalyst in many transactional relationships and it determines the nature of many businesses (Alsajjan and Dennis, 2010; Gefen et al., 2003; Morgan and Hunt, 1994; Popoola and Ibn Arshad, 2015; Wang et al., 2003; Yousafzai et al., 2007). The role of technology trust and of internet trust as a specific form of technology trust in the context of internet banking has, also, been recognized (Candra, 2013; Ezzi, 2014; Grabner-Krauter and Faullant, 2008; Rullis and Sloka, 2010; Yu et al., 2015). Definitions and conceptualization of trust diversify according to scientific areas. Psychologists view trust as a personal trait, sociologists consider it as a social construct and economists see it as an economic choice mechanism (McKnight et al., 2002a, McKnight et al., 2002b). Rousseau et al. (1998) defined trust based on social psychology as "perceptions about others' attributes and a related willingness to become vulnerable to others". Drawing on literature in social psychology and marketing, trust defined as the perceived credibility and benevolence of a target of trust by Doney and Cannon (1997). In this regard, consumers might not use e-commerce because they lack trust in internet businesses (Grewal et al., 2004). With greater trust, people can undertake their uncertainty regarding the motives, intentions and prospective actions of others on whom they depend (Kramer, 1999), as well as save money and effort, because trust reduces monitoring and legal contract costs (Fortin et al., 2002). The absence of trust represents a significant hindrance to the market penetration of e-channels in online transactions and web vendors (Liu et al., 2004). Moreover, recent research implies that trust has a crucial influence on users' willingness to engage in online exchanges of money and sensitive personal information (Agarwal et al., 2009; Alalwan et al., 2015; Al-smadi, 2012; Candra, 2013; Ezzi, 2014; Maduku, 2013; Nasri, 2011; Ozdemir et al., 2008). Mcknight et al. (2002a) and Mcknight et al. (2002b) added a third dimension of trust, which is the integrity. These dimensions of trust most often included in marketing research (Bhattacherjee, 2002; Gefen et al., 2003; Schlosser et al., 2006a, Schlosser et al., 2006b; Yousafzai et al., 2007; Yu et al., 2015). Besides that, Janouri and Gharbi (2008) added a fourth dimension, which is the orientation to resolve problems. In this book chapter, it is assumed that e-trust is composed of four dimensions: the perceived credibility, the benevolence, the perceived integrity and the orientation to resolve problems. The first dimension of trust, perceived credibility, is the extent to which one partner believes that the other partner has the required expertise to perform the job effectively and reliably (Bories, 2007; Chouk and Perrien, 2003; Gansean, 1994; Gefen, 2002; Morgan and Hunt, 1994). Then, trust based on a partner's expertise and reliability focuses on the objective credibility of an exchange partner: expectancy that the word or written statement of the partner can be relied on (Wang et al., 2003). Wang and Emurian, (2005) cited to the perceived credibility as the belief that the electronic supplier has the required knowledge to do his job effectively. The second dimension of trust, benevolence, is the extent to which one partner is genuinely interested in the welfare of the other and is motivated by the search for common gain (Bories, 2007). It reflects the belief that the supplier wishes well to the user despite its own profits (Gansean, 1994; Gefen, 2002). So, the benevolence is the belief that the electronic merchant is interested in user interests (Bhattacherjee, 2002; Chen and Dhillon, 2005; Chouk and Perrien, 2004; Wang and Emurian, 2005). The third dimension of trust, the perceived integrity, is the belief that the other party accepts the honesty and the respect of promises and commitments (Chouk and Perrien, 2004; Gefen, 2002; Isaac and Volle, 2008). The perceived integrity is the confidence that the internet banking will not exploit the vulnerability and that it will honour its commitment by promising to protect the security of transactions and the confidentiality of information (Bhattacherjee, 2002; Chouk and Perrien, 2003; Janouri and Gharbi, 2008; Wang and Emurian, 2005). The fourth dimension of trust, the orientation to resolve problems, is the will of the supplier to resolve problems that may arise during and after the purchase, problems for which a commitment was not made (Chouk and Perrien, 2003; Janouri and Gharbi, 2008). The importance of trust rises with uncertainty especially in intangible situations such as online services. In fact, the absence of trust may be one of the most important obstacles of accepting online trade (Isaac and Volle, 2008) and internet banking (Pearson et al., 2007). However, a high level of trust comforts partners to engage in innovative transactions (Guimond, 2008). The importance of trust emphasized in e-commerce acceptance by many scholars. Gefen et al. (2003) assert that technology acceptance model that features more social dimensions must include trust, especially when the relations involve social uncertainty and risk, as in internet banking. Page and Luding (2003) regard that trust issues are crucial factors of internet banking adoption, and McKnight et al. (1998) combine trust into the TRA as a belief that influences intention to engage in behavior related to a specific e-vendor. They conclude that trust is implicit to customer intentions, and at higher levels of trust, customers perceive a website as more useful. Pavlou (2003) reveals that trust influences intentions through positive attitude. Alsajjan and Dennis (2010) consider that trust has a positive effect on users' perceived usefulness and attitudinal intentions towards internet banking adoption. Thus, we can assume that trust has an impact on personal beliefs about ease of use and usefulness of internet banking and has an influence on attitude towards internet banking use, which in turn leads to behavioral intention to use internet banking. Then, it is hypothesized that:

H6: Trust has a positive effect on attitude.
H7: Trust has a positive effect on customer loyalty.
H8: Trust has a positive effect on customer satisfaction.
H9: Trust has a positive effect on intention to use.
H10: Trust has a positive effect on perceived ease of use.
H11: Trust has a positive effect on perceived usefulness.

INTERNET BANKING SERVICE QUALITY AND E-CUSTOMER SATISFACTION

In the literature, the causal relationship between banking service quality and customer satisfaction is the topic of great academic debate and no compromise has been reached on this issue (Babakus and Boller, 1992; Bahia and Nantel, 2000). Some researchers and academicians describe customer satisfaction as an antecedent of service quality (Bitner et al., 1990; Carman, 1990; Parasuraman et al., 1985), and others have argued that the service quality is an antecedent of customer satisfaction (Amin and Isa, 2008; Anderson and Sullivan, 1993; Cronin et al., 2000; Cronin and Taylor, 1992; Kashif et al., 2015; Sheng and Liu, 2010; Yap et al., 2012). In banking sector, Joseph and Stone (2003) have highlighted that the availability of internet banking services delivery and user friendliness appears to be correlated with high customer satisfaction and retention. Similarly, Rod et al. (2009) found that internet banking service quality is perceived to be high, customers are more likely to be satisfied with their online service and consequently will be more satisfied with their banks. For this reason, Anderson and Srinivasan (2003) proposes that probably website characteristics (e.g. ease of use) drives e-customer satisfaction, since the website is the principle interface between the customer and the company. Consequently, positive customer perceptions of the quality of the various e-service attributes will result in satisfaction with the e-service provided through the website (Carlson and O'Cass, 2011; Cristobal et al., 2007; Kaura et al., 2015; Raza et al., 2015; Singh and Kaur, 2013). In addition, Bressolles et al. (2014) suggest that although e-customer satisfaction is influenced by website attributes, different consumers will be affected differently. Additionally, Black et al. (2014) explain that relationship between service quality and customer satisfaction is stronger for those that are less technically complex of services. In this situation, customers who have information technology skills can easily use the internet banking service, and they will have higher-satisfaction levels than others (Herington and Weaven, 2009; Ho and Lin, 2010; Lang and Colgate, 2003; Li-hua, 2012). Consequently, internet service quality will also influence customers to revisiting the internet website and deliver positive WOM to other customers (Carlson and O'Cass, 2011). Hence, the following research hypotheses are proposed:

H12: Internet banking service quality has a positive relationship with attitude.
H13: Internet banking service quality has a positive relationship with customer loyalty.
H14: Internet banking service quality has a positive relationship with customer satisfaction.
H15: Internet banking service quality has a positive relationship with trust.
H16: Internet banking service quality has a positive relationship with intention to use.
H17: Internet banking service quality has a positive relationship with Perceived Ease of Use.
H18: Internet banking service quality has a positive relationship with Perceived Usefulness.

E-CUSTOMER SATISFACTION AND E-CUSTOMER LOYALTY

Many works have provided empirical evidence to support the statement that customer satisfaction has positive relationship on repurchase intention and customer loyalty (Aksoy, 2014; Amin et al., 2013; Baumann et al., 2011, Baumann et al., 2012; Bloemer et al., 1998; Cronin et al., 2000; Kashif et al., 2015; Kassim and Abdullah, 2010; Nguyen and Leblanc, 2002; Sharifi and Esfidani, 2014; Thaichon et al., 2014; Zeithaml et al., 1996). For example, Ramseook-Munhurrun and Naidoo (2011) found that there is a significant relationship between e-customer satisfaction and e-customer loyalty in internet banking. Accessibility is the primary sigificant side in verifying e-customer satisfaction, while security is also important determinant of behavioral intentions. Likewise, Gera (2011) found that e-customer satisfaction is a direct antecedent of positive recommendation intentions, and leads to positive WOM (Baumann et al., 2011; Kaura et al., 2015; Levy, 2014; Sotiriadis and van Zyl, 2013), and behavioral intention (Baumann et al., 2012; Cronin et al., 2000; Szymanski and Henard, 2001; Zeithaml et al., 1996). Consumers who are satisfied with internet banking are more likely to engage in a consistent relationship with internet banking in future, which demonstrates a more loyal behavior (Al-alak, 2014; Gounaris et al., 2010; Levy, 2014; Martín-Consuegra et al., 2007; Wong and Zhou, 2006). However, when dissatisfaction is felt, customers are likely to complain about the service and engage in negative WOM (Caruana, 2002; Kandampully et al., 2015; Ryngelblum et al., 2013), and they will react negatively and alter to other service providers (Amin et al., 2011; Athanassopoulos et al., 2001; Laksamana et al., 2013; Wirtz et al., 2007). In this context, if the internet banking fails to provide channels which the customers need, they will find more difficulty in developing a relationship with their customers (Amin et al., 2013; Bloemer et al., 1998; Levy, 2014). As a result, a customer who satisfied with internet banking will exhibit high loyalty toward their bank. For this reason, e-customer satisfaction is considered as causal factor of e-customer loyalty. Thus, the following research hypothesis:

H19: Customer loyalty has a positive relationship with attitude.
H20: Customer loyalty has a positive relationship with intention to use.
H21: Customer satisfaction has a positive relationship with attitude.
H22: Customer satisfaction has a positive relationship with customer loyalty.
H23: Customer satisfaction has a positive relationship with intention to use.

The conceptual model of the research is presented in Figure 1.

Figure 1. Conceptual model

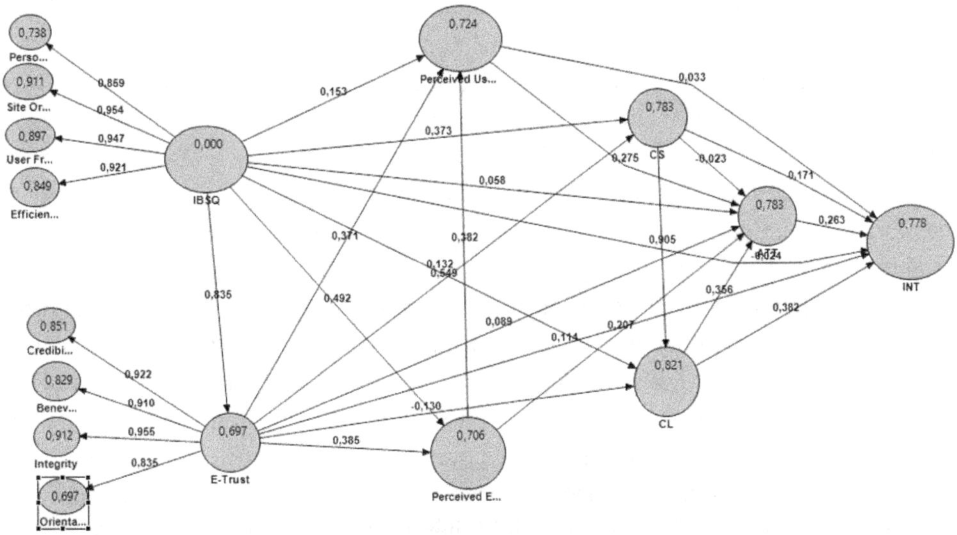

RESEARCH METHOD

Measures of the Constructs

This study a survey instrument was constructed by adapting measures from prior research to ensure the content validity of the scales. The standard TAM scales developed by Davis (1989) and validated by Taylor and Todd (1995a) and Taylor and Todd (1995b) were used to measure perceived ease of use, perceived usefulness, attitude and behavioral intention. These were previously shown to apply well to e-commerce context (Gefen et al., 2003). The measurement of trust was adapted from the scale developed by Chouk and Perrien (2005). Besides traditional three dimensions attributed to the concept of trust (credibility, integrity and benevolence), this scale introduces a new dimension "orientation to solve problems". This scale was tested and has demonstrated high reliability in e-commerce context as well (Janouri and Gharbi, 2008; Bartikowski et al., 2008).

Internet banking service quality was measured using the items from Herington and Weaven (2009) and Ho and Lin (2010). e-Customer satisfaction was measured using a five items previously used by Herington and Weaven (2009), Ramseook-Munhurrun and Naidoo (2011) and Ribbink et al. (2004) in their internet banking research. Five items to measure e-customer loyalty were adapted and modified from Amin et al. (2013), Ramseook-Munhurrun and Naidoo (2011), Ribbink et al., 2004 and Zeithaml et al. (1996). Thirty questionnaires were distributed to bank customers who have experienced using internet banking in Turkey, and minor changes to the scales were made accordingly to ensure that the questions were not repetitive. Pilot test respondents were not used in the further analysis.

The descriptive statistics of the sample showed that females account for 34.9 and males 65.1 percent of the internet banking customers. Half of the respondents were between 18-24 years of age, the percent of 27.1 was in the age group 25-34 years, 14.6 percent of respondents were between 35-44 years of age, 5.7 percent of respondents were between 45-54 years of age and 0.5 percent was in the age group 55 and + years. In this context, younger generation prefers to use internet banking services compared to older age groups (35 years old above). In addition, majority respondents claimed that they use internet for one to three hours daily (33.3 percent), followed by four to six hours (34.4 percent), for seven to eight hours daily (14.1), below one hour (4.7 percent), ten hours above with 13.5 percent. The majority of respondents were high school degree and represented 33.33%. followed by vocational school degree (27.6 percent), bachelor degree (18.8 percent), master degree (12 percent), Ph.D. degree (7.8 percent) and other education level (0.5 percent).

Finally, the survey instrument used in this study consisted of a total of 53 items related to the fourteen constructs of the research model. The items were measured using a five-point Likert-type scale for all constructs. To ensure the content validity of the instrument, all construct measures were adapted from their original source and slightly modified to fit the technology context studied. The validation of the survey instrument was done by experts from the field of information technology as well as the banking industry.

DATA COLLECTION PROCESS

To reach internet banking users, a Web-based survey was used. A snowball sampling method was applied. A total of 253 surveys were collected from the participants, of which 192 were found to be usable in the data analysis.

Data Analysis and Results

This study utilizes the Structural Equation Modeling (SEM) approach with Partial Least Square (PLS) as an analysis method. PLS has been widely used for theory testing and validation. PLS examines the psychometric properties and provides appropriate evidences on whether relationships might or might not exist (Fornell and Larcker, 1981). In this study, we performed data analysis in accordance with a two-stage methodology (Anderson and Gerbing, 1988) using SmartPLS 2.0 M3. The first step was to test the content, convergent, and discriminant validity of constructs using the measurement model, while the second step was to test the structural model and hypotheses.

Data Analysis

This book chapter utilizes the Structural Equation Modeling (SEM) approach with Partial Least Square (PLS) as an analysis method. PLS has been widely used for theory testing and validation. PLS examines the psychometric properties and provides appropriate evidences on whether relationships might or might not exist (Fornell and Larcker, 1981). In this study, we performed data analysis in accordance with a two-stage approach (Anderson & Gerbing, 1988; Henseler, Ringle, and Sinkovics, 2009) using SmartPLS 2.0 M3 (Ringle et al., 2005). The first step the measurement model is examined to test the reliability and validity of constructs using the measurement model, while the second step was to examine the structure model and hypotheses (Henseler, Ringle, and Sinkovics, 2009). When running the PLS-algorithm, however, the setting recommended by Hair et al. (2013) is used. In particular, the path-weighting scheme is selected with standardized data metric (mean = 0, variance = 1). In addition, the value of the maximum number of iterations was 300, the initial value for all outer weights was 1 and the value of stopping criteria was 1.0E-5. Based on an iterative algorithm, the PLS-SEM approach focuses on variance explanation (prediction of the constructs) that works well under conditions of small sample size and requires very few assumptions about the distribution of the variables. Moreover, the PLS-SEM approach is appropriate for research contexts that are in the early stage of development and have not been extremely examined to date (Gefen, Ridgon and Straub, 2011; Hair et al. 2013; Hair, Ringle and Sarstedt, 2011).

Evaluation of the Measurement Model

First, we assessed the reliability and validity of the measurement instrument using content, reliability, and convergent validity criteria. All the results of these criteria are presented in Table 2. Considering first the range of the factor loadings of the latent variables, the loadings are consistently above 0.791 and therefore exceed the critical level of 0.708, indicating good indicator reliability (Hair et al. 2013). In addition, the test of significance for each loading by the means of t-values shows that all values are highly significant at $p < 0.001$ (see Table 1). The internal consistency reliability was evaluated by using both Cronbach's alpha and composite reliability. According to Table 1, the internal consistency reliability is verified because each latent construct clearly exceeds the minimum recommended level of 0.70 for composite reliability (Bagozzi and Yi, 1988) and Cronbach's alpha (Nunnally and Bernstein, 1994), respectively. Convergent validity was estimated by using the

Table 1. Evaluation of the reflective measurement model.

					Inter-construct correlations[d]													
LC	**Range of factor loadings**	**CR[a]**	**A[b]**	**AVE[c]**	**CL**	**CS**	**EW**	**PN**	**SO**	**UF**	**ATT**	**BEN**	**CRD**	**INTG**	**INT**	**ORI**	**PEO**	**PU**
CL	0.914-0.957	0,97	0,96	0,88	1													
CS	0.907-0.954	0,97	0,96	0,87	0,90	1												
EW	0.894-0.924	0,94	0,9	0,83	0,71	0,77	1											
PN	0,827-0,896	0,9	0,83	0,75	0,74	0,77	0,72	1										
SO	0,859- 0,929	0,94	0,92	0,8	0,74	0,77	0,83	0,78	1									
UF	0,791- 0,906	0,92	0,89	0,75	0,7	0,77	0,85	0,73	0,88	1								
ATT	0,893- 0,926	0,94	0,9	0,83	0,83	0,83	0,74	0,71	0,72	0,69	1							
BEN	0,813- 0,912	0,92	0,88	0,74	0,68	0,78	0,71	0,70	0,64	0,67	0,72	1						
CRD	0,845-0,915	0,92	0,87	0,79	0,75	0,85	0,76	0,76	0,74	0,75	0,72	0,79	1					
INTG	0,861-0,937	0,96	0,95	0,8	0,71	0,79	0,74	0,78	0,73	0,73	0,74	0,80	0,84	1				
INT	0,868-0,897	0,91	0,86	0,78	0,85	0,84	0,69	0,73	0,67	0,66	0,82	0,68	0,71	0,75	1			
ORI	0,883-0,921	0,9	0,77	0,81	0,63	0,74	0,64	0,6	0,58	0,64	0,6	0,76	0,74	0,72	0,58	1		
PEO	0,862-0,897	0,93	0,89	0,76	0,79	0,84	0,77	0,74	0,76	0,74	0,81	0,73	0,75	0,76	0,78	0,63	1	
PU	0,896-0,915	0,95	0,93	0,82	0,79	0,83	0,76	0,72	0,72	0,68	0,82	0,75	0,78	0,74	0,77	0,67	0,80	1

[a]CR, composite reliability.
[b], Cronbach's alpha.
[c]AVE, average variance extracted.
[d]The square roots of the constructs' AVE values are shown in the diagonal line (in bold); non-diagonal elements are latent variable correlations.
[a]Average variance extracted (AVE)=(summation of the square of the factor loadings)/{(summation of the square of the factor loadings)+(summation of the error variances)}.
[b]Composite reliability (CR)=(square of the summation of the factor loadings)/{(square of the summation of the factor loadings)+(square of the summation of the error variances)}.

Table 2. Results of hypothesis testing and predictable power.

Hypotheses and path	β path coef.	t-value	f^2	R^2	q^2	Q^2
H1 Perceived Ease of Use -> ATT	0,207	1,963**	0,046		0,024	
H2 Perceived Ease of Use -> Perceived Usefulness	0,382367	4,0157***	0,155797101		0,086853219	
H3 Perceived Usefulness -> ATT	0,275338	2,7112***	0,087557604		0,044407285	
H4 Perceived Usefulness -> INT	0,033112	0,3873	0,067567568		0,024017651	
H5 ATT -> INT	0,263173	2,7671***	0,067567568	0.778	0,024017651	0,583
H6 E-Trust -> ATT	0,089372	0,9183				
H7 E-Trust -> CL	-0,130497	1,3144				
H8 E-Trust -> CS	0,549088	6,3149***		0.783		0,679390
H9 E-Trust -> INT	0,114216	1,1063				
H10 E-Trust -> Perceived Ease of Use	0,384772	3,9774***		0.706		0,532295
H11 E-Trust -> Perceived Usefulness	0,371358	3,6002***		0.724		0,589641
H12 IBSQ -> ATT	0,057768	0,5458				
H13 IBSQ -> CL	0,132392	1,6653*				
H14 IBSQ -> CS	0,373176	4,1175***				
H15 IBSQ -> E-Trust	0,834932	25,9953***		0.697		0,460065
H16 IBSQ -> INT	-0,024151	0,2319				
H17 IBSQ -> Perceived Ease of Use	0,491665	5,1333***				
H18 IBSQ -> Perceived Usefulness	0,153334	1,353				
H19 CL -> ATT	0,356351	3,0299***	0,101382488	0.783	0,050114428	0,649562
H20 CL -> INT	0,382452	3,6272***	0,103603604		0,031212154	
H21 CS -> ATT	-0,022959	0,1856	0		-0,001249864	
H22 CS -> CL	0,905404	13,7565***	0,994413408	0.821	0,551730872	0,718379
H23 CS -> INT	0,170832	1,1977	0,013513514		-0,002362195	

[a]t-values for two-tailed test:
* 1.65 (sig. level 10%).
** 1.96 (sig. level=5%).
*** t-value 2.58 (sig. level = 1%) (Hair et al., 2011).
Notes:***p<0.01, **p<0.05, *p<0.1 (two-sided).
effective size: 0 – none, 0.02 – small, 0.15 - medium, 0.35 – large (Cohen, 1988).
Effect sizes calculated using the following formulas
[1]$f^2 = R^2$included - R^2excluded / 1- R^2included (1)
[2]$q^2 = Q^2$included - Q^2excluded / 1- Q^2included (2)

average variance extracted (AVE). The values of each latent construct are above the suggested AVE value of 0.5, indicating a sufficient degree of convergent validity. Finally, the criterion by Fornell and Larcker (1981) is applied to evaluate discriminant validity. The square roots of all AVE scores are significantly larger than any other correlation coefficients among each construct, indicating good discriminant validity.

Further support for discriminant validity can be derived from the fact that all of the indicator's loadings are higher than all of its cross-loadings. Hence, all relevant criteria met the requirements for testing the structure model.

Table 3 shows weights of the first order constructs on the designated second order construct indicating that Internet banking service quality is a second order factor with four significant first order dimensions including personal need, site organization, user friendliness and efficiency of website. E-Trust is a second order factor with four significant first order dimensions including credibility, benevolence, integrity and orientation to resolve problems

EVALUATION OF THE STRUCTURAL MODEL

Similar to the previous section, the evaluation of the structural models in a PLS-SEM research context is based on specific criteria, including the level and significance of the path coefficients, the coefficients of determinates (R^2) and the predictive relevance (Q^2) of the path model (Hair et al., 2013). In addition to PLS Algorithm, the bootstrapping procedure was used and we selected 191 cases, 5000 samples, and the no sign changes option to evaluate the significance of the path coefficients (Hair, Sarstedt, Ringle, & Mena, 2012). The results of the PLS-SEM analysis show, as in Table 3, the structural model estimation and the evaluation of the formulated hypotheses.

The strength of the individual path coefficients and their statistical significance are obtained by applying the bootstrap re-sampling procedure with 5000 samples. Thus, as listed in Table 2, 14 hypothesized paths were statistically significant. Thus, H1, H2 H3, H5, H8, H10, H11, H13, H14, H15, H17, H19, H20 and H22 were supported.

Table 3. Weights of the first order constructs on the designated second-order constructs.

Second-Order Constructs	First-Order Constructs	Weight	T-Value
Internet banking service quality	Personal Need	0,858	29,973
	Site Organization	0,955	128,371
	User Friendliness	0,948	101,658
	Efficiency of Website	0,921	67,172
E-Trust	Credibility	0,922	69,443
	Benevolence	0,911	56,793
	Integrity	0,955	120,540
	Orientation to resolve problems	0,835	35,850

Notes: Critical t-values. **2.58 ($P < 0.01$).

Eight of the twenty three hypothesized paths, from perveived usefulness to intention (H4), from trust to attitude (H6), from trust to customer loyalty (H7), from trust to intention to use (H9), from internet banking service quality to attitude (H12), from internet banking service quality to intention to use (H16), from internet banking service quality to perceived usefulness (H18), from customer satisfaction to attitude (H21), from customer satisfaction to intention to use (H23) were not supported by statistically significant path coefficients. Because PLS is predictive-oriented, the R^2 measure is the essential criterion for structural model assessment, which presents the amount of explained variance of the endogenous latent variable. Although no generalities exist with regard to what is a high or low level of R^2, Chin (1998) and Henseler, Ringle, and Sinkovics (2009) describe R^2 values of 0.67, 0.33 and 0.19 as substantial, moderate and weak, respectively. While the R2 values were estimated with the bootstrapping procedure, the values of Stone–Geisser's Q^2 (Geisser 1974; Stone 1974) for assessing the research model's ability to predict were calculated by using the blindfolding procedure (Tenenhaus et al. 2005). Even though Q^2 can be measured in two forms, in this study the cross-validated redundancy measures of Q^2 are used, as suggested by Hair, Ringle, and Sarstedt (2011).

As in Table 2, The R^2 values of the endogenous latent construct were also obtained using the PLS algorithm procedure. According to the guideline by Cohen (1988), the fitted multiple regression model depends on R^2, and if R^2 value lies between 0.02 and 0.12, the model is weak, 0.13 and 0.25 is moderate and 0.26 and above is good (Gaur & Gaur, 2006; Griffith, 1996). However, Hair et al. (2012) suggested that the judgment of R2 value is high/low depends on the specific research context. To evaluate the structural models' predictive power, R squares (R^2) were calculated. R squares (R^2) indicates the amount of variance explained by the exogenous variables (Barclay et al., 1995). Using a bootstrapping technique with a re-sampling of 5000. The Stone–Geisser's Q^2 (Geisser, 1975; Stone, 1974) for the predictive sample reuse technique can be applied as a criterion for the predictive relevance besides looking at the magnitude of the R^2. According to Henseler et al. (2009), this measure is used to assess the research model's capability to predict. However, Stone–Geisser's Q^2 indicates to the interpretation that the model of our study is satisfying the predictive relevance, given that the results of R^2 versus Q^2. As the endogenous latent construct the R^2 value for customer loyalty is 0.821 with a Q^2 value of 0.718, which is relatively high. Customer satisfaction R^2=0.783, Q^2=0.679, perceived ease of use R2=0.706, Q^2 = 0.679, perceived usefulness R^2 = 0.724, Q2=0.590, attitude R^2=0.783, Q^2=0.650, intention R^2=0.778, Q^2=0.583, e-trust R^2=0.697, Q^2 =0.460 also show a large effect size. Thus, the explanatory power of all internet banking service quality variables explained %70 of the variance in e-trust is substantial, internet banking service quality variables and e-trust %71 of the variance in perceived ease of use and %72 of the variance in perceived usefulness is substantial. E-trust and internet banking service quality, perceived usefulness and perceived ease of use variables explained %82 of the variance in customer loyalty is substantial, %78 of the variance in customer satisfaction is substantial. E-trust and internet banking service quality, perceived usefulness and perceived ease of use, customer loyalty and customer satisfaction variables explained %78 of the variance in attitude can be described as substantial. All seven variables together explained %78 of the variance in intention is substantial. According to Hair et al. (2013), in SEM models, Q^2 values bigger than zero for a reflective endogenous construct imply the path model's predictive relevance for a particular construct (Using cross-validated redundancy is recommended). This study picked an omission distance (D=8). By performing blindfolding procedures, all Q^2 values are considerably above zero, which supports the model's (Figure 1) predictive relevance for the seven endogenous constructs Moreover, the path model has predictive relevance for the endogenous construct because all Q^2-values are consistently larger than zero (Henseler, Ringle, and Sinkovics, 2009). In addition, the f2 effect size, which shows the impact of a specific predictor construct on an endogenous latent construct, and the q^2 effect size for the predictive relevance are presented in Table 3. Besides measuring the values of R^2 and Q^2, the literature further suggests evaluating both effect size f^2 and q^2 for the R^2-values and Q^2-values, respectively. Whereas the effect size f^2 indicates how strongly the latent endogenous variable is influenced by individual exogenous variables, the effect size q^2 evaluates the predictive relevance of the exogenous variables for a certain endogenous variable (Hair et al., 2013; Hair, Ringle, and Sarstedt, 2011). In order to judge both of these effect sizes, Chin (1998) and Henseler, Ringle, and Sinkovics (2009) define values of 0.02, 0.15 and 0.35, which indicate small, medium and large effects, respectively.

Considering the underlying research model, the exogenous variable PEO has a medium effect size on endogenous variable PU (f^2 =0.159) and CL has a small effect size on ATT (f^2=0.101) and INT (f^2=0.104), whereas the effect size of CS on CL is very large (f^2 =0.994). Meanwhile, all other f^2-values involve a small effect size.

In terms of the prediction relevance of the individual exogenous variables, the q^2-value of 0.552 for the variable CS determines a large effect on the predictive relevance for CL. For all other q^2-values, the analysis reveals a weak contribution for the prediction relevance. Table 2 provides an overview of all relevant criteria for assessing the quality of the structure model.

Regarding model quality, Wetzels et al. (2009) provide a global fit (GoF) measure for PLS path modeling and derive a GoF measure with effect sizes for R^2 by substituting the minimum average AVE of 0.50. The GoF measures have been suggested as baseline values for validating the PLS model globally (the cut value of GoFsmall=0.1, GoFmedium=0.25, and GoFlarge=0.36) (Tenenhaus et al., 2005;

Wetzels et al., 2009). The research model obtained a GoF value of 0.78, which is greater than the cut-off value of 0.36 and is considered an adequate GoF. These results demonstrate this model performs well (see Table 4).

DISCUSSIONS AND MANAGERIAL IMPLICATIONS

The purpose of this chapter is to extend the technology acceptance model (TAM) by adding contextual variables (e-trust and internet banking service quality) and its implication on e-customer satisfaction and e-customer loyalty in the context of Turkish internet banking user's intention. Using the TAM as a theoretical framework, this study introduced trust-related dimensions (integrity, credibility, benevolence and orientation to resolve problems) and service quality dimensions (personal need, site organization, user friendliness, efficiency of website) as external variables to the original variables of the TAM (perceived ease of use, perceived usefulness, behavioral intention and attitude) to deepen people' internet banking adoption behavior. The results confirm that the all eight dimensions (personal need, site organization, user friendliness, efficiency of website, benevolence, credibility, integrity and orientation to resolve problems) are distinct constructs. The results also indicate that internet banking service quality and e-trust consisting of eight dimensions has appropriate reliability and each dimension has a significant relationship with internet banking service quality and e-trust. In this study, site organization is found to be the key driver of internet banking service quality, followed by user friendliness and efficiency of websites. And also, integrity is found to be the key driver of e-trust followed by credibility and benevolence. The findings of this study strongly support the

Table 4. Structural Model.

Fit measures	Endogenous Construct	Final Model
R^2	Customer loyalty	0.821
	Customer satisfaction	0.783
	Perceived ease of use	0.706
	Perceived usefulness	0.724
	Attitude	0.783
	Intention	0.778
	E-Trust	0.697
GoF		0,778
Note. GoF = $\sqrt{\text{Average Communality x Average } R^2}$		

relevance of using extended TAM to understand the managers' intention to adopt internet banking services. Significant effects of perceived usefulness, perceived ease of use and some service quality and trust dimensions on attitude and on behavioral intention were observed.

Our findings provide empirical support that indicates the internet banking service quality and e-trust are a second order factor with eight significant first order dimensions: personal need, site organization, user friendliness, and efficiency of website, integrity, credibility, benevolence and orientation to resolve problems. Our findings also highlight that internet banking service quality and e-trust and all eight of its significant dimensions all of which are considered by users as they evaluate a banking's web site.

As reckoned as TAM studies, the belief–attitude–intention–usage stream is still effective in predicting the people' internet banking behavior. The results show that perceived ease of use has a fairly significant effect on perceived usefulness and on attitude towards using the internet banking. Perceived usefulness almost determines attitude and intention to use the internet banking. This is consistent with the findings of previous studies in which users' intention behavior and attitude depend on their evaluation of the technology benefits. This evaluation is based on their expectation of a certain value provided by the combination of their beliefs about the main attributes of the technology in terms of usefulness and ease of use (Agarwal et al., 2009; Davis et al., 1992; Davis, 1989; Dishawa and Strong, 1999; Ezzi, 2014; Gefen, 2000; Maduku, 2013; Taylor and Todd,1995a; Taylor and Todd 1995b;Venkatesh and Davis, 2000). Furthermore, evidence exists to support that perceived ease of use has a lower effect on attitude than perceived usefulness in internet banking context. The relative importance of perceived usefulness has been widely recognized in the field of internet banking (Eriksson et al., 2005; Ezzi, 2014; Guriting and Ndubisi, 2006; Jaruwachirathanakul and Fink, 2005; Laforet and Li, 2005; Liao and Cheung, 2002; Maduku, 2013; O'Cass and Frenech, 2003; Polatoglu and Ekin, 2001; Suh and Han, 2002). Perceived ease of use was found to be a more salient determinant affecting perceived usefulness. This result seems to be consistent with the findings of prior studies and suggests that evaluation of perceived ease of use by active internet banking users increases its saliency in determining perceived usefulness or the pragmatic value of internet banking (Celik, 2008; Chang and Lu, 2004; Venkatesh and Davis, 2000). The indirect effects of perceived ease of use via perceived usefulness on attitude towards using and intention to use internet banking were also confirmed, as suggested by Celik (2008) and Suh and Han (2002).

Furthermore, we found that trust dimension has no a direct impact on behavioral intention, through perceived usefulness and attitude to adopt internet banking. Reid and Levy (2008) have also revealed a direct effect of trust on the perceptual variables of the TAM. We, also, found that internet banking service quality has no direct impact on behavioral intention, through perceived usefulness and attitude to adopt internet banking. Given that the usage of internet banking is completely voluntary and that the target user group consists of responsible with diversified backgrounds. The findings of this study suggest that to attract more users to internet banking, it is not going to be enough to make the system easy to interact with and useful. It is of paramount importance to develop internet banking systems with valuable functions and trustworthy protection of security and privacy for the users. Internet security was identified as the most important future challenge in e-banking, while customer trust,

privacy and awareness are being recognized as challenges of great importance (Gikandi and Bloor, 2010). Kim and Ahn (2006) found that the benevolence and the orientation to resolve problems dimensions of trust did not have a significant influence on the adoption of internet banking. Pavlou (2003), also, found that the multi-dimension trust including benevolence and credibility was only marginally related to the intention to transact online.

These results provide valuable insights for bank managers who should be aware of the importance of different beliefs on users' attitudes and adjust their marketing mix accordingly. Professionals should consider influencing internet banking behavior not only by developing ease of use and usefulness beliefs but also by promoting professional credibility and integrity. They should, certainly, create informative interfaces for potential adopters and provide accessible internet banking sites, fast page downloads, short transaction times, easy interfaces, a wide variety of financial products/services adapted to specific business' needs, customer-friendly website and frequent-user advantages for active users. Practitioners should, also, give extra attention to prevent internet banking users from consequently experiencing access difficulties, system crashes, drop outs, service delays and system malfunctions which consequently decrease their trust belief. Agarwal et al. (2009) note that banks should, firstly, promote elementary services to initiate customers to use internet banking and to develop their trust towards internet banking, with the hope that once customers start using internet banking for this purpose, they might continue using internet banking for other services also. Moreover, professionals should reveal their integrity and credibility by respecting their promises, by developing the features of security and privacy and by promoting their trustworthy brand. Agarwal et al. (2009) highlight that security and privacy depend on the bank's image in the minds of the users and on users' belief that organizational and technical arrangements have been made by the bank to ensure security and privacy of user data. The importance of security and privacy for the acceptance of online banking has been noted in many banking studies (Agarwal et al., 2009; Black et al., 2002; Chen and Barnes, 2007; Ezzi, 2014; Hamlet and Strube, 2000; Hernandez and Mazzon, 2007; Howcroft et al., 2002; Maduku, 2013; Polatoglu and Ekin, 2001; Sathye, 1999; Tan and Teo, 2000). Similarly, lack of privacy and security were found to be significant obstacles to the adoption of internet banking (Chen and Barnes, 2007; Sathye, 1999). Howcroft et al. (2002) noted that although consumers' confidence in their bank was strong, yet their confidence in the technology was weak. Agarwal et al. (2009) highlighted that the variables under security and trust include transparency, reliability, safety and privacy, which depict the degree of importance users attach with risk-free operations while doing their banking transactions online. As user satisfaction with these aspects is a reflection of users' belief that an organizational and technical infrastructure exists to support use of the system, dissatisfaction regarding these parameters indicates that users do not really believe that the banks have necessary measures in place to ensure safety and privacy of user data. The spat of scams related to online fraud have instilled more fear in the minds of users, thereby increasing user demand for security and safety in their online banking transactions. To combat this idea, internet banking professionals can organize training courses "one to one" and promotion approaches on various internet banking applications to increase the users' familiarity and to promote transparency and integrity. Internet banking professionals should consider alleviating these concerns by implementing measures to safeguard client assets and information such as authentication processes and passwords, and by supplying sufficient information about site security, giving unconditional loss guarantees, providing accessible customer service and educating customers. As safety and security are of higher concern to users while banking online, bank professionals should, also, try to attract users by focusing on and communicating the integrity of the bank and highlight the ways in which safety and security of the data have been ensured by the bank. Banks should work to build a safe and secure innovative reputation and obtain endorsements and positive word-of-mouth that should improve perceptions of banks' trustworthiness and will, also, enhance the perceptions of potential users about the website's usefulness. Popoola and Ibn Arshad, 2015 as well as Alsajjan and Dennis (2010) suggest that besides classical communication actions, promoting internet banking requires contemporary marketing communications, such as e-mails, blogs and social networks.

The internet banking service quality dimensions identified in this study are associated to web based-banking. Therefore, in order to maintain a high level of internet banking service quality, the internet banks should pay attention to all four dimensions identified. In this study, the site organization is found to be the key driver of internet banking service quality followed by user friendliness, efficiency of website, and personal need, respectively. It means that customers are looking for banks that provide easy to use user interfaces and lay outs and easily accessible. In other words, customers are paying more emphasize on the site organization dimensions than other dimensions as key factor in establishing relationship with their banks. Therefore, internet banking is required to focus on strategic choice in increasing consumer awareness and acceptance of new technology (banking system) in order to gain competitive advantage in a rigid competitive environment. This finding was in not consistent with Herington and Weaven (2009) and Sohail and Shaikh (2008) who stated that efficiency of website is the most influencing factor in user's evaluation of internet banking service quality and the internet banking customers are more focussing on download speed and completing a transaction quickly. Findings of the present study not support the results of research by Ho and Lin (2010) on internet banking, customers expect to complete their transactions correctly, on time, and have their e-mails response quickly. Similarly, Thaichon et al. (2014) suggests that by enhancing service quality, firms can influence customers' satisfaction, value, trust and commitment, which are important for firms' long-term sustainability. Therefore, the internet banking has to make sure that these expectations are met accordingly. For example, if a customer has a bad experience in regards to the interactivity of functionality of the internet banking website, then a customer might make overall internet banking service quality judgment before a transaction took place. In order to provide a good quality of internet services, internet banking should provide their customers with effective and efficient website in a suitably presented environment and update the technology development (Chemingui, 2013; Chen and Teng, 2013; Collier and Bienstock, 2006; Herington and Weaven, 2009; Ho et al., 2012; Jayawardhena, 2004). For this reason, Ho et al. (2012) emphasizes more the importance of physical (system functions and interface) and psychological attributes (services, information, attitudes) in understanding the dynamics of online customer behavior. Additionally, Ho et al. (2012) discover that quality plays an important part on internet searching behavior in two ways: improving the effectiveness and efficiency of websites, and users' acceptance of the website systems. Nowadays, customers are easily to find a large number of alternative banking sites, therefore, the main drivers of internet banking are depend on banking websites functionality and user interface (Vatanasombut et al., 2008).

Internet banking service quality has a positive relationship with customer satisfaction. As a result supports that the higher levels of internet banking service quality will significant effect to have a high level of e-customer satisfaction. In addition to relationships between internet banking service quality and e-customer satisfaction, prior research suggested that internet banking service quality has a positive relationship with e-customer satisfaction (Carlson and O'Cass, 2011; Herington and Weaven, 2009; Ho et al., 2012; Kaura et al., 2015). For example, Herington and Weaven (2009) highlights that internet banking service quality dimensions, such as efficiency of website is plays significant role in enhancing e-customer satisfaction. Additionally, Bressolles et al. (2014) suggests that managers should pay attention in upgrading usability aspects of their navigation interface and focus on security or privacy aspects in order to maintain the relationship with their customers. Internet banking service quality has a positive relationship with e-customer loyalty, and e-customer satisfaction has positive relationship with e-customer loyalty. Although, the results support H22, however, internet banking service quality has positive relationship with e-customer loyalty, thus H13 is supported. This finding is not consistent with the results found in the previous study (Al-Hawari, 2014, Al-Hawari, 2015; Casaló et al., 2008; Ganguli and Roy, 2011; Kashif et al., 2015; Levy, 2014; Ranjan et al., 2015). This may indicate that bank customers feel that having excellent internet bank service quality will have positive impact on e-customer loyalty. The results report that internet service quality has a direct effect on e-customer loyalty via e-customer satisfaction. This direct effect of internet banking service quality on e-customer loyalty discloses a new contribution to the existing literature that internet banking service quality has not been measured as a predictor factor of e-customer loyalty. In this context, Casaló et al. (2008) suggests that the ability of internet banking in terms of manageability of internet service quality will enhance e-customer satisfaction and e-loyalty. In fact, Carlson and O'Cass (2011) concluded that if e-service is delivered and evaluated as being of sufficient quality, then satisfaction with the service delivered will be a result. Similarly, Casaló et al. (2008) shows that satisfaction with previous interactions with the bank's website have a positive effect on customer loyalty and positive WOM. Additionally, Baumann et al. (2007) highlight that when customers are satisfied with a particular bank and they are willing to recommend their bank to others. As a result, the more satisfied customers with their banks, the more loyal and committed to the bank they developed. As a result, when customers are dissatisfied and they are willing to switch their banks to others (Amin et al., 2011; Collier and Bienstock, 2006; Fang et al., 2011; Laksamana et al., 2013; Ranaweera and Prabhu, 2003).

This work suggests that the higher level of e-customer satisfaction leads to a lower intention to leave the relationship and has the potential to increase e-customer loyalty, however, the intention of consumers to maintain the relationship with the internet banking largely depends on the satisfaction or dissatisfaction with their present services provider. Therefore, internet banking needs to improve the relationship between bank and customers, and –will raise the customer trust toward internet banking transactions. However, the correlation between e-satisfaction and e-loyalty becomes less important as risk perception increased. For this reason, Järvinen (2014) highlights that bank managers should take actions by reducing the levels of perceived risk in high-complexity banking services and maintaining customer service at high priority. If internet banking successful in transmitting a low perceived risk image, a risk-averse consumer will prefer to stay loyal, and this situation, the relationship between e-customer satisfaction and e-customer loyalty is stronger when perceived risk is significantly reduced (Aldas-Manzano et al., 2011; Dahlstrom et al., 2014). For this reason, the customers should perceive a feeling that the operations of internet banking system are secured and the legitimate of internet banking operation are backed up by the Central Bank of Turkey. Thus, it will create confidence that Turkish government rules are sufficiently enough to monitor internet banking. This has significant impact on consumer behavior decisions and influences their perceptions toward internet banks. This are consistent with Laforet and Li (2005) and Zhou (2011) who highlighted that consumers emphasize more on security aspects, and customers will hesitate to deal with their banks if they do not trust the internet banking transactions (Kshetri, 2013). If trust for internet banking is established accordingly, then the customer is more likely to accept it as a complementary in doing their online transactions (Dahlstrom et al., 2014; McNeish, 2015; Yap et al., 2010). In fact, if internet banking consumers sense that their transactions are unsafe, they will evacuate from the website even before completing the transactions. Therefore, it is fundamental for internet banking to increase consumer trust, as the perceived risk is correlated to possible losses transaction is greater than in offline banking transactions (Aldas-Manzano et al., 2011; Mukherjee and Nath, 2003, 2007). For this reason, Chemingui (2013), Jan and Kalthom (2014), Hurley et al. (2014), Yu et al. (2015), Zhao et al. (2010) and Zhou (2011) suggest that trust is plays important role in enhancing relationship between internet banks and customers. In addition, bank should proactively notify their customers that their personal data and all banking transactions are guarantee safely and kept confidentially. By providing numerous methods for customers to contact the online bank to get assistance is essential in improving the quality of the bank's online service operations. In fact, banks should provide online transaction procedures, information on how to deal with security problems, and instructions on how to use internet banking services securely. In addition, Kassim and Abdullah (2010), Rajaobelina et al. (2014) and Yu et al. (2015) suggests that to encourage e-customer satisfaction and e-customer loyalty, the banks should develop online systems which are trustworthy, secured, responsive, personalized for their users, perceived security/privacy, and information quality. In this context, banks should provide online transaction procedures, information on how to deal with security problems, and instructions on how to use internet banking services securely.

A revised TAM was proposed to investigate the impact of internet banking service quality and trust on the original variables of the TAM. A multidimensional trust variable (benevolence, integrity, credibility, orientation to resolve problems) and internet banking service quality (personal need, site organization, user friendliness, efficiency of website) was proposed to have significant influence on the intention to use internet banking directly and indirectly through perceived usefulness, perceived ease of use and attitude towards using internet banking. Thus, the contributions of trust on internet banking usage in Turkish internet banking users context were investigated. It was shown that mainly integrity and credibility dimensions of trust and site organization and user friendliness dimensions of internet banking service quality determine the usefulness of internet banking, the attitude to adopt e-banking services and the behavioral intention. The contributions of this study to customers internet banking adoption are three fold. First, it successfully applied the extended TAM in a new information context. Second, the trust variable was found to be an important determinant of perceived ease of use and perceived usefulness, attitude towards internet banking and behavioral intention. Three, internet banking service quality variable was found to be an important determinant of perceived ease of use, attitude towards internet banking and behavioral intention. Users' perception that e-transactions are easy and controllable is essential for customers' internet banking adoption, because ease of use and trust create controllability and usefulness perceptions. And also, site organization is essential for customers' internet banking adoption, because easy to use, well organized and user friendly user interfaces.

The findings of this study have implications for developing usable internet banking systems. Considering the important amounts that have been invested in internet banking systems throughout the world, it is of paramount importance to ensure that business users will actually use them. To achieve this goal, attention must be given to designing easy-to-use, useful and trustworthy systems. Internet banking professionals need to develop the beliefs of usefulness, ease of use, credibility and integrity of business users regarding internet banking.

LIMITATIONS AND FUTURE RESEARCH

This empirical study has some limitations. The total number of internet banking, sample size and covered areas of the study should be increased in order to achieve a proper result. The study findings and implications are obtained from a single study that examined a particular technology context and targeted a specific user group. Thus, additional research efforts, replicating the study in other working environments or for other emerging technologies in the banking industry, are needed to evaluate the validity of the investigated model and our findings. This study was conducted with a snapshot research approach. This reduces the ability of this study to reflect the temporal changes in the research constructs, especially when internet banking experiences increase. Longitudinal evidence might enhance our understanding of the causality and the interrelationships between variables that are important to the acceptance of internet banking. Self-selection bias could be a problem because only internet businesses with an active e-mail account were used in the data collection process. Therefore, offline surveys should be performed complementarily in conjunction with online surveys to collect representative samples. The intentions of the research participants to use internet banking were measured through self-reports in this study. Although some prior studies on technology acceptance provide evidence about the strong correlation between self-reports and actual system usage (Taylor and Todd, 1995a, Taylor and Todd, 1995b), future research efforts should be conducted to validate usage externally. Finally, there may be a need to extend the TAM by adding other service quality-related concepts whose importance is increasing for internet banking. The concept of perceived risk, which concerns perceived risk in parties' technologies and actions (Hoecht, 2004; Lim, 2003), can be a more comprehensive concept (Ozdemir et al., 2008). Additional internet service quality dimensions should be investigated such as interactivity and website services ability, that may improve our ability to predict customer usage intention more accurately.

REFERENCES

Adams, D. A., Nelson, R. R., & Todd, P. A. (1992). Perceived usefulness, ease of use, and usage of information technology: A replication. *Management Information Systems Quarterly*, *16*(2), 227–247. doi:10.2307/249577

Agarwal, R., Rastogi, S., & Mehrotra, A. (2009). Customers' perspectives regarding e-banking in an emerging economy. *Journal of Retailing and Consumer Services*, *16*(5), 340–351. doi:10.1016/j.jretconser.2009.03.002

Ajzen, I. (2006). *Theory of planned behavior*. Available at: http://people.umass.edu/aizen/tpb.diag.html#null-link

Al-Ajam, A. S., & Md Nor, K. (2015). Challenges of adoption of internet banking service in Yemen. *International Journal of Bank Marketing*, *33*(2), 178–194. doi:10.1108/IJBM-01-2013-0001

Al-alak, B. A. (2014). Impact of marketing activities on relationship quality in the Malaysian banking sector. *Journal of Retailing and Consumer Services*, *21*(3), 347–356. doi:10.1016/j.jretconser.2013.07.001

Al-hawari, M. A. (2015). How the personality of retail bank customers interferes with the relationship between service quality and loyalty. *International Journal of Bank Marketing*, *33*(1), 41–57. doi:10.1108/IJBM-09-2013-0096

Al-Hawari, M. A. A. (2014). Does customer sociability matter? Differences in e-quality, e-satisfaction, and e-loyalty between introvert and extravert online banking users. *Journal of Services Marketing*, *28*(7), 538–546. doi:10.1108/JSM-02-2013-0036

Al-Jabri, I. M., & Sohail, M. S. (2012). Mobile banking adoption: Application of diffusion of innovation theory. *Journal of Electronic Commerce Research*, *13*(4), 379–391.

Al-smadi, M. O. (2012). Factors affecting adoption of electronic banking: An analysis of the perspectives of banks' customers. *International Journal of Business and Social Science*, *3*(17).

Alalwan, A.A., Dwivedi, Y.K., Rana, N.P., Lal, B., & Williams, M.D. (2015). Consumer adoption of internet banking in Jordan: examining the role of hedonic motivation, habit, self-efficacy and trust. *Journal of Financial Services Marketing*, *20*(2), 145-157.

Aldás-Manzano, J., Lassala-Navarré, C., Ruiz-Mafé, C., & Sanz-Blas, S. (2009). The role of consumer innovativeness and perceived risk in online banking usage. *International Journal of Bank Marketing*, *27*(1), 53–75. doi:10.1108/02652320910928245

Aldas-Manzano, J., Ruiz-Mafe, C., Sanz-Blas, S., & Lassala-Navarré, C. (2011). Internet banking loyalty: Evaluating the role of trust, satisfaction, perceived risk and frequency of use. *Service Industries Journal*, *31*(7), 1165–1190. doi:10.1080/02642060903433997

Alsajjan, B., & Dennis, C. (2010). Internet banking acceptance model: Cross-market examination. *Journal of Business Research*, *63*(9), 957–963. doi:10.1016/j.jbusres.2008.12.014

Amin, M., Isa, Z., & Fontaine, R. (2011). The role of customer satisfaction in enhancing customer loyalty inMalaysian Islamic banks. *Service Industries Journal*, *31*(9), 1519–1532. doi:10.1080/02642060903576076

Amin, M., Isa, Z., & Fontaine, R. (2013). Islamic banks: Contrasting the drivers of customer satisfaction on image, trust, and loyalty of Muslim and non-Muslim customers in Malaysia. *International Journal of Bank Marketing*, *31*(2), 79–97. doi:10.1108/02652321311298627

Anderson, J. C., & Gerbing, D. (1988). Structural equation modeling in practice: A review and recommended two-step approach. *Psychological Bulletin*, *103*(3), 411–423. doi:10.1037/0033-2909.103.3.411

Angelakopoulos, G., & Mihiotis, A. (2011). E-banking: Challenges and opportunities in the Greek banking sector. *Electronic Commerce Research*, *11*(3), 297–319. doi:10.1007/s10660-011-9076-2

Bagozzi, R. P., & Yi, Y. (1988). On the Evaluation of Structural Equation Models. *Journal of the Academy of Marketing Science*, *16*(1), 74–94. doi:10.1007/BF02723327

Bajaj, A., & Nidumolu, S. R. (1998). A feedback model to understand information system usage. *Information & Management*, *33*(4), 213–224. doi:10.1016/S0378-7206(98)00026-3

Barclay, D., Higgins, C., & Thompson, R. (1995). The partial least squares (PLS) approach to causal modeling: Personal computer adoption and use as an illustration. *Technology Studies*, *2*(2), 285–309.

Bartikowski, B., Chandon, J. L., & Muller, B. (2008). *Mesurer la confiance des internautes dans les sites web marchands: adaptation de l'échelle de Mcknight, Kacmar et Choudhury*. Communication au 7ème congrès sur les tendances du marketing en Europe, Venise.

Baumann, C., Burton, S., Elliott, G., & Kehr, H. M. (2007). Prediction of attitude and behavioural intentions in retail banking. *International Journal of Bank Marketing*, *25*(2), 102–116. doi:10.1108/02652320710728438

Black, N. J., Lockett, A., Ennew, C., Winklhofer, H., & McKechnie, S. (2002). Modelling consumer choice of distribution channels: An illustration from financial services. *International Journal of Bank Marketing*, *20*(4), 161–173. doi:10.1108/02652320210432945

Bobbitt, L. M., & Dabholkar, P. A. (2001). Integrating attitudinal theories to understand and predict use of technology-based self-service. *International Journal of Service Industry Management*, *12*(5), 423–450. doi:10.1108/EUM0000000006092

Bodet, G. (2008). Customer satisfaction and loyalty in service: Two concepts, four constructs, several relationships. *Journal of Retailing and Consumer Services*, *15*(3), 156–162. doi:10.1016/j.jretconser.2007.11.004

Bressolles, G., Durrieu, F., & Senecal, S. (2014). A consumer typology based on e-service quality and e-satisfaction. *Journal of Retailing and Consumer Services*, *21*(6), 889–896. doi:10.1016/j.jretconser.2014.07.004

Calisir, F., & Gumussoy, C. A. (2008). Internet banking versus other banking channels: Young consumers' view. *International Journal of Information Management*, *28*(3), 215–221. doi:10.1016/j.ijinfomgt.2008.02.009

Candra, S. (2013). Revisit technology acceptance model for internet banking (Case Study: Public Banking in Indonesia). *International Conference on Computer, Networks and Communication Engineering (ICCNCE 2013)*.

Cao, X., & Mokhtarian, P. (2005). *The intended and actual adoption of online purchasing: a brief review of recent literature*. University of California Transportation.

Carlson, J., & O'Cass, A. (2011). Developing a framework for understanding e-service quality, its antecedents, consequences, and mediators. *Managing Service Quality*, *21*(3), 264–286. doi:10.1108/09604521111127965

Casaló, L. V., Flavián, C., & Guinalíu, M. (2008). The role of satisfaction and website usability in developing customer loyalty and positive word-of-mouth in the e-banking services. *International Journal of Bank Marketing*, *26*(6), 399–417. doi:10.1108/02652320810902433

Celik, H. (2008). What determines Turkish consumers' acceptance of internet banking? *International Journal of Bank Marketing*, *26*(5), 353–370. doi:10.1108/02652320810894406

Chang, S., & Lu, M. (2004). Understanding internet adoption and use behavior: A Hong Kong perspective. *Journal of Global Information Management*, *12*(3), 21–43. doi:10.4018/jgim.2004070102

Charfeddine, L., & Nasri, W. (2013). The behavior intention of Tunisian banks' customers on using internet banking. *International Journal of Innovation in the Digital Economy*, *4*(1), 16–30. doi:10.4018/jide.2013010102

Chemingui, H., & Ben lallouna, H. (2013). Resistance, motivations, trust and intention to use mobile financial services. *International Journal of Bank Marketing*, *31*(7), 574–592. doi:10.1108/IJBM-12-2012-0124

Chen, C. (2013). Perceived risk, usage frequency of mobile banking services. *Managing Service Quality: An International Journal*, *23*(5), 410–436. doi:10.1108/MSQ-10-2012-0137

Chen, M.-Y., & Teng, C.-I. (2013). A comprehensive model of the effects of online store image on purchase intention in an e-commerce environment. *Electronic Commerce Research*, *13*(1), 1–23. doi:10.1007/s10660-013-9104-5

Chen, Q., & Wells, W. D. (2002). Attitude toward the site II: New information. *Journal of Advertising Research*, *42*(2), 33–45. doi:10.2501/JAR-42-2-33-45

Chen, R.-F., Hsiao, J.-L., & Hwang, H.-G. (2012). Measuring customer satisfaction of Internet banking in Taiwan: Scale development and validation. *Total Quality Management & Business Excellence*, *23*(7/8), 749–767. doi:10.1080/14783363.2012.704284

Chen, Y. H., & Barnes, S. (2007). Initial trust and online buyer behaviour. *Industrial Management & Data Systems, 107*(1), 21–36. doi:10.1108/02635570710719034

Cheng, T. C., Lam, D. Y. C., & Yeung, A. C. L. (2006). Adoption of internet banking: An empirical study in Hong Kong. *Decision Support Systems, 42*(3), 1558–1572. doi:10.1016/j.dss.2006.01.002

Cheong, J. H., & Park, M. C. (2005). Mobile internet acceptance in Korea. *Internet Research, 15*(2), 125–140. doi:10.1108/10662240510590324

Chin, W. W. (1998). The Partial Least Squares Approach to Structural Equation Modeling. In G. A. Marcoulides (Ed.), *Modern Methods for Business Research* (pp. 295–358). London: Lawrence Erlbaum Associates.

Chiou, J.-S., & Shen, C.-C. (2012). The antecedents of online financial service adoption: The impact of physical banking services on Internet banking acceptance. *Behaviour & Information Technology, 31*(9), 859–871. doi:10.1080/0144929X.2010.549509

Chouk, I., & Perrien, J. (2005). La confiance du consommateur vis-a'-vis d'un marchand internet: Proposition d'une échelle de mesure. *Revue Française du Marketing, 205*(1), 5–20.

Cohen, J. (1988). *Statistical Power Analysis for the Behavioral Sciences*. Hillsdale, NJ: Lawrence Erlbaum.

Collier, J. E., & Bienstock, C. C. (2006). Measuring service quality in e-retailing. *Journal of Service Research, 8*(3), 260–275. doi:10.1177/1094670505278867

Cronin, J. J. Jr, & Taylor, S. A. (1992). Measuring service quality: A reexamination and extension. *Journal of Marketing, 56*(3), 55–68. doi:10.2307/1252296

Dahlstrom, R., Nygaard, A., Kimasheva, M., & Ulvnes, A. M. (2014). How to recover trust in the banking industry? A game theory approach to empirical analyses of bank and corporate customer relationships. *International Journal of Bank Marketing, 32*(4), 268–278. doi:10.1108/IJBM-03-2014-0042

Davis, F. D. (1989). Perceived usefulness, perceived ease of use and user acceptance of computer technology. *Management Information Systems Quarterly, 13*(3), 319–337. doi:10.2307/249008

Davis, F. D. (1993). User acceptance of information technology: System characteristics, user perceptions and behavioral impacts. *International Journal of Man-Machine Studies, 38*(3), 475–487. doi:10.1006/imms.1993.1022

Davis, F. D., Bagozzi, R., & Warshaw, P. R. (1992). Extrinsic and intrinsic motivation to use computers in the workplace. *Journal of Applied Social Psychology*, *22*(14), 1109–1130. doi:10.1111/j.1559-1816.1992.tb00945.x

Davis, F. D., Bagozzi, R. P., & Warshaw, P. R. (1989). User acceptance of computer technology: A comparison of two theoretical models. *Management Science*, *35*(8), 982–1003. doi:10.1287/mnsc.35.8.982

DeYoung, R., Lang, W. W., & Nolle, D. L. (2007). How the internet affects output and performance at community banks. *Journal of Banking & Finance*, *31*(4), 1033–1060. doi:10.1016/j.jbankfin.2006.10.003

Dick, A. S., & Basu, K. (1994). Customer loyalty, toward an integrated conceptual framework. *Journal of the Academy of Marketing Science*, *22*(2), 99–113. doi:10.1177/0092070394222001

Dishawa, M. T., & Strong, D. M. (1999). Extending the technology acceptance model with task-technology fit constructs. *Information & Management*, *36*(1), 9–21. doi:10.1016/S0378-7206(98)00101-3

Eriksson, K., Kerem, K., & Nilsson, D. (2005). Customer acceptance of internet banking in commerce: Using and extension of the technology acceptance model. *Journal of Internet Banking and Commerce*, *12*(3), 1–26.

Ezzi, S. Z. (2014). A theoretical model for internet banking: Beyond perceived usefulness and ease of use. *Archives of Business Research*, *2*(2), 31–46. doi:10.14738/abr.22.184

Fang, Y.-H., Chiu, C.-M., & Wang, E. T. (2011). Understanding customers' satisfaction and repurchase intentions: An integration of IS success model, trust, and justice. *Internet Research*, *21*(4), 479–503. doi:10.1108/10662241111158335

Fishbein, M., & Ajzen, I. (1975). *Belief, Attitude, Intention and Behavior: An Introduction to Theory and Research*. Reading, MA: Addison-Wesley.

Fonseca, J. R. (2014). e-banking culture: A comparison of EU 27 countries and Portuguese case in the EU 27 retail banking context. *Journal of Retailing and Consumer Services*, *21*(5), 708–716. doi:10.1016/j.jretconser.2014.05.006

Fornell, C., & Larcker, D. F. (1981). Evaluation Structural Equation Models with Unobservable Variables and Measurement Error. *JMR, Journal of Marketing Research*, *18*(1), 39–50. doi:10.2307/3151312

Fusilier, M., & Durlabhji, S. (2005). An exploration of student Internet use in India. *Campus-Wide Information Systems*, *22*(4), 233–246. doi:10.1108/10650740510617539

Ganguli, S., & Roy, S. K. (2011). Generic technology-based service quality dimensions in banking: Impact on customer satisfaction and loyalty. *International Journal of Bank Marketing*, *29*(2), 168–189. doi:10.1108/02652321111107648

Gaur, A. S., & Gaur, S. S. (2006). Statistical methods for practice and research: A guide to data analysis using SPSS. *Sage (Atlanta, Ga.).*

Gefen, D. (2000). E-commerce: The role of familiarity and trust. *Omega*, *28*(6), 725–737. doi:10.1016/S0305-0483(00)00021-9

Gefen, D., Karahanna, E., & Straub, D. W. (2003). Trust and TAM in online shopping: An integrated model. *Management Information Systems Quarterly*, *27*(1).

Gefen, D., Ridgon, E. E., & Straub, D. W. (2011). Editor's Comments: An Updated and Extension to SEM Guidelines for Administrative and Social Science Research. *Management Information Systems Quarterly*, *35*(2), iii–xiv.

Gefen, D., & Straub, D. (2003). Managing user trust in b to c e- service. *e-Service Journal, 2*(2), 7-24.

Geisser, S. (1974). A Predictive Approach to the Random Effect Model. *Biometrika*, *61*(1), 101–107. doi:10.1093/biomet/61.1.101

Geisser, S. (1975). The predictive sample reuse method with applications. *Journal of the American Statistical Association*, *70*(350), 320–328. doi:10.1080/01621459.1975.10479865

Gerrard, P., & Cunningham, J. B. (2003). The diffusion of internet banking among Singapore consumers. *International Journal of Bank Marketing*, *21*(1), 16–28. doi:10.1108/02652320310457776

Gikandi, J., & Bloor, C. (2010). Adoption and effectiveness of electronic banking in Kenya. *Electronic Commerce Research and Applications*, *9*(1), 277–282. doi:10.1016/j.elerap.2009.12.003

Giordani, G., Floros, C., & Judge, G. (2014). Econometric investigation of internet banking adoption in Greece. *Journal of Economic Studies (Glasgow, Scotland)*, *41*(4), 586–600. doi:10.1108/JES-04-2011-0042

Grabner–Krauter, S., & Faullant, R. (2008). Consumer acceptance of internet banking: The influence of internet trust. *International Journal of Bank Marketing*, *16*(7), 483–504. doi:10.1108/02652320810913855

Grewal, D., Iyer, G., & Levy, M. (2004). Internet retailing: Enablers, limiters and market consequences. *Journal of Business Research*, *57*(7), 703–7013. doi:10.1016/S0148-2963(02)00348-X

Griffith, D. A. (1996). *Some guidelines for specifying the geographic weights matrix contained in spatial statistical models*. Practical Handbook of Spatial Statistics.

Gu, J. C., Lee, S. C., & Suh, Y. H. (2009). Determinants of behavioral intention to mobile banking. *Expert Systems with Applications*, *36*(9), 11605–11616. doi:10.1016/j.eswa.2009.03.024

Guriting, P., & Ndubisi, N. O. (2006). Borneo online banking: Evaluating customer perceptions and behavioural intention. *Management Research News*, *29*(1/2), 6–15. doi:10.1108/01409170610645402

Ha, S., & Stoel, L. (2008). Promoting customer-retailer relationship building: Influence of customer trustworthiness of customer loyalty programme marketing. *Journal of Customer Behaviour*, *7*(3), 215–229. doi:10.1362/147539208X369652

Hair, J. F., Hult, G. T. M., Ringle, C. M., & Sarstedt, M. (2013). *A Primer on Partial Least Squares Structural Equation Modeling (PLS-SEM)* (1st ed.). Thousand Oaks, CA: Sage.

Hair, J. F., Ringle, C. M., & Sarstedt, M. (2011). PLS-SEM: Indeed a Silver Bullet. *Journal of Marketing Theory and Practice*, *18*(2), 139–152. doi:10.2753/MTP1069-6679190202

Hair, J. F., Sarstedt, M., Ringle, C. M., & Mena, J. A. (2012). An Assessment of the Use of Partial Least Squares Structural Equation Modeling in Marketing Research. *Journal of the Academy of Marketing Science*, *40*(3), 414–433. doi:10.1007/s11747-011-0261-6

Hamlet, C., & Strube, M. (2000). Community banks go online. *ABA Banking Journal, 92*(3), 61-65.

Hanafizadeh, P., Keating, B. W., & Khedmatgozar, H. R. (2014). A systematic review of internet banking adoption. *Telematics and Informatics*, *31*(3), 492–510. doi:10.1016/j.tele.2013.04.003

Harrison, T. S., Onyia, O. P., & Tagg, S. K. (2014). Towards a universal model of internet banking adoption: Initial conceptualization. *International Journal of Bank Marketing*, *32*(7), 647–687. doi:10.1108/IJBM-06-2013-0056

Hasim, M. S., & Salman, A. (2010). Factors affecting sustainability of internet usage among youth. *The Electronic Library*, *28*(2), 300–313. doi:10.1108/02640471011033657

Hausman, A., & Siekpe, J. (2008). The effect of web interface features on consumer online purchase intentions. *Journal of Business Research, 8*(1), 1-9.

Henseler, J., Ringle, C. M., & Sinkovics, R. R. (2009). The Use of Partial Least Squares Path Modeling in International Marketing. In R. R. Sinkovics & P. N. Ghauri (Eds.), *Advances in International Marketing 20* (pp. 277–320). Bingley, UK: Emerald. doi:10.1108/S1474-7979(2009)0000020014

Henseler, J., Ringle, C. M., & Sinkovics, R. R. (2009). The Use of Partial Least Squares Path Modeling inInternational Marketing. In R. R. Sinkovics & P. N. Ghauri (Eds.), *Advances in International Marketing* (Vol. 20, pp. 277–320). doi:10.1108/S1474-7979(2009)0000020014

Herington, C., & Weaven, S. (2009). E-retailing by banks: E-service quality and its importance to customer satisfaction. *European Journal of Marketing*, *43*(9/10), 1220–1231. doi:10.1108/03090560910976456

Hernandez, J. M., & Mazzon, J. A. (2007). Adoption of internet banking: Proposition and implementation of an integrated methodology approach. *International Journal of Bank Marketing*, *25*(2), 72–88. doi:10.1108/02652320710728410

Ho, C.-T. B., & Lin, W.-C. (2010). Measuring the service quality of internet banking: Scale development and validation. *European Business Review*, *22*(1), 5–24. doi:10.1108/09555341011008981

Ho, L.-A., Kuo, T.-H., & Lin, B. (2012). The mediating effect of website quality on Internet searching behavior. *Computers in Human Behavior*, *28*(3), 840–848. doi:10.1016/j.chb.2011.11.024

Hoecht, A. (2004). Control in collaborative research and technology development: A case study in the chemical industry. *Journal of Managerial Psychology*, *19*(3), 218–234. doi:10.1108/02683940410527720

Howcroft, B., Hamilton, R., & Hewer, P. (2002). Consumer attitude and the usage and adoption of home-based banking in the United Kingdom. *International Journal of Bank Marketing*, *20*(3), 111–121. doi:10.1108/02652320210424205

Hu, P. J., Chau, P. Y. K., Sheng, O. R. L., & Tam, K. Y. (1999). Examining the technology acceptance model using physician acceptance of telemedicine technology. *Journal of Management Information Systems*, *16*(2), 91–112. doi:10.1080/07421222.1999.11518247

Hurley, R., Tahir Jan, M., & Abdullah, K. (2014). The impact of technology CSFs on customer satisfaction and the role of trust. *International Journal of Bank Marketing*, *32*(5), 429–447. doi:10.1108/IJBM-11-2013-0139

Igbaria, M., Parasuraman, S., & Baroudi, J. J. (1996). A motivational model of microcomputer usage. *Journal of Management Information Systems*, *13*(1), 127–144. doi:10.1080/07421222.1996.11518115

Jan, M. T., & Kalthom, A. (2014). The impact of technology CSFs on customer satisfaction and the role of trust. *International Journal of Bank Marketing*, *32*(5), 429–447. doi:10.1108/IJBM-11-2013-0139

Janouri, F., & Gharbi, J. E. (2008). *Affrontement entre le risque perçu et le capital marque dans la formation de la confiance dans un site marchand*. Available at: www.esscpeap. net/conferences/marketing/2008-cp/Materiali/Paper/Fr/Janouri-Gharbi.pdf

Jaruwachirathanakul, B., & Fink, D. (2005). Internet banking adoption strategies for a developing country: The case of Thailand. *Internet Research*, *15*(3), 295–311. doi:10.1108/10662240510602708

Jaruwachirathanakul, B., & Fink, D. (2005). Internet banking adoption strategies for a developing country: The case of Thailand. *Internet Research*, *15*(3), 295–311. doi:10.1108/10662240510602708

Järvinen, R. A. (2014). Consumer trust in banking relationships in Europe. *International Journal of Bank Marketing*, *32*(6), 551–566. doi:10.1108/IJBM-08-2013-0086

Karahanna, E., Straub, D. W., & Chervany, N. L. (1999). Information technology adoption across time: A cross-sectional comparison of pre-adoption and post-adoption beliefs. *Management Information Systems Quarterly*, *23*(2), 183–213. doi:10.2307/249751

Karjaluoto, H., Mattila, M., & Pento, T. (2002). Factors underlying attitude formation towards online banking in Finland. *International Journal of Bank Marketing*, *20*(6), 261–272. doi:10.1108/02652320210446724

Kashif, M., Wan Shukran, S. S., Rehman, M. A., Sarifuddin, S., Estelami, H., & Heinonen, K. (2015). Customer satisfaction and loyalty in Malaysian Islamic banks: A PAKSERV investigation. *International Journal of Bank Marketing*, *33*(1), 23–40. doi:10.1108/IJBM-08-2013-0084

Kassim, N., & Abdullah, N. A. (2010). The effect of perceived service quality dimensions on customer satisfaction, trust, and loyalty in e-commerce settings: A cross cultural analysis. *Asia Pacific Journal of Marketing and Logistics*, *22*(3), 351–371. doi:10.1108/13555851011062269

Kaura, V., Prasad, C. S. D., & Sharma, S. (2015). Service quality, service convenience, price and fairness, customer loyalty, and the mediating role of customer satisfaction. *International Journal of Bank Marketing*, *33*(4), 404–422. doi:10.1108/IJBM-04-2014-0048

Khalil, N., Sutanonpaiboon, J., & Mastor, N. H. (2010). Malay, Chinese, and internet banking. *Chinese Management Studies*, *4*(2), 141–153. doi:10.1108/17506141011053069

Kim, M. S., & Ahn, J. H. (2006). Comparison of trust sources of an online market-maker in the e-marketplace: Buyer's and seller's perspectives. *Journal of Computer Information Systems*, *47*(1), 84–94.

Kshetri, N. (2013). Cybercrime and cyber-security issues associated with China: Some economic and institutional considerations. *Electronic Commerce Research*, *13*(1), 41–69. doi:10.1007/s10660-013-9105-4

Laforet, S., & Li, X. (2005). Consumers: Attitudes towards online and mobile banking in China. *International Journal of Bank Marketing*, *23*(5), 362–380. doi:10.1108/02652320510629250

Laksamana, P., Wong, D., Kingshott, R. P., & Muchtar, F. (2013). The role of interaction quality and switching costs in premium banking services. *Marketing Intelligence & Planning*, *31*(3), 229–249. doi:10.1108/02634501311324591

Levy, S. (2014). Does usage level of online services matter to customers' bank loyalty? *Journal of Services Marketing*, *28*(4), 292–299. doi:10.1108/JSM-09-2012-0162

Li-hua, Y. (2012). Customer satisfaction antecedents within service recovery context: Evidences from 'Big 4' banks in China. *Nankai Business Review International*, *3*(3), 284–301. doi:10.1108/20408741211264585

Liao, Z., & Cheung, M. T. (2002). Internet-based e-banking and consumer attitudes: An empirical study. *Information & Management*, *39*(4), 283–295. doi:10.1016/S0378-7206(01)00097-0

Lim, N. (2003). Consumers' perceived risk: Sources versus consequences. *Electronic Commerce Research and Applications*, *2*(3), 216–228. doi:10.1016/S1567-4223(03)00025-5

Lin, C. H., Shih, H. Y., & Sher, P. J. (2007). Integrating technology readiness into technology acceptance: The TRAM model. *Psychology and Marketing*, *24*(7), 641–657. doi:10.1002/mar.20177

Lin, H. F. (2011). An empirical investigation of mobile banking adoption: The effect of innovation attributes and knowledge-based trust. *International Journal of Information Management*, *31*(3), 252–260. doi:10.1016/j.ijinfomgt.2010.07.006

Lin, J., & Lu, H. (2000). Towards an understanding of the behavioural intention to use a web site. *International Journal of Information Management*, *20*(1), 197–208.

Luarn, P., & Lin, H. H. (2005). Toward an understanding of the behavioral intention to use mobile banking. *Computers in Human Behavior*, *21*(6), 873–891. doi:10.1016/j.chb.2004.03.003

Maduku, D. K. (2013). Predicting retail banking consumers' attitude towards internet banking services in South Africa. *Southern African Business Review*, *17*(3).

McFarland, D., & Hamilton, D. (2004). Adding contextual specificity to the technology acceptance model. *Computers in Human Behavior*, *22*(3), 427–447. doi:10.1016/j.chb.2004.09.009

McKechnie, S., Winklhofer, H., & Ennew, C. (2006). Applying the technology acceptance model to the online retailing of financial services. *International Journal of Retail & Distribution Management*, *34*(4/5), 388–410. doi:10.1108/09590550610660297

Mcknight, D. H., Choudhury, V., & Kacmar, C. (2002a). Developing and validating trust measures for ecommerce: An integrative typology. *Information Systems Research*, *13*(3), 334–359. doi:10.1287/isre.13.3.334.81

McKnight, D., Choudhury, V., & Kacmar, C. (2002b). The impact of initial consumer trust on intentions to transact with a web site: A trust building model. *The Journal of Strategic Information Systems*, *11*(1), 297–323. doi:10.1016/S0963-8687(02)00020-3

McNeish, J. (2015). Consumer trust and distrust: Retaining paper bills in online banking. *International Journal of Bank Marketing*, *33*(1), 5–22. doi:10.1108/IJBM-08-2013-0088

Moon, J., & Kim, Y. (2001). Extending the tam for a word wide web context. *Information & Management*, *38*(4), 217–230. doi:10.1016/S0378-7206(00)00061-6

Mukherjee, A., & Nath, P. (2003). A model of trust in online relationship banking. *International Journal of Bank Marketing*, *21*(1), 5–15. doi:10.1108/02652320310457767

Mukherjee, A., & Nath, P. (2007). Role of electronic trust in online retailing: A re-examination of the commitment-trust theory. *European Journal of Marketing*, *41*(9/10), 1173–1202. doi:10.1108/03090560710773390

Narayanasamy, K., Rasiah, D., & Tan, T. M. (2011). The adoption and concerns of e-finance in Malaysia. *Electronic Commerce Research*, *11*(4), 383–400. doi:10.1007/s10660-011-9081-5

Nasri, W. (2011). Factors influencing the adoption of internet banking in Tunisia. *International Journal of Business and Management*, *6*(8). doi:10.5539/ijbm.v6n8p143

Nor, K. M., & Pearson, J. M. (2008). An exploratory study into the adoption of internet banking in a developing country: Malaysia. *Journal of Internet Commerce*, *7*(1), 29–73. doi:10.1080/15332860802004162

Norazah, M. S., Ramayah, T., & Norbayah, M. S. (2008). Internet shopping acceptance: Examining the influence of intrinsic versus extrinsic motivations. *Direct Marketing: An International Journal*, *2*(2), 97–110. doi:10.1108/17505930810881752

Nunnally, J. C., & Bernstein, I. H. (1994). *Psychometric Theory* (3rd ed.). New York: McGraw-Hill.

Nysveen, H., Pedersen, P. E., & Thorbjørnsen, H. (2005). Intentions to use mobile services: Antecedents and cross-service comparisons. *Journal of the Academy of Marketing Science*, *33*(3), 330–346. doi:10.1177/0092070305276149

O'cass, A., & Frenech, T. (2003). Web retailing adoption: Exploring the nature of Internet users web retailing behavior. *Journal of Retailing and Consumer Services*, *10*(2).

Oliveira, T., Faria, M., Thomas, M. A., & Popovič, A. (2014). Extending the understanding of mobile banking adoption: When UTAUT meets TTF and ITM. *International Journal of Information Management*, *34*(5), 689–703. doi:10.1016/j.ijinfomgt.2014.06.004

Ozdemir, S., Trott, P., & Hoecht, A. (2008). Segmenting internet banking adopter and non-adopters in the Turkish retail banking sector. *International Journal of Bank Marketing*, *26*(4), 212–236. doi:10.1108/02652320810884777

Pavlou, P. A. (2003). Consumer acceptance of electronic commerce: Integrating trust and risk with technology acceptance model. *International Journal of Electronic Commerce*, *7*(3).

Pikkarainen, T., Pikkarainen, K., Karjaluoto, H., & Pahnila, S. (2004). Consumer acceptance of online banking: An extension of the technology acceptance model. *Internet Research*, *14*(3), 224–235. doi:10.1108/10662240410542652

Polatoglu, V. N., & Ekin, S. (2001). An empirical investigation of the Turkish consumers' acceptance of internet banking services. *International Journal of Bank Marketing*, *19*(4), 156–165. doi:10.1108/02652320110392527

Popoola, N. F., & Ibn Arshad, M. D. R. (2015). Strategic approach to build consumers trust in adoption of internet banking in Nigeria. *Journal of Internet Banking and Commerce*, *20*(1).

Rajaobelina, L., Ricard, L., Bergeron, J., & Toufaily, É. (2014). An integrative model of installed online trust in the financial services industry. *Journal of Financial Services Marketing*, *19*(3), 186–197. doi:10.1057/fsm.2014.18

Ramseook-Munhurrun, P., & Naidoo, P. (2011). Customers' perspectives of service quality in internet banking. *Services Marketing Quarterly*, *32*(4), 247–264. doi:10.1080/15332969.2011.606753

Ranaweera, C., & Prabhu, J. (2003). The influence of satisfaction, trust and switching barriers on customer retention in a continuous purchasing setting. *International Journal of Service Industry Management*, *14*(4), 374–395. doi:10.1108/09564230310489231

Ranjan, K. R., Sugathan, P., Rossmann, A., & Martin, C. (2015). A narrative review and metaanalysis of service interaction quality: New research directions and implications. *Journal of Services Marketing*, *29*(1), 3–14. doi:10.1108/JSM-01-2014-0029

Reid, M., & Levy, Y. (2008). Integrating trust and computer self-efficacy with TAM: An empirical assessment of customers' acceptance of Banking Information Systems (BIS) in Jamaica. *Journal of Internet Banking and Commerce*, *13*(3), 1–19.

Rezaei, S., Amin, M., & Wan Ismail, W. K. (2014). Online repatronage intention: An empirical study among Malaysian experienced online shoppers. *International Journal of Retail & Distribution Management*, *42*(5), 390–421. doi:10.1108/IJRDM-03-2012-0026

Ribbink, D., Van Riel, A. C., Liljander, V., & Streukens, S. (2004). Comfort your online customer: Quality, trust and loyalty on the internet. *Managing Service Quality*, *14*(6), 446–456. doi:10.1108/09604520410569784

Ringle, C. M., Sarstedt, M., & Straub, D. M. (2012). Editor's Comments: A Critical Look at the Use of PLS-SEM in MIS Quarterly. *Management Information Systems Quarterly*, *36*(1), iii–xiv.

Ringle, C. M., Wende, S., & Will, A. (2005). *SmartPLS 2.0 (M3) Beta*. Retrieved from http://www.smartpls.de

Rod, M., & Ashill, N. J. (2010). Management commitment to service quality and service recovery performance: A study of frontline employees in public and private hospitals. *International Journal of Pharmaceutical and Healthcare Marketing*, *4*(1), 84–103. doi:10.1108/17506121011036042

Rod, M., Ashill, N. J., Shao, J., & Carruthers, J. (2009). An examination of the relationship between service quality dimensions, overall internet banking service quality and customer satisfaction: A New Zealand study. *Marketing Intelligence & Planning*, *27*(1), 103–126. doi:10.1108/02634500910928344

Rousseau, D., Sitkin, S., Burt, R., & Camerer, C. (1998). Not so different after all: A cross-discipline view of trust. *Academy of Management Review*, *23*(3), 393–404. doi:10.5465/AMR.1998.926617

Rullis, H., & Sloka, B. (2010). Marketing within Internet banking in Latvia: Possibilities and challenges in commercial banks. *Economics and Management*, *15*(1), 705–710.

Sathye, M. (1999). Adoption of internet banking by Australian consumers: An empirical investigation. *International Journal of Bank Marketing*, *17*(7), 324–334. doi:10.1108/02652329910305689

Sayar, C., & Wolfe, S. (2007). Internet banking market performance: Turkey versus the UK. *International Journal of Bank Marketing*, *25*(3), 122–141. doi:10.1108/02652320710739841

Shih, H. (2004). Extended technology acceptance model of Internet utilization behavior. *Information & Management*, *41*(6), 719–729. doi:10.1016/j.im.2003.08.009

Shih, K.-H., Hung, H.-F., & Lin, B. (2010). Assessing user experiences and usage intentions of m-banking service. *International Journal of Mobile Communications*, *8*(3), 257–277. doi:10.1504/IJMC.2010.032974

Sohail, M. S., & Shaikh, N. M. (2008). Internet banking and quality of service: Perspectives from a developing nation in the Middle East. *Online Information Review*, *32*(1), 58–72. doi:10.1108/14684520810865985

Stone, M. (1974). Cross-Validatory Choice and Assessment of Statistical Predictions. *Journal of the Royal Statistical Society. Series A (General)*, *36*(2), 111–147.

Suh, B., & Han, I. (2002). Effect of trust on customer acceptance of internet banking. *Electronic Commerce Research and Applications*, *1*(1).

Tan, M., & Teo, T. S. H. (2000). Factors influencing the adoption of internet banking. *Journal of the Association for Information Systems*, *1*(5), 1–38.

Taylor, S., & Todd, A. (1995a). Assessing information technology usage: The role of prior experience. *Management Information Systems Quarterly*, *16*(4).

Taylor, S., & Todd, P. A. (1995b). Understanding information technology usage: A test of competing models. *Information Systems Research*, *6*(2), 144–176. doi:10.1287/isre.6.2.144

Teller, C., & Gittenberger, E. (2011). Patronage Behaviour of Elderly Supermarket Shoppers -Antecedents and Unobserved Heterogeneity. *International Review of Retail, Distribution and Consumer Research*, *21*(5), 483–499. doi:10.1080/09593969.2011.619701

Tenenhaus, M., Esposito Vinzi, V., Chatelin, Y.-M., & Lauro, C. (2005). PLS Path Modeling. *Computational Statistics & Data Analysis*, *48*(1), 159–205. doi:10.1016/j.csda.2004.03.005

Teo, T. S. H., Lim, V. K. G., & Lai, R. Y. C. (1999). Intrinsic and extrinsic motivation in internet usage. *International Journal of Management Sciences*, *27*(1), 25–37.

Thaichon, P., Lobo, A., & Mitsis, A. (2014). An empirical model of home internet services quality in Thailand. *Asia Pacific Journal of Marketing and Logistics*, *26*(2), 190–210. doi:10.1108/APJML-05-2013-0059

Triandis, H. C. (1979). Values, attitudes and interpersonal behavioural. *Nebraska Symposium on Motivation*, *27*(1), 195–259. PMID:7242748

Vatanasombut, B., Igbaria, M., Stylianou, A. C., & Rodgers, W. (2008). Information systems continuance intention of web-based applications customers: The case of online banking. *Information & Management*, *45*(7), 419–428. doi:10.1016/j.im.2008.03.005

Venkatesh, V. (2000). Determinants of perceived ease of use: Integrating control, intrinsic motivation and emotion into the technology acceptance model. *Information Systems Research*, *11*(4), 342–365. doi:10.1287/isre.11.4.342.11872

Venkatesh, V., & Davis, F. D. (1996). A model of the antecedents of perceived ease of use: Development and test. *Decision Sciences*, *27*(3), 451–481. doi:10.1111/j.1540-5915.1996.tb01822.x

Venkatesh, V., & Davis, F. D. (2000). A theoretical extension of the technology acceptance model: Four longitudinal field studies. *Management Science*, *46*(2), 186–204. doi:10.1287/mnsc.46.2.186.11926

Venkatesh, V., Morris, M. G., Davis, G. B., & Davis, F. D. (2003). User acceptance of information technology: Toward a unified view. *Management Information Systems Quarterly*, *27*(3), 425–478.

Wagner, G., Schramm-Klein, H., & Steinmann, S. (2013). Effects of Cross-Channel Synergies and Complementarity in a Multichannel E-commerce System — An Investigation of the Interrelation of E-Commerce, M-commerce and IETV-Commerce. *International Review of Retail, Distribution and Consumer Research*, *25*(5), 571–581. doi:10.1080/09593969.2013.832697

Wang, Y. S., Wang, Y. M., Lin, H. H., & Tang, T. I. (2003). Determinants of user acceptance of internet banking: An empirical study. *International Journal of Service Industry Management*, *14*(5), 501–519. doi:10.1108/09564230310500192

Weir, C. S., Anderson, J. N., & Jack, M. A. (2006). On the role of metaphor and language in design of third party payments in e-banking: Usability and quality. *International Journal of Human-Computer Studies*, *64*(8), 770–784. doi:10.1016/j.ijhcs.2006.03.003

Wetzels, M., Odekerken-Schroder, G., & Oppen, C. V. (2009). Using PLS path modeling for assessing hierarchical construct models: Guidelines and empirical illustration. *Management Information Systems Quarterly*, *33*(1), 177–195.

Williams, M. D., Dwivedi, Y. K., Lal, B., & Schwarz, A. (2009). Contemporary trends and issues in IT adoption and diffusion research. *Journal of Information Technology*, *24*(1), 1–10. doi:10.1057/jit.2008.30

Wold, H. O. (1974). Causal Flows with Latent Variables: Partings of the Ways in the Light of NIPALS Modeling. *European Economic Review*, *5*(1), 67–86. doi:10.1016/0014-2921(74)90008-7

Yang, H., & Yoo, Y. (2004). It's all about attitude: Revisiting the technology acceptance model. *Decision Support Systems*, *38*(1), 19–31. doi:10.1016/S0167-9236(03)00062-9

Yap, K. B., Wong, D. H., Loh, C., & Bak, R. (2010). Offline and online banking-where to draw the line when building trust in e-banking? *International Journal of Bank Marketing*, *28*(1), 27–46. doi:10.1108/02652321011013571

Yoon, H. S., & Steege, L. M. B. (2013). Development of a quantitative model of the impact of customers' personality and perceptions on internet banking use. *Computers in Human Behavior*, *29*(3), 1133–1141. doi:10.1016/j.chb.2012.10.005

Yousafzai, S. Y., Pallister, J. G., & Foxall, G. R. (2009). Multi-dimensional role of trust in internet banking adoption. *Service Industries Journal*, *29*(5), 591–605. doi:10.1080/02642060902719958

Yu, C. S. (2012). Factors affecting individuals to adopt mobile banking: Empirical evidence from the UTAUT model. *Journal of Electronic Commerce Research*, *13*(2), 104–121.

Yu, P. L., Balaji, M., & Khong, K. W. (2015). Building trust in internet banking: A trustworthiness perspective. *Industrial Management & Data Systems*, *115*(2), 235–252. doi:10.1108/IMDS-09-2014-0262

Zeithaml, V. A., Berry, L. L., & Parasuraman, A. (1996). The behavioral consequences of service quality. *Journal of Marketing*, *60*(2), 31–46. doi:10.2307/1251929

Zhao, A. L., Koenig-Lewis, N., Hanmer-Lloyd, S., & Ward, P. (2010). Adoption of internet banking services in China: Is it all about trust? *International Journal of Bank Marketing*, *28*(1), 7–26. doi:10.1108/02652321011013562

Zhou, T. (2011). An empirical examination of initial trust in mobile banking. *Internet Research*, *21*(5), 527–540. doi:10.1108/10662241111176353

Zhou, T., Lu, Y., & Wang, B. (2010). Integrating TTF and UTAUT to explain mobile banking user adoption. *Computers in Human Behavior*, *26*(4), 760–767. doi:10.1016/j.chb.2010.01.013

Chapter 8
Towards a Customer-Centric Framework for Evaluation of E-Health Service Quality

Michael Mutingi
Namibia University of Science and Technology, Namibia

ABSTRACT

As the awareness of the importance of healthcare service quality and pressures from stakeholders continue to grow, healthcare service providers have no option except to develop appropriate service quality evaluation procedures. Patient satisfaction is imperative, and has become a critical issue especially in e-health services. Although healthcare service providers have become aware of the need for improving customer experience through provision of customer-centric services, virtually none of the existing e-health quality evaluation frameworks are grounded on customer-centric metrics. In this chapter, a critical analysis of existing evaluation initiatives is presented from the context of e-health services. Critical areas of e-health service are investigated to determine quality dimensions that influence customer experience. From this analysis, a customer-centric evaluation framework is proposed, comprising four e-health service quality scales. The framework provides a platform for continuous improvement in e-health service.

INTRODUCTION

The healthcare service sector forms the largest percentage of the overall service sector in almost every economy across the globe (Hamid and Sarmad, 2008). Governments, non-governmental organizations and other stakeholders continue to increase their focus on healthcare service industry. This is evidenced by increasing investments

DOI: 10.4018/978-1-5225-3628-4.ch008

and regulatory controls on the sector (Hamid and Sarmad, 2008; Mitchell, 2000). In addition, the ever-increasing rate of adoption and usage of electronic service (e-service) technologies in the healthcare demonstrates the increasing attention of stakeholders on the sector (Glasgow, 2007; Gustafson and Wyatt, 2004; Mantzana et al., 2006). Researchers have argued that the use of e-service technologies can potentially revolutionize the sector in terms of quality of service, efficiency, reduced costs, and customer experience (Wickramasinghe and Misra, 2004; Brender, 2006; Gustafson and Wyatt, 2004).

The concept of e-service quality basically defines the extent to which a web site facilitates efficient and effective shopping, purchasing and delivery (Zemblyte, 2015; Akinci et al., 2010). This phenomenon has become a central issue in e-service research. According to Zemblyte (2015), it is crucial to develop quality frameworks for measurement and evaluation of service quality of e-services, stretching from pre-purchase through to the post-purchase phase. This necessitates development of instruments or frameworks for e-service quality evaluation, preferably from a multi-dimensional perspective. Though a significant number of existing researches has focused on developing frameworks that show how e-service quality can be created, negligible attention has been focused on developing frameworks for evaluating e-service quality, covering the whole customer experience in e-service delivery. Customer experience should encompass information search, order placement, customer service, order delivery and customer satisfaction. This idea can be applied in specific service industry, most importantly the e-health service industry. E-health leverages the information and communication technology to connect healthcare service provider, patients and governments, to educate and inform healthcare professionals, stimulating innovation and improvement in the healthcare system (Oh et al., 2005; Brender, 2006).

The purpose of this chapter is to present a multi-dimensional framework for evaluating e-service quality from the customer's perspective, incorporating the entire customer experience. By assessing past case studies, conceptual and literature review articles, a set of e-service quality evaluation criteria is developed from the context of e-health service. From these criteria, appropriate quality dimensions are dimensions are developed. Finally, quality scales are then synthesized. The proposed instrument offers several advantages over conventional e-service evaluation frameworks when applied to e-health services.

The rest of the chapter is organized as follows: The next section provides a background to the e-health, e-health service quality, and e-health service quality evaluation issues. This is followed by a description of the research methodology applied in this study. Literature search findings are then evaluated and presented, detailing the evaluation criteria pertinent to e-healthcare service. A customer-centric framework for evaluation of e-health services is proposed. Managerial insights are then derived and outlined. The chapter ends with a summary.

BACKGROUND

The concept of e-service quality differs from the traditional service quality (Parasuraman et al., 2005; Zeithaml 1998). Unlike traditional service quality, e-service quality focuses on the concept on internet-based customer experience (Li and Suomi, 2009; Li et al., 2009). Basically, service quality is determined from the variation between the expected service and perceived service from the service provider (Zeithaml 1998). Based on empirical studies in various service industries, Parasuraman et al. (2005) developed a generic instrument, called SERQUAL, for assessing and measuring service quality. The SERVQUAL instrument consists of five fundamental dimensions, outlined as follows (Parasuraman et al., 2005):

1. Tangibles, which defines the appearance of physical facilities, equipment, personnel and communication materials;
2. Reliability, which explains the ability to perform the promised service dependably and accurately;
3. Responsiveness, which signifies the willingness to help customer and provide prompt services;
4. Assurance, which defines the knowledge and courtesy of employees and their ability to convey trust and confidence;
5. Empathy, which relates to the care and individualized attention provided to customers.

The SERVQUAL instrument has been used widely in the literature (Rowley, 2006; Santos, 2003). However, its application in the context of e-services is limited (Li et al., 2009). This is mainly attributable to three main aspects of e-services: (i) the absence of direct face-to-face human contact, (ii) the absence of a traditional tangible product, and (iii) the customer self-service. In this view, researchers have suggested improved e-service quality frameworks to cater for these issues. For instance, Zeithaml et al. (2002) proposed an E-SERVQUAL framework to measure e-service quality. Parasuraman et al. (2005) transformed the framework into E-S-QUAL and E-RecS-QUAL. E-S-QUAL contains the core dimensions of e-service quality, that is, efficiency, system availability, fulfillment, and privacy. The E-RecS-

QUAL considers responsiveness, compensation, and contact dimensions. Though widely accepted, these frameworks may not be applicable in specific contexts, such as e-health services.

For ease and clarity of deliberation, we provide a brief overview of the concepts of e-health and e-health service quality, together with some background on the issues relating to the evaluation e-health service systems.

E-Health: An Overview

Broadly speaking, the term "e-health" refers to the use of information and communication technologies in the provision of healthcare services. As such, e-health encompasses a wide range of aspects in the field. Not surprisingly, the term Healthcare IT has also been used as a synonym for e-health.

According to the World Health Organization (WHO), e-health is defined as the "*... cost-effective and secure use of information and communication technologies in support of the health and health-related fields including healthcare, health surveillance and health education, knowledge and research.*" (Eysenbach, 2001). It is a rapidly emerging field that integrates the aspects of medical informatics, public health as well as business. Health services and information are delivered and enhanced through the use of internet and the associated technologies" (Eysenbach, 2001). E-health has widely been known for enhancing innovation in the delivery of healthcare services and in cost-effective health system management (Oh et al., 2005).

To date, e-health technologies and innovations have been implemented extensively in several countries across the world. This is attributed to the extensive permeation of information and communications technologies in the service industries. Some of the applications of e-health service technologies include the following:

- Consumer informatics, including thousands of health-oriented websites and mobile health technologies (Mutingi, 2015);
- Online health information technology systems, such as health portals, which are widely used for disseminating important health information to promote healthy life styles in patients and the society at large (Adibi, 2015; Bower, 2005: Brender, 2006);
- Telemedicine and telehealth systems for transmission of medical information for cost-effective diagnosis enhanced patient access and improved utilization of digital health assets (Mutingi, 2015; Adibi, 2015; Yang et al., 2001).
- Electronic health records (EHR), including patient records, e-prescriptions, health administration, e-booking, as well as digital imaging and archiving (Perera and Chakrabarti, 2015; Mutingi, 2015); and,

- E-health systems used to store public health data which researchers can collect and analyze, for instance, disease modelling and biostatics for research in infectious diseases (Connell, 2007).

To effectively implement and realize the usefulness of an e-health system, decision makers should have a service quality evaluation framework. Such a framework can be used to make informed decisions at the planning, development or operation stage of the e-healthcare system (Sarmad and Hamid, 2008; Brender, 2006). In practice, several stakeholders are involved in e-health service evaluation decisions, which potentially complicates the process.

E-Health Service Evaluation Challenges

Several challenges are associated with research into the evaluation and application of e-health service systems. This can be attributed to the fact that e-health lies at the intersection of complex systems, namely, information systems, health services, and service quality evaluation frameworks.

Information system evaluation is a research area with its own complex challenges (Sarmad and Hamid, 2008). This is further complicated by the fact that system evaluation includes both social and technical aspects of the system (Eng, 2002). When information system research is combined with e-health system research, the implication is that both social science and information system science aspects are to be considered simultaneously.

Healthcare services in general, and e-health in particular, are sensitive and critical areas controlled by strict regulations pertaining to patient data and healthcare delivery information. In a rapidly evolving environment, this becomes even more challenging (Friedman and Wyatt, 2000). Since e-health systems include the social aspects, this poses strong challenges on the evaluation and success of the system.

The process of developing e-service quality evaluation framework is generally a challenging task (Sarmad and Hamid, 2008). In light of the complexities in the information system and health system evaluation domains, developing effective quality evaluation frameworks is a difficult task. In practice, decision makers must familiarize themselves with several quality evaluation criteria and quality dimensions from which suitable quality scales can be derived.

The next section describes the research methodology that was followed in developing a customer-focused for e-health service.

RESEARCH METHODOLOGY

To develop an effective customer-centric framework for evaluation of e-health services, it is important to carry out an intensive literature search in order to learn from pertinent case studies, as well as conceptual and review articles in the literature. The research methodology used in this research consists of four major steps, as presented in Figure 1. In the first place, an initial literature search survey was carried out to explore the relevant literature in online databases, including EBSCO Inspec, ISI Web of Science, Ei Compendex, ScienceDirect, Google Scholar and Research Gate. To limit the range of articles, appropriate key words were used, including e-service quality, e-health service, quality evaluation, customer experience, E-S-QUAL, and E-RecS-QUAL.

In the second step, articles obtained from the literature search survey were analyzed further and streamlined to ensure that all the selected articles were strictly about e-service quality, with a special focus on e-health service quality. Articles published in reputable peer-reviewed journals, conference proceedings, book chapters, and books were preferred.

The third and final step involved content analysis of selected articles to identify appropriate critical e-service quality evaluation criteria, while considering the

Figure 1. Research methodology

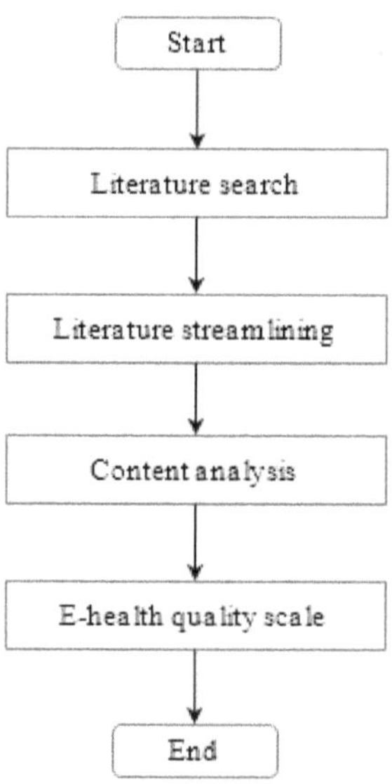

challenges associated with e-health services and their evaluation. Fourth and finally, suitable quality dimensions were synthesized from evaluation criteria, based on the customer's perspective in the healthcare context. Furthermore, the identified quality dimensions were synthesized into e-health quality scales.

LITERATURE SEARCH FINDINGS

A total of 28 past empirical studies were identified from the literature search survey. In addition to these, literature reviews and conceptual articles were also examined to investigate and identify essential e-service quality dimensions that influence the customer's perception, adoption, utilization and satisfaction levels of e-health services.

E-Health Service Quality Evaluation Criteria

As indicated earlier, the health sector is inundated with sensitive and critical issues that are controlled by regulatory systems. These regulations and controls should be considered when developing service quality evaluation criteria, quality dimensions and quality scales. We develop a set of evaluation criteria, learning from a several of empirical studies in the literature, alongside conceptual and literature review articles. A total of 9 sets of evaluation criteria were obtained, as listed in Table 1.

User Satisfaction

The first set of evaluation criteria is concerned with user satisfaction, a widely regarded measure of service systems. User satisfaction can be examined based on attributes such as fulfillment, reliability, efficiency, flexibility, privacy or security, and individualization. Fulfillment relates to the perceived usefulness and completeness of a given e-service Loiacono et al. (2002). Reliability, on the other hand, focuses on the ability of an e-health service to provide the correct technical function over time (Zeithaml and Parasuraman, 2000; Smaglik et al., 1998). Efficiency refers to the time of execution of an e-health service, or the time spent to complete the requested service in a satisfactory manner. For example, a website is expected to provide for quick navigation and search (Eistano, 2014). However, since the healthcare sector handles sensitive personal information, privacy and security are an important consideration when providing e-health services (Samad and Hamid, 2008). E-service must ensure that information is protected and is not shared. An interesting addition to user satisfaction is individualization or customization (Zemblyte, 2015). This signifies the degree to which the e-service can be adjusted to satisfy the individual user's specific needs.

Table 1. A collection of evaluation criteria and their descriptions

Evaluation Criteria	Measured Attributes	Brief Description	References
1. User satisfaction	Fulfillment	Usefulness and completeness of the service	Loiacono et al.(2002)
	Reliability	Ability to always provide appropriate service functions	Zeithaml and Parasuraman (2000)
	Efficiency	Time spent to complete or provide the required satisfactory service	Wang et al. (2005)
	Flexibility	Ability of the system or service to give choices and dynamic information	Zeithaml and Parasuraman (2000)
	Privacy/Security	Protection of information; information not shared	
	Individualization	The degree the service satisfies the individual user's specific needs	Burgess (2004)
2. Inner satisfaction	Inner care	Service satisfaction touching the inner being of the customer	Mutingi and Mbohwa (2016).
	Inner value	Inward feeling of improved personal worth due to service quality	Mutingi and Mbohwa, 2016).
	Inner trust	Inward belief built due to the goodness of the service provided	Mutingi and Mbohwa (2016). Li et al. (2009)
3. Functionality	Availability	Sustained ability of the system to provide service, along with alternatives choices	Hamid and Sarmad (2008)
	Accuracy	The degree to which the service provided is free of errors	Melander-Wikman et al. (2005)
	Speed	The response time, e.g., downloading time, web page loading time	Einasto (2015)
	Validity	Clarity of information and regular update of the information	Hamid and Sarmad (2008)
	Robustness	Technical functionality in the presence of possible threats to the system	Melander-Wikman et al. (2005)
4. Compatibility	Compatibility	The ease of fitting the e-service system into the healthcare system	Chau and Hu (2001).
5. Accessibility	Content accessibility	Compliance and access level in comparison to other alternative service with the same quality	Santouridis et al. (2012)
	User interface	Availability of options of user interface options, e.g., multiple-window interface	Santouridis et al. (2012).
	Disability access	Availability of disability access and language translation facilities	Caldwell et al. (2007)
6. User-friendliness	Easy to learn	The time needed by the average user to understand and use the system	Mutesi et al. (2017)
	Easy to use	The simplicity of the service and its various functions	Hamid and Sarmad (2008); Li et al. (2009)
	Aesthetics/design	Website design and its visual attractiveness or stylish design	Mutesi et al. (2017)
7. Responsiveness	Responsiveness	Quick response to call for user requirements/assistance	
	Compensation	Helpfulness in handling and solving problems and returns as reported by the customer	Einasto (2014)
8. Communication	Contact	Ease of contact and dialogue; availability of a variety of communication channels	Nath and Singh (2010)
	Feedback	Confirmation from service provider concerning customer requests	Einasto (2014)
9. Benefits	Effort saving	The degree of convenience in using e-health features	
	Value	Value added to citizens knowledge about their health conditions, treatment and healthcare facilities	Stroetmann et al. (2006)
	Cost saving	The cost or money saved by using the e-service provided	Hamid and Sarmad (2008)
	Time saving	The amount of time saved by using the e-health service provided	Hamid and Sarmad (2008)

Inner Satisfaction

An additional set of criteria to be considered in e-health is inner satisfaction of the e-service user. This concept agrees with the idea of holistic customer experience (Mutingi and Mbohwa, 2016). Given that the sector deals with patients who are social beings in need of special attention, we argue that the focus of attention should seek not only to satisfy the basic health of the individual, but also to satisfy the inner being of the user, which includes the emotional, mental or inward health. Thus, we define inner health in terms of three dimensions: (i) inner care, which defines the service satisfaction touching the inner being of the user, (ii) inner value, which refers to the internal feeling of improved personal worth or personal state of mind driven by the quality of e-service, and, (iii) inner trust which defines the internal belief enhanced by the perceived quality of the service.

Functionality

Functionality is a set of criteria that can be defined in terms of availability, accuracy, speed, validity, and robustness of an e-service (Melander-Wikman et al., 2005). Availability measures the sustained ability of the e-service system to provide the required service, along with its alternatives choices. On the other hand, measures the clarity of information, which should be updated regularly. The user expects the e-service to function with accuracy (that is, free of errors), and speed (that is, with minimum response time). In addition, the user also expects the system to provide technical functionality despite the possible threats to the e-service systems.

Compatibility

Compatibility is a crucial evaluation criterion related to usability of an e-service system. It is a measure of the degree to which the e-service technology is perceived to be consistent with existing values, culture, experiences, and needs of the potential users (Chau and Hu, 2001). This tends to affect the user's perception towards accepting new e-health technologies. Therefore, potential users are likely to adopt and use new technologies that are compatible with current practices and experiences.

Accessibility

Accessibility also relates to usability (Bevan and Macleod, 1993). This criterion is necessitated by the fact that most countries across the globe have a significant percentage of people with disabilities. Researchers have indicated that users with disabilities are more likely to use available e-health technologies than others (Lenhart et al., 2003; Bevan et l., 2007). Terry Ma and Zaphiris (2003) postulated that an e-health system must be accessible and usable by various users suffering from different disabilities and illness. This implies that the system should be accessible and effective despite the varying needs of the users (Mont, 2007; Zeng and Parmanto, 2004). Therefore, accessibility is an important criterion that influences the user's perception on the e-health service system.

User-Friendliness

A user-friendly e-health service is one that is easy to learn and easy to use. For instance, a website is regarded as easy to learn if the time needed by the average user to understand and use the system is relatively short (Sarmad and Hamid, 2008; Einasto, 2014). On the other hand, ease of use implies that the website is characterized by simplicity of the service and its various functions. This implies that the website is logically structured and it enhances quick access and ease of navigation by an average user (Einasto, 2014).

Responsiveness

Responsiveness can be defined in terms of speed of response to user requirements (Zemblyte, 2015). The e-health service provider should provide accurate services according to promise, and should be equipped with alternative options for offering recovery services. On the other hand, compensation refers to the helpfulness in handling and solving problems faced by the user. The user expects to be compensated for the problems that are created by the service provider. The service provider must compensate the user whenever the service provided fails to meet the user requirements.

Communication

Communication can be described in terms of contact and feedback (Einstano, 2014). Since e-health deals with delicate customers, parameters such as ease of contact, availability of choice of language, dialogue, and user participation are highly expected. This calls for enhanced availability of a variety of open communication channels that are always available for the customers. Feedback pertains to confirmation from service providers regarding customer requests (Einasto, 2014; Akinci et al., 2010).

Benefits

When assessing the overall potential impact of an e-health service, the user evaluates the benefits associated with the adoption of an e-health service. This entails evaluation of the risks and benefits associated with the design and implementation of an e-health initiative (Gustafson and Wyatt, 2004). The benefits or savings of e-health services can be viewed from four perspectives, namely: (i) effort saving, which refers to the degree of convenience in using the e-health features, (ii) value, which defines the value added to citizens knowledge about their health conditions, treatment and healthcare facilities, (iii) cost saving, which refers to the cost saved by adopting and using the e-service, and (iv) time saving, which defines the amount of time saved by using the e-health service.

The next section presents a customer-centric framework for evaluating e-service quality in the context of e-health services.

E-Service Quality Evaluation Frameworks

Zeithaml et al. (2000, 2002) suggested an approach to measuring e-service quality, which is explained in terms of the e-SERVQUAL model. The model was later transformed into two separate scales: E-S-QUAL and E-RecS-QUAL (Parasuraman et al., 2005). The E-S-QUAL contains the core dimensions of e-service quality, namely, efficiency, system availability, fulfillment, and privacy. On the contrary, the E-RecS-QUAL focuses on responsiveness, compensation, and contact. According to Zemblyte (2015), the two quality scales can be used as a basis for analyzing service quality in specific contexts. The quality scales provide a means for conceptualizing e-service quality (Brady and Cronin, 2001; Collier and Bienstock, 2006; Fassnacht and Koese, 2006).

According to Van Ossel (1998), a service quality scale or model should include a complete set of quality dimensions that covers all the desired aspects of quality evaluation, particularly from the user's perspective (Mutingi and Mbohwa, 2016; Sarmad and Hamid, 2008). Furthermore, the quality dimensions should be holistic, and should clearly explain all the various conceivable perceptions of service quality. This ensures that the quality scale will clearly evaluate the required aspects of quality, without possible ambiguities. However, care must be taken to include too many dimensions into the model as this may result in a complex quality scale. Given the importance of quality evaluation in e-health services, the criticality of the customers (patients), and all the associated stakeholders, it is quite difficult to identify and select appropriate dimensions for e-health services. In this study, suitable evaluation criteria and the ensuing dimensions are developed from the customer's perspective.

A CUSTOMER-CENTRIC FRAMEWORK FOR E-HEALTH SERVICE QUALITY EVALUATION

From a total of 28 past empirical studies identified from the literature search survey, we identified and outlined 9 e-health service evaluation criteria, from the customer's perspective (as presented in the in the previous section). Some of them were previously analyzed in Rowley (2006) and Kim et al (2006). Addition literature review and conceptual articles provided a valuable source of information on key e-service quality dimensions that influence the customer's perception on the quality of e-health services.

After the identification of 9 major e-health evaluation criteria, further literature analysis identified a total of 22 dimensions, deriving from the information obtained in Table 1. These, in turn, were synthesized into four e-health service quality scales. The four scales are presented as follows:

1. Website e-service quality scale
2. Core e-service quality scale
3. E-service quality recovery scale
4. E-service inner satisfaction scale

Figure 2 presents a framework for e-health service quality evaluation. We provide further discussions on these four e-health service quality scales.

Figure 2. A framework for e-health service quality evaluation

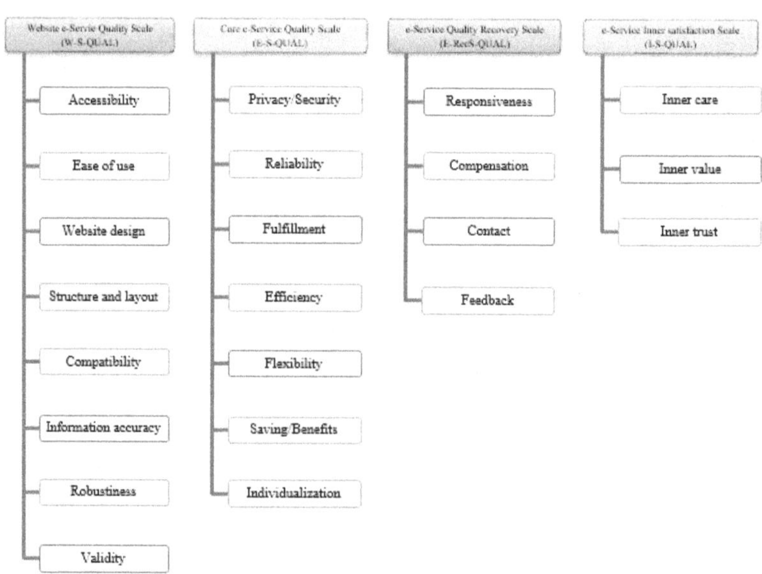

4.1 Website E-Service Quality Scale

The W-S-QUAL is the e-health quality scale that encompasses e-health service wherein the customer primarily interacts with the website. This involves retrieval of pertinent information whose content must satisfy the patient or customer. In the absence of face-to-face interaction, much is expected of website e-service quality (Zemblyte, 2015; Yoo and Donthu, 2001). This is further complicated by the fact that customer perception of website e-service quality will vary considerably. Similar supporting arguments have been presented in the literature (Santouridis et al. 2012; Yang and Jun, 2002; Cai and Jun, 2003). Moreover, e-service quality should incorporate incubative and active perspectives, that is, before and after the use of the website (Santos, 2003).

In light of the above challenges, e-health service quality should include the W-S-QUAL scale, to take care of the website quality dimensions (Zemblyte, 2015).

4.2 Core E-Service Quality Scale

The E-S-QUAL is the core e-health quality scale which primarily contains the core dimensions of service quality. In addition to the basic four dimensions identified by Zeithaml et al. (2000, 2002), three more dimensions were identified and added to the core e-health service quality scale: (i) individualized attention, (iii) saving, and (iii) flexibility. As suggested by Zemblyte (2015), individualization is about providing service considering the specific needs of an individual. Individual attention is related to the empathy dimension in the SERVQUAL model (Zeithaml, Parasuraman & Malhotra, 2002). In e-health services, individualized attention plays a crucial role due to the criticality of the service and the variation of the needs of the health service required by individual customers.

When customers choose e-health services over conventional services, they often expect to gain some benefits (savings) in form of effort saving, value addition, cost saving, and time saving. This dimension has an important influence on the adoption, usage and perception of e-services (Lapointe et al., 2002). In this view, the savings dimension is an important addition to the core e-service quality scale. On the other hand, flexibility is also an important aspect for measuring quality in e-health services, as suggested by Zeithaml (2002). Thus, flexibility is an important addition, specifically to the core e-service quality scale.

4.3 E-Service Quality Recovery Scale

The E-RecS-QUAL defines the e-service quality recovery scale (Zemblyte, 2015; Akinci et al., 2010; Parasuraman et al., 2005), which essentially consists of four dimensions in all: (i) responsiveness, (ii) compensation, (iii) contact, and (iv) feedback. Here, the first three are the original dimensions suggested by Parasuraman et al. (2005), while the fourth one is an additional dimension identified in this study. As originally stated in Parasuraman et al. (2005), responsiveness focuses on handling of returns through the e-service means. Therefore, we argue that, in e-health services, responsiveness is often complemented by individualized services, where specific customer needs are attended to.

4.4 E-Service Inner Satisfaction Scale

Defined as the E-I-QUAL, the e-service inner satisfaction scale is quality scale introduces a more social approach to service quality, particularly important to modern patient care in the society. This development assumes that patient care is expected to include not only the physiological health of the patient, but also such aspects as mental health, psychological health, social health, and the health of the inner being. In addition, patient care seeks to provide satisfaction to the patient's close family members who interact with the patient. For instance, e-counselling (online counselling), mobile health (mHealth), and e-drug delivery and tele-health emphasize on social touch (Mutingi, 2015), apart from the conventional core health service.

MANAGERIAL INSIGHTS

The proposed customer-centric quality evaluation framework has a wide range of managerial implications for practicing decision makers in the field as well as researchers in e-service quality. First, the study provides a more comprehensive perspective of e-health service quality, which is a useful development towards holistic evaluation of e-service quality based on four e-service quality scales. Second, the study suggests a customer-centric approach to service quality evaluation, providing the manager with a renewed view on e-health service quality which enables evaluation of quality over the entire customer experience. In turn, enhanced customer experience will have a long-lasting positive impact on the inner satisfaction of the patient, the family, and the society at large. Inner satisfaction has a significant influence on customer loyalty and corporate image, which enhances market share and business competitiveness in the long run. Third, the new dimensions, particularly those in the inner satisfaction quality scale will provide a more focused set of criteria for designing e-health services from the perspective of the patient. Quality evaluation metrics can be developed based on these new dimensions, which are derived from critical areas that are most important to those in need of the service. Fourth, this research introduces a more social approach to service quality, which is especially important when it comes to modern patient care. Patient care goes beyond caring for biological health, to include satisfaction of the inner being.

Apart from e-healthcare service managers, other related e-service managers will find this framework useful for effective evaluation of their e-services. These include library e-services, electronic retailing (or e-tailing) services (Wolfinbarger and Gilly, 2003), among others.

SUMMARY

Given the ever-growing awareness of the pressure for healthcare service quality, healthcare service providers need appropriate service quality evaluation frameworks. Patient satisfaction has become a critical issue in e-health services. For improved customer experience, it is important to provide customer-centric services through the application of e-health quality evaluation frameworks that are grounded on customer-centric metrics. This chapter provided a critical analysis of existing literature on service quality evaluation initiatives, with a special focus on developing customer-centric e-health quality services. Literature search findings enabled the analysis of critical areas of e-health service quality which influence customer experience. By analyzing and categorizing these critical areas, a set of e-health quality criteria were identified, covering the entire customer experience from the context of e-healthcare services. The proposed framework consists of four quality scales, namely, the website e-service quality scale, the core e-service quality scale, the e-service quality recovery scale, the e-service inner satisfaction scale. New dimensions were added to the quality scales, based on the identified critical e-health service quality criteria.

The proposed framework for is a useful guide towards effective evaluation and verification of e-health service quality. The study may be a useful instrument for managing e-healthcare services. Therefore, the framework provides a platform for continuous improvement in e-health services. This research can be extended to other related e-service organizations, such as e-tailing service providers who intend to offer high quality e-services.

ACKNOWLEDGMENT

This research was partly supported by the Namibia University of Science and Technology, Windhoek, Namibia.

REFERENCES

Adibi, H. (2015). Social perspectives on health: mHealth implications. In mHealth Multidisciplinary Verticals (pp. 219 – 236). Taylor & Francis Group, LLC.

Akinci, S., Atilgan-Inan, E., & Aksoy, S. (2010). Re-assessment of E-S-Qual and E-RecS-Qual in a pure service setting. *Journal of Business Research, 63*(3), 232–240. doi:10.1016/j.jbusres.2009.02.018

Bevan, N., & Macleod, M. (1993). Usability assessment and measurement. In The Management and Measurement of Software Quality. Ashgate Technical/Gower Press.

Bevan, N., Petrie, H., & Claridge, N. (2007). Strategies for providing guidance on usability and accessibility. *Proceedings HCI International*. doi:10.1007/978-3-540-73283-9_3

Bower, A. (2005). *The diffusion and value of healthcare information technology*. RAND Corporation.

Brady, M. K., & Cronin, J. J. Jr. (2001). Some new thoughts on conceptualizing perceived service quality: A hierarchical approach. *Journal of Marketing, 65*(3), 34–49. doi:10.1509/jmkg.65.3.34.18334

Brender, J. (2006). *Handbook of Evaluation Methods for Health Informatics*. Burlington, MA: Elsevier Academic Press Publications.

Burgess, L. (2004). A conceptual framework for understanding and measuring perceived service quality in net-based customer support systems. *CollECTeR LatAm Conference*.

Cai, S., & Jun, M. (2003). Internet user's perceptions of online service quality: A comparison of online buyers and information searchers. *Managing Service Quality, 13*(6), 504–519. doi:10.1108/09604520310506568

Caldwell, B., Cooper, M., Reid, L., & Vanderheiden, G. (2007). *Web Content Accessibility Guidelines 2.0, W3C Working Draft*. Biot.

Chau, P., & Hu, P. (2001). Information technology acceptance by individual professionals: A model comparison approach. *Decision Sciences, 32*(4), 699–719. doi:10.1111/j.1540-5915.2001.tb00978.x

Collier, J. E., & Bienstock, C. C. (2006). Measuring service quality in e-retailing. *Journal of Service Research, 8*(3), 260–275. doi:10.1177/1094670505278867

Connell, N., & Young, T. (2007). Evaluating healthcare information systems through an 'enterprise' perspective. *Information & Management*, *44*(4), 433–440. doi:10.1016/j.im.2007.04.002

Einasto, O. (2014). E-service quality criteria in university library: A focus group study. *Procedia: Social and Behavioral Sciences*, *147*, 561–566. doi:10.1016/j.sbspro.2014.07.163

Eng, T. R. (2002). E-health research and evaluation: Challenges and opportunities. *Journal of Health Communication*, *7*(4), 267–272. doi:10.1080/10810730290001747 PMID:12356287

Eysenbach, G. (2001). What is e-health? *Journal of Medical Internet Research*, *3*(2), e20. doi:10.2196/jmir.3.2.e20 PMID:11720962

Fassnacht, M., & Koese, I. (2006). Quality of electronic services – conceptualizing and testing a hierarchical model. *Journal of Service Research*, *9*(1), 19–37. doi:10.1177/1094670506289531

Friedman, C., & Wyatt, J. (2000). *Evaluation Methods in Medical Informatics*. New York, NY: Springer-Verlag.

Glasgow, R. E. (2007). E-health evaluation and dissemination research. *American Journal of Preventive Medicine*, *32*(5), S119–S26. doi:10.1016/j.amepre.2007.01.023 PMID:17466816

Gustafson, D., & Wyatt, J. (2004). Evaluation of e-health systems and services. *British Medical Journal*, *328*(7449), 1150. doi:10.1136/bmj.328.7449.1150 PMID:15142895

Hamid, A., & Sarmad, A. (2008). Evaluation of e-health services: user's perspective criteria. *Transforming Government: People. Process and Policy*, 2(4), 243–255.

Kim, M., Kim, J. H., & Lennon, Sh. J. (2006). Online service attributes available on apparel retail web sites: An E-S-Qual approach. *Managing Service Quality*, *16*(1), 51–77. doi:10.1108/09604520610639964

Kim, M., Kim, J.-H., & Lennon, S. J. (2006). Online service attributes available on apparel retail web sites: An E-S-QUAL approach. *Managing Service Quality: An International Journal*, *16*(1), 51–77. doi:10.1108/09604520610639964

Lapointe, L., Lamothe, L., & Fortin, J. P. (2002). The dynamics of IT adoption in a major change process in healthcare delivery. *Proceedings of the 35th HICSS Conference*. doi:10.1109/HICSS.2002.994117

Lenhart, A., Horrigan, J., Rainie, L., Allen, K., Boyce, A., Madden, M., & O'Grady, E. (2003). *The Ever-shifting Internet Population. A New Look at Internet Access and the Digital Divide*. Washington, DC: Pew Internet & American Life Project.

Li, H., Liu, Y., & Suomi, R. (2009). Measurement of e-service quality: An empirical study on online travel service. *Proceedings of 17th European Conference on Information Systems.*

Li, H., & Suomi, R. (2009). A Proposed Scale for Measuring E-Service Quality. *International Journal of u- and e-service. Science and Technology*, *2*(1), 1–10.

Loiacono, E. T., Watson, R. T., & Goodhue, D. L. (2002). Webqual: A web site quality instrument. *Proceedings of the AMA Winter Marketing Educators' Conference.*

Mantzana, V., & Themistocleous, M. (2006). A method for the identification of actors involved in the adoption of innovations in healthcare organizations. *Proceedings of the 39th Hawaii International Conference on System Sciences.*

Melander-Wikman, A., Jansson, M., & Gard, G. (2005). The mobihealth usability evaluation questionnaire. *E-health. International Journal (Toronto, Ont.)*, *2*(1), 9–14.

Mitchell, J. (2000). Increasing the cost-effectiveness of telemedicine by embracing e-health. *Journal of Telemedicine and Telecare, 6*, S16-S19.

Mont, D. (2007). *Measuring disability prevalence. SP Discussion Paper No. 0706.* Retrieved 22 July 2017: http://siteresources.worldbank.org/DISABILITY/Resources/Data/MontPrevalence.pdf

Mutesi, J., Mutingi, M., & Charkraborty, A. (2017). Service quality assessment for internet banking: Empirical evidences from Namibia. *e-Service Journal, 10*(1), 42–65. doi:10.2979/eservicej.10.1.03

Mutingi, M. (2015). Enhancing Decision Support in Health-Care Systems through mHealth. In mHealth Multidisciplinary Verticals (pp. 511 – 526). Taylor & Francis Group, LLC.

Mutingi, M., & Mbohwa, C. 2016. Sustainable Process Innovation for Customer Experience: A Conceptual Framework. IEEE Industrial Engineering and Engineering Management, 2182-2187.

Nath, A.K., & Singh, R. (2010). Evaluating the Performance and Quality of Web Services in Electronic Marketplaces. *e-Service Journal*, *7*(1), 43-59.

Oh, H., Rizo, C., Enkin, M., & Jadad, A. (2005). What is e-health? A systematic review of published definitions. *Journal of Medical Internet Research, 24.*

Parasuraman, A., Zeithaml, V. A., & Malhotra, A. (2005). E-S-QUAL – A multiple-item scale for assessing electronic service quality. *Journal of Service Research*, *7*, 213–233. doi:10.1177/1094670504271156

Perera, C., & Chakrabarti, R. (2015). mHealth and Medical Imaging. In mHealth Multidisciplinary Verticals (pp. 651 – 668). Taylor & Francis Group, LLC.

Rowley, J. (2006). An analysis of the e-service literature: Towards a research agenda. *Internet Research*, *16*(3), 339–359. doi:10.1108/10662240610673736

Santos, J. (2003). E-service quality: A model of virtual service quality dimensions. *Managing Service Quality*, *3*(3), 233–246. doi:10.1108/09604520310476490

Santouridis, I., Trivellas, P., & Tsimonis, G. (2012). Using E-S-QUAL to measure internet service quality of e-commerce web sites in Greece. *International Journal of Quality and Service Sciences*, *4*(1), 86–98. doi:10.1108/17566691211219751

Smaglik, P., Hawkins, R., Pingree, S., & Gustafson, D. (1998). The quality of interactive computer use among HIV infected individuals. *Journal of Health Communication*, *3*(1), 53–68. doi:10.1080/108107398127508 PMID:10947374

Stroetmann, K., Jones, T., & Dobrev, A. (2006). Evaluating the economic impact of e-health applications approach and method. *E-health Conference 2006*.

Terry Ma, H., & Zaphiris, P. (2003). *The Usability and Content Accessibility of the E-government in the UK, Centre for Human-Computer Interaction Design*. City University, London. Retrieved 20 July 2017 from www.soi.city.ac.uk/zaphiri/Papers/HCII2003/HCII2003-Accessibility.pdf

Van Ossel, G. (1998). Service guarantees and service-level agreements. In *Services Management. An Integrated Approach*. Financial Times Pitman Publishing.

Wang, L., Bretschneider, S., & Gant, J. (2005). Web-based e-government services with a citizen-centric approach. *Proceedings of 38th Annual Hawaii International Conference on Systems Sciences*. doi:10.1109/HICSS.2005.252

Wickramasinghe, N., & Goldberg, S. (2004). How M 1/4 EC2 in healthcare. *International Journal of Mobile Commerce*, *2*(1), 22–47.

Wickramasinghe, N., & Misra, S. (2004). A wireless trust model for healthcare. *International Journal e-Health*, *1*(1), 60-77.

Wolfinbarger, M., & Gilly, M. C. (2003). eTailQ: Dimensionalizing, measuring and predicting etail quality. *Journal of Retailing*, *79*(3), 183–198. doi:10.1016/S0022-4359(03)00034-4

Yang, Z., & Jun, M. (2002). Consumer perceptions of e-service quality: From internet purchaser and non-purchaser perspectives. *The Journal of Business Strategy*, *1*, 19–41.

Yang, Z., Peterson, T., & Huang, L. (2001, Summer). Taking the pulse of internet pharmacies. *Marketing Health Services*, 5–10. PMID:11406905

Yoo, B., & Donthu, N. (2001). Developing a scale to measure the perceived quality of an internet shopping site (Sitequal). *Quarterly Journal of Electronic Commerce*, *2*, 31–46.

Zeithaml, V. A., Parasuraman, A., & Malhotra, A. (2000). *A Conceptual Framework for Understanding e-Service Quality: Implications for Future Research and Managerial Practice*. Marketing Science Institute, Cambridge, MA, Report No. 00-115.

Zeithaml, V. A., Parasuraman, A., & Malhotra, A. (2002). Service quality delivery through web sites: A critical review of extant knowledge. *Academy of Marketing Science Journal*, *30*(4), 362–375. doi:10.1177/009207002236911

Zemblyte, J. (2015). The Instrument for Evaluating E-Service Quality. *Procedia: Social and Behavioral Sciences*, *213*, 801–806. doi:10.1016/j.sbspro.2015.11.478

Zeng, X., & Parmanto, B. (2004). Web content accessibility of consumer health information web sites for people with disabilities: A cross sectional evaluation. *Journal of Medical Internet Research*, *6*(3), e19. doi:10.2196/jmir.6.2.e19 PMID:15249268

KEY TERMS AND DEFINITIONS

Customer-Centric: Focused towards satisfying all the customer needs and specifications.

E-Health: Leveraging on information and communication technology to healthcare provider and patients and governments, to improve the healthcare system.

Individualization: The act of providing individualized service, or customization.

Reliability: Correct technical functioning of the site; the assurance that the website is always in good working condition.

Responsiveness: The desire to solve the customer's problems and provide support.

Security: Protection of information; information not shared.

Speed: Quick navigation, search, and downloading.

User-Friendliness: Easy to learn and easy to use; the possibility to find easily the information that I need.

Chapter 9
Measuring the Service Quality of Services: TRADONIC SERVQUAL Model

Abhishek Vashishth
Indian Institute of Management Tiruchirappalli, India

Ayon Chakraborty
Indian Institute of Management Tiruchirappalli, India

ABSTRACT

SERVQUAL and e-SERVQUAL have been considered the most effective and powerful approaches in evaluating the quality and gaps in the service delivered in traditional and electronic services, respectively, but neither SERVQUAL nor e-SERVQUAL can measure the overall service quality of the firm. Therefore, this chapter aims to propose and test a new scale that can measure the overall service quality of the firm.

INTRODUCTION

Services form a considerable part of the world economy and the growth of the service sector has long been considered as an indicator of a country's economic progress. The service sector accounts for a significant proportion of GDP in most countries, including low-income and developing countries, where it frequently generates over 50 percent of GDP. For instance, service sector is the largest sector in India accounting for 57.9% of the total GDP whereas manufacturing and agriculture sector contributed for 24.2% and 17.9% respectively (Statistical Year Book India, 2015). Thus, it becomes very important for the firms to pay more attention on the service

DOI: 10.4018/978-1-5225-3628-4.ch009

and more precisely, the quality of service to be able to differentiate themselves from their competitors.

Given the importance of quality of service, a lot of attention has been paid on the ways and methods of measuring the quality of service and customer satisfaction. Service quality has its roots in the business and management field. Marketers realized that to retain customers, and to support market growth, they must provide high quality of service (Dabholkar, Shephard & Thorpe, 2000; Zeithaml, 2002). It is said that service quality is an important antecedent of consumer assessment of value, which in turn influences customer satisfaction, which then motivates loyalty (Babakus & Boller, 1992). There has been much debate as to what constitute service quality and how its measures can be operationalized in various service industries, yet no consensus has been reached (Chowdhary & Prakash, 2007). Since service itself is a complex phenomenon, efforts to define service quality and its dimensions have been subjected to academic debate. One of the most cited and applied concept of service quality is by Parasuraman, Zeithaml and Berry (1985) who simple put it as: the overall evaluation of a specific service firm that results from comparing the firm's performance with the customer's general expectations of how firms in that industry should perform. They introduced SERVQUAL that has been considered as the most effective and powerful approach in evaluating the quality and gaps in the service delivered and the customer expectations in traditional services. The approach defines the service quality as "the discrepancy between a customer's expectations of a service and the customer's perception of the service offering" (Parasuraman, 1998). The SERVQUAL helps managers in determining the service quality gaps and prioritizing these gaps that needs to be focused on and allocating essential resources to fulfill them.

Though various grounds have been identified for criticizing the SERVQUAL methodology but as per the articles published in recent decade the process has gained great importance for identifying various quality gaps and thus helps in preparing various strategies for efficiently allocating resources and terminologies for attaining better customer satisfaction along with profit maximization. Over the years, there have been many adaptations of the original SERVQUAL model to meet the specific operational characteristics of the service industry (Dowell and Long, 1998; Walker, 1996; Getty and Thompson, 1994; Knutson et al., 1991). This has led to the evolution of terms such as LODGQUAL, LODGSERV, DINESERV and GROVQUAL. All of them individually have sought to develop and build over the more generic SERVQUAL methodology.

Last decade has seen numerous researchers developing and applying service quality models across different industries and countries (for example see Amin and Nasharuddin, 2013; 239). SERVQUAL has been applied in many sectors such as healthcare (Kilbourne, 2004; Lam, 1997; Headley and Miller, 1993; Carman, 1990);

banking (Lam, 2002; Zhou, 2002; Mels et al., 1997); telecommunications (van der Wal, 2002); retail (Parasuraman et al., 1988); information technology (Jiang et al., 2000); and library services (Cook and Thompson, 2000).

Also, with the advancement in the technology services now can be offered electronically. These electronic services can be defined as, "interactive services that are delivered on the internet using advanced telecommunications, information, and multimedia technologies" (Boyer et al., 2002). Online services, including online banking services, are becoming an attractive alternative to visiting service outlets or phoning call centers for increasing number of customers (HR-Focus 2000). Some of the reasons for customers to prefer online services (as online banking services) are: convenience (Meuter et al., 2000; Szymanski & Hise, 2000), feeling more in control of the service process (Bateson, 2000; Dabholkar, 1996) and avoiding human contact and saving time (Dabholkar 1996; Meuter et Al., 2000). As far as online services are concerned, it is quite easier for customers to evaluate and compare the benefits of competing services (Santos, 2003). In addition, the switching costs are very low, that is why retaining the customer in the Internet space is of vital importance (Reichheld & Schefter, 2000). With the introduction of electronic services methodology like E-S-QUAL and E-RecS-QUAL (Zeithaml et al. 2000) have been implemented to measure the service quality of the e-services. The SERVQUAL model measures the service quality of the traditional services offered at the service premises whereas the E-SERVQUAL model measures the service quality of the electronic services offered by the firm.

The business scenario is changing due to factors such as opening-up of markets, increased used of information technology, increased customer knowledge, use of electronic services, etc. (Sangeetha and Mahalingam, 2011). With these advancements it has become increasingly difficult to define the services offered in modern industries such as banking, retail, healthcare, etc. as purely electronic or traditional. The boundary between the traditional and electronic services has vanished and an overlap can be seen in both the services. Owing to this growing body of knowledge in service quality research (definition, modeling, measurement, data collection procedure, data analysis, etc.) and competitive business environment have necessitated a fresh understanding and outlook on SERVQUAL model (Amin and Nasharuddin, 2013; Guiry et al., 2013; Sangeetha and Mahalingam, 2011). This is increasingly visible in case of finance (banking), retail, hospitality, and healthcare sectors where the customers can avail the services electronically, traditionally or sometimes using both the channels.

Zeithaml, Parasuraman, and Malhotra (2002) stated that "A comparison of the way consumers evaluate service quality and electronic service quality reveals differences in the role of expectations, number and nature of dimensions, and cognitive-emotional content. In addition, there seems to be a greater degree of consumer trade-offs and hence curvilinearity along e-SQ dimensions than in the case of SQ". Thus it can be said that SERVQUAL can only be used to measure the service quality of traditional services whereas E-SERVQUAL can only be used to measure the service quality of electronic services. Therefore, none of them is said to measure the overall service quality of the firm which goes against Parasuraman, Zeithaml and Berry (1985) who put service quality as: the overall evaluation of a specific service firm that results from comparing the firm's performance with the customer's general expectations of how firms in that industry should perform. This reflects the need of a model which is capable of measuring the overall service quality of the firms representing the modern industries (where services cannot be classified as purely traditional or electronic) such as Banking, Retail etc.

The objective of this article is to propose and test a new scale to help the managers measure the overall service quality of the modern firms. The rest of the article is organized as follows: The next session provides a literature review of the SERVQUAL and E-SERVQUAL scale and the differences in the dimensions of both the scale with the motivation behind a new scale. The following section discusses the constructs and dimensions of the proposed scale followed by the methodology and the future analysis part. The last section discusses the findings of the study followed by conclusion.

LITERATURE BACKGROUND

The SERVQUAL Model

The SERVQUAL model was first published in 1985 by A. Parasuraman, Valarie A. Zeithaml and Leonard L. Berry for measuring and managing service quality across a broad range of service categories. It was based on their definition of quality as 'difference between the expected and perceived performance'. The authors further conducted a qualitative study involving consumers views on service quality and elicited ten determinants of the service quality that focused more on the 'process' of service delivery and not on the 'output or technical' as defined by Gronross (1984). The scale consisted of 22 pairs of statements – measuring expectations of customers by asking each respondent to rate, on a 7 point scale, how essential each item is for an excellent service. The second set of 22 identical statements ascertains the respondent's perception to the level of service given by the service provider. The difference between the ranked perception and the ranked expectations is calculated: the average score is the SERVQUAL overall service quality score. This model has been vigorously tested and improved upon (Parasuraman et al., 1985, 1988, 1990, 1991, 1993, 1994, 2004; Zeithaml, et al., 1996; Zeithaml, Parasuraman & Malhotra,

2002; Parasuraman, et al., 2005) as shown in Table 1. In 1988, the ten factors were collapsed to five dimensions as shown in Figure 1: Reliability, Assurance (competence, courtesy, credibility, and security), Tangibles, Empathy (access, communication, knowing the customer) and Responsiveness, better known as the R.A.T.E.R. dimensions, defined as:

1. **Reliability:** The ability to perform the promised service dependably and accurately.
2. **Assurance:** The knowledge and courtesy of employees and their ability to convey trust and confidence.
3. **Tangibles:** The appearance of physical facilities, equipment, personnel and communication materials.
4. **Empathy:** Care and individualized attention provided to customers.
5. **Responsiveness:** The willingness to help customers and provide prompt services.

Some of the notable attempts with SERVQUAL model to develop new quality measurement models and/or scales in various sectors are by Calabrese and Scoglio, 2012; Durvasula, et al., 2011; Baccarani et al., 2010; Urban, 2009; Chatterjee and Chatterjee, 2005; Gupta et al., 2005; Chiu and Lin, 2004; Getty and Getty,

Table 1. Evolution of SERVQUAL

SERVQUAL	Dimensions
Conceptual model of SQ: the Gap Theory Model 1985	10 determinants of service quality • Reliability, Responsiveness, Competence, Access, Courtesy, Communication, Credibility, Security, Knowing customer needs, Tangibles
SERVQUAL, 1988	5 R.A.T.E.R. dimensions were developed through factor analysis • Reliability, Assurance, Tangibles, Empathy, Responsiveness, • Used a 7 point Likert type scale
SERVQUAL, 1991	• Dropped negative wording • Dropped the normative 'should' and replaced it with 'would' • Allocated 100 points among the 5 dimensions on a 10 point scale
SERVQUAL, 1993	Expectation component interpreted as: • Adequate service • Desired service • Predicted service
SERVQUAL, 1994	• Reformatted 22 items to 21 items on a 9 point Likert-type scale • Included use of 'no opinion' • Perception next to desired and adequate separately
SERVQUAL, 1996	• Developed a conceptual framework of both financial and behavioural consequences of service quality

Figure 1. Dimensions of SERVQUAL

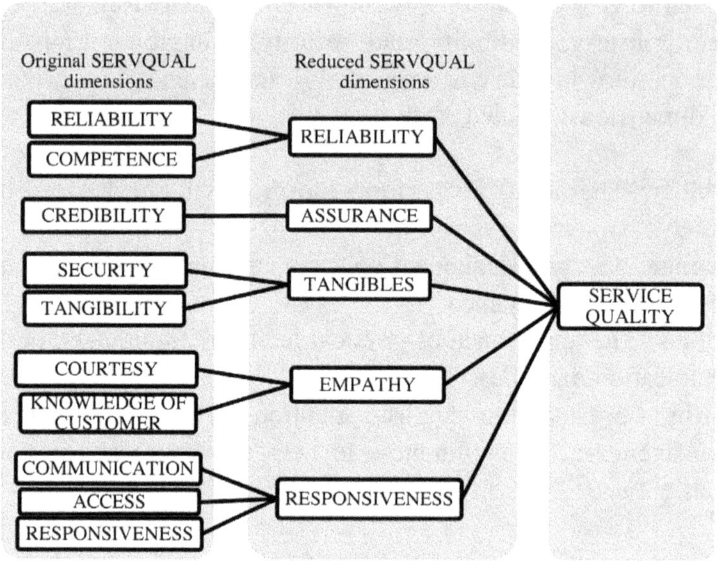

2003; Luk and Layton, 2002; Brady and Cronin, 2001; Candido and Morris, 2000; Bennington and Cummane, 1998; and Lovelock, 1984. The SERVQUAL model has its own strengths and weaknesses. Strengths lie in its simplicity, its linearity and in the rationality of firm actions. It can be used on a regular basis to track customer perceptions of service quality of a particular firm compared to its competitors. Once data have been analyzed they can be visually presented so that it is easy to identify strengths and weaknesses relative to competition. The model limits itself in neglecting the systemic nature of firm and its environment (Mauri et al., 2013). Also, there is little evidence that customers access service quality in terms of Perception minus Expectation gaps. This because they found that the dominant contributor to the gap score was the perception score because of a generalized response tendency to rate expectations high.

Electronic Service Quality Measurement

Electronic services can be defined as "those services that can be delivered electronically," (Javalgi, Martin, & Toddl, 2004) and similarly as "provision of services over electronic networks" (Rust & Kannan, 2003). Boyer, Hallowell & Roth (2002) use the definition, "interactive services that are delivered on the Internet using advanced telecommunications, information, and multimedia technologies." E-service has been regarded as having the potential not only to deliver strategic benefits, but also to enhance operational efficiency and profitability (J.J. Cronin, 2003). The first formal definition of Web site service quality, or e-SQ, was provided by Zeithaml, Parasuraman, and Malhotra (2000). In their terms, e-SQ can be defined as the extent to which a Web site facilitates efficient and effective shopping, purchasing, and delivery of products and services (Zeithaml et al. 2000). As can be observed in this definition, the meaning of service is comprehensive and includes both pre– and post–Web site service aspects.

E-SQ Instrument is an instrument similar to the SERVQUAL scale, developed specifically for measuring online services (e-services) quality. The model has been developed in 2000 and tested and revised in 2002 by Parasuraman, Zeithaml and Malhotra who made an exploratory study on quality perceptions of customers as far as online shopping is concerned. During the research Parasuraman et al. observed that there was missing data on some items. After an analysis of these items they concluded that they were all related to service recovery. That is why they separated those items to develop a separate e service recovery scale (E-RecS-QUAL). The rest of the items formed an e-core service quality scale (E-S-QUAL). The E-S-QUAL scale consists of 4 dimensions with 22 attributes and the E-RecS-QUAL consists of 3 dimensions with 11 attributes. After the development of these scales, they were empirically tested by using questionnaires distributed to sample of users of the most visited at that time web sites in the USA – amazon.com and walmart.com (Parasuraman et Al., 2005).

Table 2. E-S-QUAL SCALE (Parasuraman, A., Zeithaml, V. and Malhotra A., 2005)

E-S-QUAL scale	
DIMENSION	**DEFINITION**
Efficiency	The ease and speed of accessing and using the website
Fulfilment	The extent to which the site's promises about order delivery and item availability are fulfilled
System Availability	The correct technical functioning of the site
Privacy	The degree to which the site is safe and protects customer information

Table 3. E-RecS-QUAL SCALE (Parasuraman, A., Zeithaml, V. and Malhotra A., 2005)

E-RecS-QUAL	
DIMENSION	**DEFINITION**
Responsiveness	Effective handling of problems and returns through the site
Compensation	The degree to which the site compensates the customers for problems
Contact	The availability of assistance through telephone or online representatives

Better e-service quality will enhance the relationship with customers and their satisfaction. So the measurement of e-service quality is very important but it is a complex process due to the complex nature of services (Vivek Agrawal et al. 2014). A major contribution in measuring the e-service quality has been made by Parasuraman (2005) to which other researchers have done some modifications to come up with similar models with different dimensions (Table 4). Similar to SERVQUAL, e-SQ's

Table 4. E-SERVQUAL Dimensions

Author(s)	Dimensions
Dabholkar (1996)	Website designs, reliability, delivery, ease of use, enjoyment and control.
Zeithaml et al. (2000)	Efficiency, reliability, fulfillment, privacy, responsiveness, compensation, and contact.
Yoo and Douthu (2001)	Ease of use, aesthetic design, processing speed, and security.
Cox and Dale (2001)	Website appearance, communication, accessibility, credibility, understanding and availability.
Jun and Cai (2001)	Website designs, information, ease of use, access, courtesy, responsiveness, and reliability.
Yang (2001)	Website design, security and information.
Wolfinbarger and Gilly(2002, 2003)	Website design, reliability, security, and customer service.
Zeithaml et al. (2002)	Security, communication, reliability, responsiveness and delivery.
Madu and Madu (2002)	Performance, features, structure, aesthetics, reliability, serviceability, security and system integrity, trust, responsiveness, service differentiation and customization, Web store police, reputation, assurance and empathy.
Loiacono et al. (2002)	Information, interactivity, trust, response time, website design, intuitiveness, flow, innovativeness, integrated communication, business process and substitutability.
Yang and Jun (2002)	Website design, security, reliability, responsiveness, accessibility and customization.
Surjadaja et al. (2003)	Security, interaction, responsiveness, information, reliability, delivery, and customization.
Santos (2003)	Ease of use, appearance, linkage, structure, content, efficiency, reliability, communication, security, incentive and customer support.
Yang et al. (2003)	Responsiveness, credibility, ease of use, reliability, convenience, communication, access, competence, courtesy, personalization, collaboration, security and aesthetics.
Yang et al. (2004)	Reliability, responsiveness, competence, ease of use, security and product portfolio.
Field et al. (2004)	Website design, reliability, security, and customer service.
Kim and Stoel (2004)	Web appearance, entertainment, information, transaction capability, responsiveness and trust.
Yang and Fang (2004)	Responsiveness, reliability, credibility, competence, access, courtesy, communication, information, responsiveness and website design.
Gounaris et al. (2005)	Website design, information, trust, responsiveness and reputation.
Parasuraman et al. (2005)	Efficiency, availability, fulfillment, privacy, responsiveness, compensation and contact.
Lee and Lin (2005)	Website design, reliability, responsiveness, trust and personalization.
Kim et al. (2006)	Efficiency, fulfillment, system availability, privacy, responsiveness, compensation, contact, information and graphic style.
Fassnacht and Koese (2006)	Graphic quality, layout, attractiveness of selection, information, ease of use, technical quality, reliability, functional benefit and emotional benefit.
Cristobal et al. (2007)	Website design, customer service, assurance and order management.
Sohn and Tadisina (2008)	Trust, speed of delivery, reliability, ease of use, customized communication, website content and functionality

strength also lies in its simplicity and linearity but only several e-SQ dimensions are relevant across several industries, whereas others are more or less specific to particular online service industries. Any attempt to promote a 'global' (or 'generic') measure of e-SQ could be subject to similar criticisms to those directed at 'generic' measurement instruments in traditional service quality (Ladhari, 2010).

Difference in Traditional and Electronic Services

Traditional services and electronic services differ from each other in varying aspects (Surjadjaja, Ghosh & Anthony, 2003) such as, in a traditional service, only people —the employees— are involved in the interface of the service encounter whereas in e-service, ICT or employees are involved in the interface, mediated by the Internet. Moreover, during an e-service encounter, the customers are restricted to hearing and viewing whereas, in traditional services, customers' can- experience the service by using all their senses. Furthermore, traditional service is restricted by distance and opening hours, whereas e-service has substantially removed these barriers. Also, given the importance and advancements in information technology it has become increasingly difficult to define the services offered in modern industries such as banking, retail, healthcare, etc. as electronic or traditional. With these advancements the customer has more control and freedom over the way of getting served. In the last decade, the boundary between the traditional and electronic services has vanished and an overlap can be seen in both the services.

Zeithaml, Parasuraman, and Malhotra (2002) stated that "A comparison of the way consumers evaluate service quality and electronic service quality reveals differences in the role of expectations, number and nature of dimensions, and cognitive-emotional content. In addition, there seems to be a greater degree of consumer trade-offs and hence curvilinearity along e-SQ dimensions than in the case of SQ". Thus it can be said that SERVQUAL can only be used to measure the service quality of traditional services whereas E-SERVQUAL can only be used to measure the service quality of electronic services. Therefore, none of them is said to measure the overall service quality of the firm which goes against Parasuraman, Zeithaml and Berry (1985) who put service quality as: the overall evaluation of a specific service firm that results from comparing the firm's performance with the customer's general expectations of how firms in that industry should perform. This reflects the need of a model which is capable of measuring the overall service quality of the firms representing the modern industries (where services cannot be classified as purely traditional or electronic) such as Banking, Retail etc.

METHODOLOGY

The present study adopted eight steps of scale development paradigm suggested by Churchill (1979). These include specifying the domain of the construct, generating the sample of items, collecting data, purifying the measure, collecting further data, assessing scale reliability, assessing validity, and developing norms (Figure 2). The first four steps mentioned above are particularly relevant for conducting a pilot study which is the focus of this chapter.

Specifying Domain of the Construct

The literature shows that even after almost thirty years from its presentation, the SERVQUAL model still raises interest among scholars. It is also interesting to note that the wide-ranging literature on marketing covers the model in its original and

Figure 2. Suggested procedure for developing better measures (Churchill, 1979)

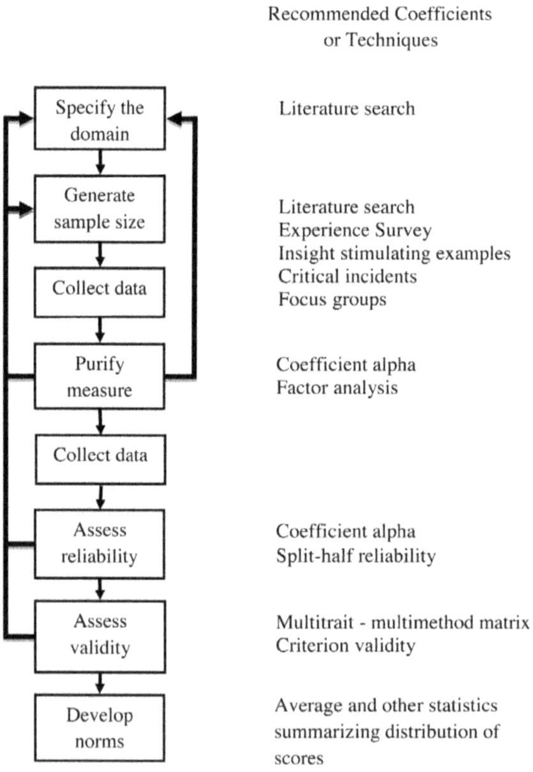

revised versions. The versions are either adjustments or applications of the model but without invalidating its theoretical principles. Also, the E-SERVQUAL model has been widely accepted and its variants are just either addition or modification the dimensions. This shows the robustness of both the models. A rigorous literature review was done to cover all the dimensions mentioned in the literature to measure the service quality of both the traditional services and electronic services (Table 1 and 4). The proposed model integrates the dimensions of SERVQUAL and E-SERVQUAL and other dominant dimensions in the literature, with the relevant modifications and provides a methodology to measure the quality of services offered in present business scenario where both traditional and electronic services are offered at the same time. Also, some dimensions which are only attributed to electronic services are kept separate to measure the service quality efficiently.

Given the differences in the traditional and electronic services (Surjadjaja, Ghosh & Anthony, 2003) it is evident that personalization cannot be achieved in traditional services. Similarly the efficiency of the site can only be attributed to the electronic services. But, all the other dimensions such as responsiveness, reliability, aesthetics, contact, recognition and assurance can be linked to both traditional and electronic services.

Generate Sample of Items

The emphasis at the early stage of item generation is to develop a set of items which tap each of the dimension of the construct at issue (Churchill, 1979). For this purpose a questionnaire was developed after reviewing the prior research in the area of service quality measurement and electronic service quality measurement. All the dimensions and attributes mentioned in the questionnaire have been included in the literature listed in Table 1 and 4. The questionnaire was designed to measure the divergence between the expectations and perceptions (perceived quality) of customers who use the online as well as traditional services along the quality dimensions found in the E-SQ instrument, SERVQUAL instrument and the other modifications of these instruments. This is expected to help the managers evaluate the overall satisfaction of customers with the services of the firm and to give an insight into the important quality dimensions that can be used to measure quality of services in general. For this purpose, the researchers have used a seven-point Likert Scale, ranging from "1="Very dissatisfied" to "7=Very satisfied". The questionnaire contains 47 questions which measure the perceived quality of the customer. The questions are related to the banking industry as it offers both the traditional and electronic services. The banking sector is marked with increasing use of information technology, increased customer knowledge and extensive use of electronic services. Also, the services offered in banks cannot be solely defined as traditional or electronic as there is high overlap and no clear boundary exists between the traditional and electronic services offered. The customers use both the channels to complete the transactions depending on the nature of transaction. For example simple transactions like transferring the money can be done online whereas a bit complex transactions such as applying for loan maybe done traditionally. No double- barreled statements were noticed in the final questionnaire. The detailed questionnaire is provided in appendix.

Collect the Data

The questionnaire contained 47 questions that measured the perceived quality of the customer. The pilot data was collected online through random sampling from 44 respondents on the various dimensions of the proposed model. Respondents were selected based on their relationship with a bank and experience with its net banking services. Relationship with the bank varied from less than one year to more than 5 years. Experience with net banking varied from less than 1 to more than 3 years. In all there were 7 female and 33 male respondent. 18 respondents had an experience of 1 to 3 years with the net banking whereas 3 were having less than one year. Others 19 respondent had experience of more than 3 years with banking. There was no missing data as all the questions were star (compulsory) marked.

Purify the Measure

Factor analysis can be used to suggest dimensions. Factor analysis has the tendency to produce many more dimensions than can be conceptually identified, when applied before the purification stage as in this case. This is partly due to the garbage items which do not have a common core but which do produce additional dimensions in the factor analysis (Churchill, 1979). In this case there were 8 dimensions identified based on Eigen value using exploratory factor analysis (EFA) with a total explained variance up to 81.28% with principal component extraction method and Promax rotation. Also, the component correlation matrix suggested that no two dimensions had high correlation.

FINDINGS

An exploratory pilot survey was conducted as the objective was to gain insights about new scales applicability in service organizations. This kind of survey helps to uncover or provide preliminary evidence of association among concepts. Further, it can help to explore the valid boundary of a theory (Forza, 2002).

The study used Structural Equation Modeling (SEM) in order to calculate the path linkages and check the fit of the model using AMOS SPSS 20. A second order scale was developed to measure the latent constructs (proposed dimensions). We present the results from our pilot study.

The factor loadings are greater than .7 which shows strong association between the first order and second order constructs. Personalization was found to have a relatively

Figure 3. Proposed Integrated SERVQUAL model

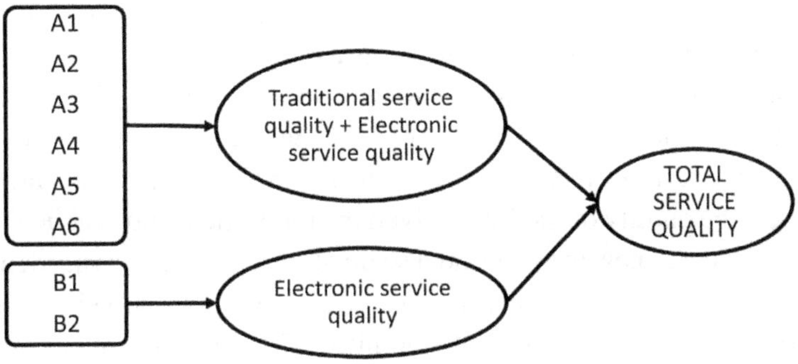

Table 5. Dimensions of Integrated SERVQUAL model

INTEGRATED SERVQUAL MODEL			
Abbreviation	**Dimensions**	**Definition**	**Factor loading**
A1	Reliability	The ability to perform the promised service dependably (Traditional) and correct technical functioning of site (Electronic).	.99
A2	Responsiveness	Willingness to serve customer (Traditional) and effective handling of problems through site (Electronic).	.97
A3	Assurance	Ability of employee to convey trust and confidence (Traditional) and privacy and security of customer information (Electronic).	.93
A4	Contact	Courtesy of the employees (Traditional) and availability online assistance through site (Electronic).	.86
A5	Recognition	Knowledge of employee about customer (Traditional) and recognition on site (Electronic).	.81
A6	Aesthetics	Appearance of physical appearances (Traditional) and site aesthetics (Electronic).	.80
B1	Efficiency	Ease and speed of accessing and using the website (Electronic).	.77
B2	Personalisation	The extent to which user can personalise the site according to his needs and fulfilment (Electronic).	.66

weaker factor loading which can be attributed to small sample size and large number of variables. The CMIN/DF value was found to be 1.78 which indicates good model fit. The loose fit of other indices such as RMR (root mean square residual) and GFI (goodness of fit indices) can be attributed to the small sample size.

CONCLUSION AND FUTURE RESEARCH

Service quality is a key differentiator between firms, and needs careful management. Precise measurement can help firms manage service quality better. Here to fore, SERVQUAL has been the key instrument to measure service quality. The SERVQUAL model measures the service quality of the traditional services offered at the service premises whereas the E-SERVQUAL model measures the service quality of the electronic services offered by the firm. However, none of them can be said to measure the overall service quality of the firm. Given the theoretical and practical deviations observed in services, a new measurement scale of service quality is needed, and is the key research goal of this study. We propose and test a model, following a highly cited scale development paradigm given by Churchill (1979) which is capable of measuring the overall service quality of the firms representing the modern industries (where services cannot be differentiated purely traditional or electronic) such as Banking, Retail, hospitality etc. The model integrates the dimensions of SERVQUAL and E-SERVQUAL and other dominant dimensions in the literature, with the relevant modifications and provides a methodology to measure the quality of services offered in modern industries. We test the scale in banking scenario where both traditional and electronic services are offered at the same time.

The pilot test suggests that the proposed scale is capable of measuring the overall service quality of a modern firm whose services cannot be strictly classified as purely electronic or traditional. The model emerging from pilot survey supports our findings from literature and the relations that appeared weak such as the impact of personalization on total quality (Factor loading of 0.66) will be further tested though large-scale survey. Qualitative cases will be done to include practitioner involvement in developing a scale having practical implications. The dimensions and the items will be refined according to the suggestions of the model. Data will be collected for the improved tool. The dimensions will be redefined on the basis of the new factor analysis with large data. Also, reliability and validity check will be done to make the model robust.

REFERENCES

Amin, M., & Nasharuddin, S. Z. (2013). Hospital service quality and its effects on patient satisfaction and behavioural intention. *Clinical Governance: An International Journal, 18*(3), 238–254. doi:10.1108/CGIJ-05-2012-0016

Babakus, E., & Boller, G. W. (1992). An empirical assessment of the SERVQUAL scale. *Journal of Business Research, 24*(3), 253–268. doi:10.1016/0148-2963(92)90022-4

Baccarani, C., Ugolini, M., & Bonfanti, A. (2010). A conceptual service quality map: The value of a wide opened perspective. *Proceedings of 13th Toulon-Verona Conference "Excellence in Services"*.

Bateson, J. E. G. (2000). Perceived Control and the Service Experience. In Handbook of Service Marketing and Management. Thousand Oaks, CA: SAGE. doi:10.4135/9781452231327.n11

Bennington, L., & Cummane. (1998). Measuring service quality: A hybrid Methodology. *Total Quality Management, 9*(6), 395-405.

Boyer, K. K., Hallowell, R., & Roth, A. V. (2002). E-services: Operating strategy—a case study and a method for analyzing operational benefits. *Journal of Operations Management, 20*(2), 175–188. doi:10.1016/S0272-6963(01)00093-6

Brady, K. M., & Cronin, J. Jr. (2001, July). Some new thoughts on conceptualizing perceived service quality: A hierarchical approach. *Journal of Marketing, 65*, 34–49. doi:10.1509/jmkg.65.3.34.18334

Calabrese, A., & Scoglio, F. (2012). Reframing the past: A new approach in service quality assessment. *Total Quality Management & Business Excellence, 23*(11), 1329–1343. doi:10.1080/14783363.2012.733259

Candido, C. J. F., & Morris, D. S. (2000). Charting service quality gaps. *Total Quality Management, 11*(4-6), 463–472. doi:10.1080/09544120050007779

Carman, J. M. (1990). Consumer perceptions of service quality: An assessment of the SERVQUAL dimensions. *Journal of Retailing, 66*(l), 33–55.

Chatterjee, S., & Chatterjee, A. (2005). Prioritization of service quality parameters based on ordinal responses. *Total Quality Management & Business Excellence, 16*(4), 477–489. doi:10.1080/14783360500078482

Chiu, H. C., & Lin, N. P. (2004). A service quality measurement derived from the theory of needs. *Service Industries Journal, 24*(1), 187–204. doi:10.1080/02642060412331301202

Chowdhary, N., & Prakash, M. (2007). Prioritizing service quality dimensions. *Managing Service Quality: An International Journal*, *17*(5), 493–509. doi:10.1108/09604520710817325

Chowdhary, N., & Prakash, M. (2007). Prioritizing service quality dimensions. *Managing Service Quality: An International Journal*, *17*(5), 493–509. doi:10.1108/09604520710817325

Cook, C., & Thompson, B. (2000). Reliability e-service quality in e-retailing. *Journal of Service Research*, *8*(3), 260–275.

Cronin, J. J. Jr. (2003). Looking back to see forward in services marketing: Some ideas to consider. *Managing Service Quality*, *13*(5), 332–337. doi:10.1108/09604520310495813

Dabholkar, P. A. (1996). Consumer evaluations of new technology-based self-service options: An investigation of alternative models of service quality. *International Journal of Research in Marketing*, *13*(1), 29–51. doi:10.1016/0167-8116(95)00027-5

Dabholkar, P. A., Shepherd, C. D., & Thorpe, D. I. (2000). A comprehensive framework for service quality: An investigation of critical conceptual and measurement issues through a longitudinal study. *Journal of Retailing*, *76*(2), 139–173. doi:10.1016/S0022-4359(00)00029-4

Dowell, J., & Long, J. B. (1998). Conception of the cognitive engineering design problem. *Ergonomics*, *41*(2), 126–139. doi:10.1080/001401398187125

Durvasula, S., Lysonshi, S., & Madhavi, A. D. (2011). Beyond service attributes: Do personal values matter? *Journal of Services Marketing*, *25*(1), 33–45. doi:10.1108/08876041111107041

Forza, C. (2002). Survey research in operations management: A process-based perspective. *International Journal of Operations & Production Management*, *22*(2), 152–194. doi:10.1108/01443570210414310

Getty, J. M., & Getty, R. L. (2003). Lodging quality index (LQI): Assessing customers' perceptions of quality delivery. *International Journal of Contemporary Hospitality Management*, *15*(2), 94–104. doi:10.1108/09596110310462940

Getty, J. M., & Thompson, K. N. (1994). A procedure for scaling perceptions of lodging quality. *Journal of Hospitality Research*, *18*(2), 75–96.

Grönroos, C. (1984). A service quality model and its marketing implications. *European Journal of Marketing*, *18*(4), 36–44. doi:10.1108/EUM0000000004784

Guiry, M., Scott, J. J., & Vequist, D. G. IV. (2013). Experienced and potential medical tourists' service quality expectations. *International Journal of Health Care Quality Assurance*, *26*(5), 433–446. doi:10.1108/IJHCQA-05-2011-0034 PMID:23905303

Gupta, A., McDaniel, J. C., & Herath, S. K. (2005). Quality management in service firms: Sustaining structures of total quality service. *Managing Service Quality*, *15*(4), 389–402. doi:10.1108/09604520510606853

Headley, D. E., & Miller, S. J. (1993). Measuring service quality and its relationship to future consumer behavior. *Journal of Health Care Marketing*, *13*(4), 32–41. PMID:10131732

HR-Focus. (2000). Online recruiting: what works, what doesn't. *HR Focus, 00*(3), 11-15.

Javalgi, R. G., Martin, C. L., & Todd, P. R. (2004). The export of e-services in the age of technology transformation: Challenges and implications for international service providers. *Journal of Services Marketing*, *18*(7), 560–573. doi:10.1108/08876040410561884

Jiang, J. J., Klein, G., & Crampton, S. M. (2000). A note on SERVQUAL, reliability and validity in information system service quality measurement. *Decision Sciences*, *31*(3), 725–734. doi:10.1111/j.1540-5915.2000.tb00940.x

Kilbourne, W. E., Duffy, J. A., Duffy, M., & Giarchi, G. (2004). The applicability of SERVQUAL in cross-national measurements of health-care quality. *Journal of Services Marketing*, *18*(6/7), 524–533. doi:10.1108/08876040410561857

Knutson, B., Stevens, P., Wullaert, C., & Patton, M. (1991). Lodgserv: A service quality index for the lodging industry. *Hospitality Research Journal*, *14*(7), 277–284.

Ladhari, R. (2010). Developing e-service quality scales: A literature review. *Journal of Retailing and Consumer Services*, *17*(6), 464–477. doi:10.1016/j.jretconser.2010.06.003

Lam, S. K. S. (1997). SERVQUAL: A tool for measuring patients' opinions of hospital service quality in Hong Kong. *Total Quality Management*, *8*(4), 145–152. doi:10.1080/0954412979587

Lam, T. K. P. (2002). Making sense of SERVQUAL's dimensions to the Chinese customers in Macau. *Journal of Market-Focused Management*, *5*(10), 43–58. doi:10.1023/A:1012575412058

Lovelock, C. H. (1984). *Services Marketing*. Englewood Cliffs, NJ: Prentice Hall, Inc.

Luk, S. T. K., & Layton, R. (2002). Perception gaps in customer expectations: Managers versus service providers and customers. *Service Industries Journal*, *22*(2), 109–128. doi:10.1080/714005073

Mauri, A. G., Minazzi, R., & Muccio, S. (2013). A review of literature on the gaps model on service quality: a 3-decades period: 1985–2013. *International Business Research*, *6*(12), 134. doi:10.5539/ibr.v6n12p134

Mels, G., Christo, B., & Deon, N. (1997). The dimensions of service quality: The original European perspective revisited. *Service Industries Journal*, *17*(1), 173–189. doi:10.1080/02642069700000009

Meuter, M. L., Ostrom, A. L., Roundtree, R. I., & Bitner, M. J. (2000). Self-service technologies: Understanding customer satisfaction with technology-based service encounters. *Journal of Marketing*, *64*(3), 50–64. doi:10.1509/jmkg.64.3.50.18024

Meuter, M. L., Ostrom, A. L., Roundtree, R. I., & Bitner, M. J. (2000). Self-Service Technologies: Understanding Customer Satisfaction with Technology-Based Service Encounters. *Journal of Marketing*, *64*(July), 50–64. doi:10.1509/jmkg.64.3.50.18024

Parasuraman, A. (1998). Customer service in business-to-business markets: An agenda for research. *Journal of Business and Industrial Marketing*, *13*(4/5), 309–321. doi:10.1108/08858629810226636

Parasuraman, A. (2004). Assessing and improving service performancefor maximum impact: Insights from a two-decade-long research journey. *Performance Measurement and Metrics*, *5*(2), 45–52. doi:10.1108/14678040410546064

Parasuraman, A., Berry, L. L., & Zeithaml, V. A. (1991). Refinement and reassessment of the SERVQUAL scale. *Journal of Retailing*, *67*(4), 420.

Parasuraman, A., Berry, L. L., & Zeithaml, V. A. (1993). More on improving service quality measurement. *Journal of Retailing*, *69*(1), 140–147. doi:10.1016/S0022-4359(05)80007-7

Parasuraman, A., Zeithaml, V. A., & Berry, L. L. (1985). A conceptual model of service quality and its implications for future research. *The Journal of Marketing*, 41-50.

Parasuraman, A., Zeithaml, V. A., & Berry, L. L. (1988). Servqual. *Journal of Retailing*, *64*(1), 12–40.

Parasuraman, A., Zeithaml, V. A., & Berry, L. L. (1990). *Delivering quality service*. Balancing Customer.

Parasuraman, A., Zeithaml, V. A., & Berry, L. L. (1994). Alternative scales for measuring service quality: A comparative assessment based on psychometric and diagnostic criteria. *Journal of Retailing*, *70*(3), 201–230. doi:10.1016/0022-4359(94)90033-7

Parasuraman, A., Zeithaml, V. A., & Malhotra, A. (2005). ES-QUAL a multiple-item scale for assessing electronic service quality. *Journal of Service Research*, *7*(3), 213–233. doi:10.1177/1094670504271156

Reichheld, F. F., & Schefter, P. (2000, July). E-loyalty: Your secret weapon on the Web. *Harvard Business Review*, 105–113.

Rust, R. T., & Kannan, P. K. (2003). E-service: A new paradigm for business in the electronic environment. *Communications of the ACM*, *46*(6), 36–42. doi:10.1145/777313.777336

Sangeetha, J., & Mahalingam, S. (2011). Service quality models in banking: A review. *International Journal of Islamic and Middle Eastern Finance and Management*, *4*(1), 83–103. doi:10.1108/17538391111122221

Santos, J. (2003). E-service quality: A model of virtual service quality dimensions. *Managing Service Quality*, *13*(3), 233–246. doi:10.1108/09604520310476490

Statistical Year Book India. (2015). Ministry of Statistics and Programme Implementation, Government of India. Retrieved from www.mospi.nic.in/statistical-year-book-india/2015

Surjadjaja, H., Ghosh, S., & Antony, J. (2003). Determining and assessing the determinants of e-service operations. *Managing Service Quality: An International Journal*, *13*(1), 39–53. doi:10.1108/09604520310456708

Szymanski, D. M., & Hise, R. T. (2000). E-satisfaction: An initial examination. *Journal of Retailing*, *76*(3), 309–322. doi:10.1016/S0022-4359(00)00035-X

Urban, W. (2009). Service quality gaps and their role in service enterprises development. *Technological and Economic Development of Economy*, *15*(4), 631–645. doi:10.3846/1392-8619.2009.15.631-645

Van der Wal, R. W. E., Pampalis, A., & Bond, C. (2002). Service quality in a cellular telecommunications company: A South African experience. *Managing Service Quality*, *12*(5), 323–335. doi:10.1108/09604520210442119

Walker, R. H. (1996). Towards identifying how visitors to Tasmania define and assess service quality in the hospitality industry. *Australian Journal of Hospitality Management*, *3*(2), 27–39.

Zeithaml, V. A. (2000). Service quality, profitability and the economic worth of customers: What we know and what we need to learn. *Journal of the Academy of Marketing Science*, *28*(1), 67–85. doi:10.1177/0092070300281007

Zeithaml, V. A., Berry, L. L., & Parasuraman, A. (1996). The behavioral consequences of service quality. *The Journal of Marketing*, 31-46.

Zeithaml, V. A., Parasuraman, A., & Malhotra, A. (2002). Service quality delivery through web sites: A critical review of extant knowledge. *Journal of the Academy of Marketing Science*, *30*(4), 362–375. doi:10.1177/009207002236911

Zhou, L., Zhang, Y., & Xu, J. (2002). *A critical assessment of SERVQUAL's applicability in the banking context of China*. Valdosta, GA: Asia Pacific Advances in Consumer Research.

APPENDIX

Questionnaire

Based on your experiences as a consumer of online and traditional banking services, please provide information on how you perceive the quality of the banking services you use in comparison to your expectations. Please answer with the number of your choice using a scale of 1-7 as explained below.

If you do not use online banking services, please do not fill out the questionnaire!

Scale explanation:

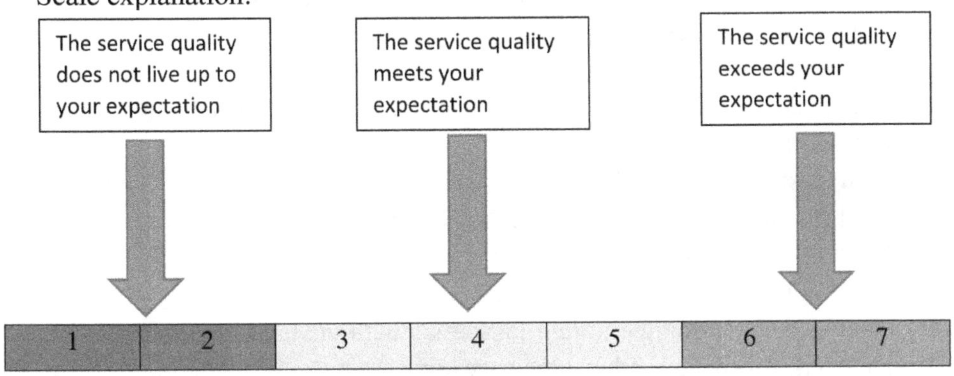

Please provide the following information:

NameofyourBank……………………………………………………………………

Gender: Male Female

Age: ……………………………………………………………

Occupation: ……………………………………………………

Length of Relationship with Bank:

<1 year

1-5 years

>5 years

Length of Internet Banking Use:

< 1 year

1-3 years

> 3 years

Frequency of visiting the bank: ………………………………………………

Frequency of Internet Banking:

……………………………………………………………

Transactions: Per month:

…………………………………………………………………..

1. Employees of bank are always willing to help me.
2. Bank does tells me exactly when services will be performed.
3. Employees of the bank are always available on their desk to help the customers.
4. The bank gives prompt responses to my request by e-mail or other means.
5. The bank quickly resolves problems I encounter with my online transactions.
6. The bank take customer feedback seriously and is always up to resolve the complaints.
7. When bank promises to do something by a certain time, it does so.
8. The bank is dependable.
9. The bank delivers the services in promised time period.
10. The site launches and runs right away.
11. The site is always available for business.
12. The bank's website is always functional and I am able to finish my transaction without failure every time.
13. I can trust employees of the bank.
14. I feel safe in my transactions with bank's employees.
15. The bank does not misuse my personal information.
16. I have confidence in the bank's service.
17. I feel safe while doing online transactions with the bank.
18. The bank's website has high security level.
19. The bank's website is capable of keeping my details secure.
20. Employees of bank are polite.
21. I get support from the bank employees.
22. Employees of the bank always wear a smile on their face.
23. The bank is easily accessible by telephone.
24. The bank's website has customer service representatives' available Online.
25. I can always contact to the bank representatives for online assistance on website.
26. Employees of the bank know me.
27. Employees of bank give me personal attention.
28. Employees of bank know what my needs are.
29. The bank's website is able to recognize and remember my preferences.
30. The bank's website gives me suggestions on the basis of my previous transactions.
31. The bank's website remember my last login details and previous transactions.
32. Bank's physical facilities are visually appealing.
33. Bank has up to date equipments.
34. The banks' employee wear tidy clothes and are well groomed.
35. The website design is aesthetically attractive.
36. This bank's Web site design is innovative.
37. This bank's Web site does not have fine print that is difficult to read.
38. I am able to get on the site quickly.

39. It is quick to complete a transaction through the bank's website.
40. Using the bank's website does not require a lot of effort.
41. The organization and structure of online content is easy to follow.
42. I don't get lost on this bank's Web site.
43. I am able to customize the bank's website according to my need.
44. I have the option of changing the color and font size of the bank's website.
45. I am able to do all transactions through bank's website without going to bank.
46. I am able to add and remove the services offered on bank's website according to my use.
47. The bank's website provides all the services offered by bank.

Chapter 10
Design of Cloud-Based CAPP System to Aid Process Planning for Sheet Metal Products

Norman Gwangwava
Botswana International University of Science and Technology, Botswana

ABSTRACT

In the design of new products, it is usual for different planners to specify different processing routes for the same part, which do not always incorporate the optimal solution to the problem. Computerization of process planning has evolved as a solution to speed up planning for new products. The essence of the chapter is to explore machine capabilities, parameters, and constrains inherent in sheet metal forming processes and then illustrate CAPP software design for process sequencing for new products to support standardisation of production routes. The verification of the designed CAPP system has realised significant savings of 40% reduction in process planning effort, 5% in material, 15% in scrap and 10% in tooling. In order to design the cloud-based CAPP system, data was gathered on the machinery capabilities for the sheet metal operations in a case study company that manufactures sheet metal accessories for the construction industry. A computer-based system was developed to store the machinery tool capabilities and their functional parameters and to facilitate sharing of information across the system modules. Autodesk Inventor® was used as the source of input for the CAD models into the system. Visual Basic. NET programming language was used to design a module for feature recognition (FR) to capture characteristics from a CAD model drawing of the new sheet metal product. A module was then developed to sequence the process operations for the CAD model based on the part features. The process sequencing system helps designers and process planners to link the new model design with the production path, the tools, and equipment required in order to improve their design and planning in the early stages of the product life cycle.

DOI: 10.4018/978-1-5225-3628-4.ch010

INTRODUCTION

Process planning is an important task in the manufacturing cycle; it determines how a product is to be manufactured. Traditionally process planning was carried out manually by experienced engineers who would interpret engineering drawings of products and determine the `best' route and manufacturing methods based on available materials, processes, and the time scale and cost restraints. Computer-aided Process Planning (CAPP) is an important element of production planning. However, existing methodologies failed to meet the requirements for CAPP systems in the cloud, which is a distributed, collaborative, and web-based environment (Xu and Lu, 2015).

The sheet metal industry is ever growing due to the increasing demand for white goods and the ever rising demand in the built environment, but one of the limitations that it faces is the manual planning of production processes, thus causing time inefficiency in meeting the customer requirements. Manual planning is time consuming, labour intensive and may involve human errors. More time is spent attempting various iterations in order to establish the optimum method. This leads to many hours being lost as idle time in the production system. Experienced process planners have no difficulties in managing the complexity of process planning compared to the novice who is facing significantly bigger challenges. Grasping the scope and understand the type of decisions made in process planning is difficult (Lundgren et al., 2014). Large volumes of scrap accumulate at production lines as a result of the non-optimal process sequencing done manually. There is rework that looms in some instances due to some product batches that has to be either trimmed or cropped.

Traditionally companies have been buying complex standalone application software to be installed and hosted on company premises, which has proved to be an expensive investment. Many small-to-medium enterprises (SMEs) cannot afford the high initial investment. The approach also makes it difficult to manage ICT costs. Such a model is viewed as a one-fits-all approach, where companies would pay for certain number of users for their software licence for certain fixed periods. Many users would have to be trapped in the old software version, according to their licence renewal period. Cloud Computing (CC) is revolutionizing business approach across many enterprises, with its pay-as-you-go model and the SOA architecture. The service oriented model in which "Everything is treated as a Service" (XaaS) is helping manufacturers, among many other businesses, to align production efficiency with business strategy as well as promote collaboration through intelligent factory networks where resources of various kinds are shared. The huge success of cloud computing brings the opportunity to the manufacturing systems with high scalability, productivity and agility. It also provides practical solutions to the small and medium-sized enterprises with elastic investment solutions (Wang et al., 2017).

The concept of Cloud Manufacturing (CM) emanated from the adoption of CC. CM requires implementation of each manufacturing application in the manufacturing cloud. Modern manufacturing practice is a collaborative environment supported by various Computer-Aided software applications (CAx) and networks. Modern product design and process planning systems require the application of new concepts and models that include the dynamic coordination of all activities and stages with the purpose of improving the quality of the product design and process planning (Milošević et al., 2017). Collaborative environments allow engineers and designers to gain better understanding of products, increasing the quality and providing design for manufacture from the very beginning, thus reducing the need for expensive re-work during later stages of development (Milošević et al., 2011). The knowledge implemented into a typical CAPP system is subject to frequent updates with newly acquired expert knowledge. Market conditions are constantly changing, while the prices, terms of delivery, and production volumes require modifications of current process plans (Siller et al., 2008). Xu and Lu (2015) discusses the requirements of CAPP systems in a cloud environment and propose a feasible system framework for next-generation CAPP systems. The proposed system is built upon Service Oriented Architecture (SOA) architecture, which enables smooth system integration among different manufacturers. A feasible mechanism to incorporate shared engineering practices and knowledge from each manufacturer is also presented as part of the proposed system. Semantic knowledge representation, management, sharing, access, and re-use approaches can support collaborative production planning in a flexible and efficient as well as an effective way (Gernhardt et al., 2015).

In order to solve the problems associated with process sequencing, the chapter reports on research work done on a case study company that performs sheet metal processes to manufacture pre-punched, light pressed metal bracing straps, and nail plate connectors for the construction of roof trusses, anti-split nail plates for use in treated timber pole production, hurricane clips/purlin clips and truss hangers. In a typical sheet metal forming company, large volumes of raw material, in the form of metal sheets and metal coils are stored in-house earlier-in-time to processing. The processing tool-dies for the discrete production tasks occupy a substantial section of the factory floor, and the finished goods are stored in batches prior to purchase and/ or dispatch. During the manufacturing stage there are several factors contributing to delays, thus fallback in customer-requirements satisfaction. The vast numbers of tooling-dies are inherently moved from one work station to another, whenever there is need for a production line to be run. This process involves frequent and lengthy adjustments of tool-die accessories (bolt-on): further-more the operators are susceptible to operation delay as a result of the operational techniques employed. Ultimately there is more idle-time and down-time in the production system. Another factor contributing to the production time delay is the transportation of work-in-progress from one workstation to another.

The increasing number of product variants necessitate adaptive and flexible process planning methods (Mourtzis et al., 2016). Due to the dynamic nature of modern enterprises, particularly the SMEs, product designs are changing regularly. Mass customised products are also on the increase, particularly because of the high capabilities of the modern day architectural practices, as well as demands from the educated customers. Traditional CAPP systems cannot cope with such dynamism.

Due to this background, a cloud based CAPP system becomes imperative for the sheet metal companies in these ever dynamic market changes: a system that will facilitate collaborative and agile process sequencing for new products in a cost effective and time efficient manner.

BACKGROUND

There has been an increase in the number of SMEs entering the construction industry due to the increasing demand in the built environment. Manufacturers servicing the construction industry perform sheet metal processes to manufacture HVAC units, pre-punched light pressed metal bracing straps, nail plate connectors for construction of roof trusses, anti-split nail-plates for use in treated timber pole production, hurricane clips/purlin clips and truss hangers, among many other products. Large volumes of raw material, metal sheets, (L=2.4m by W=1.2m) and metal coils, (weight range = 0.5-1.2tons per coil), are stored in-house earlier-in-time to processing. The processing tool-dies for the discreet production tasks occupy a substantial section of the factory floor, and the finished goods are stored in batches prior to purchase and/or dispatch. These factors contribute to quite a significant amount of inventory in the system that can be controlled or minimised.

Taiichi Ohno (1912-1990) said "Inventory is wicked" and he also said "The more inventories a company has the less likely they will have what they need."

During the manufacturing stage there are several factors contributing to delays, thus fall-back in customer-requirement satisfaction. The vast numbers of tooling-dies are inherently moved from work station to another, whenever there is a need for a production-line to be run. This process involves frequent and lengthy adjustments of tool-die accessories (bolt-on): further-more the operators are susceptible to operation delay as a result of the operational techniques employed. Summarily there is a lot of idle-time and down-time on the production system in the organizations. Also contributing to the production time delay is the transportation of work-in-progress from one workstation to another.

Scrap is a very common feature sheet metal production systems as there is always significant amounts of cut-offs and scrap that is disposed of after certain production lines. For example the Hurricane clip (BPS) line yields scrap to material ratio of 1:4, i.e. 25kg scrap from every 100kg of material, based on the case study company. There is rework that looms in other instances such that there are some product batches that have to be either trimmed or cropped to substitute for a for a depleted product batch.

The solution to the challenges faced by SMEs in sheet metal forming and many other manufacturing industries is a cloud based CAPP system to help industries adapt in these ever dynamic market changes: a system that will facilitate collaborative and agile process sequencing for new products in a cost effective and time efficient manner. The system facilitates the pre-conception, design and process planning for new and existing product families. For already existing plants, it is coessential to consider the suitability of insertion of such software product into already existing systems without the need of an expansive interface—for example, between CAD/CAM system and CAPP system and between CAPP system and the software for economical, wage, and storage records. One of the other demands for choosing a suitable CAPP system or another software application is its flexibility and possibility to adapt to the user's requirements (Monkova and Monka, 2014). Rigid, traditional business philosophy in many industries results in lack of competitiveness.

There has been intensive research in CAPP systems, particularly in machining operations. Several research projects focusing on machining and inspection led to the development of STEP-NC. Overall research activities in specific areas of STEP-NC based on manufacturing technology and processes focus more on milling rather than turning activities, due to ISO 14649, Part 11 for milling operation being established before Part 12 (Turning) (Yusof et al., 2009). Literature survey by Xie and Xu (2006) pointed out that no platform or system was available that can efficiently integrate the CAPP and CAM and computer numerical control (CNC systems) in sheet metal product development processes. A prototype STEP-compliant system for designing, nesting, and process planning for sheet metal parts was then developed with a number of modules integrated (Xie and Xu, 2006). Although sheet metal forming is a major manufacturing process to transform sheet metal blanks into complex 3D parts, the state of the art in CNC sheet metal still is still recognized as a separate and vague island of automation (Zahid et al., 2015). Discreet sheet metal parts manufacturing methodology differs inherently from conventional material removal manufacturing processes.

LITERATURE REVIEW

This section gives a literature review of states-of-the-art in sheet metal forming and process planning. The important characteristics and the forming processes employed to produce a wide variety of products are discussed.

Metal Forming

Metal forming processes can be classified into two basic categories: bulk deformation processes and sheet metalworking processes (Serope and Schmid, 2009). Each category includes several major classes of shaping operations. The article focuses on sheet metal working. Sheet metalworking processes are forming and cutting operations performed on metal sheets, strips, and coils. A part (less than about 6.25mm or 0.2inches) produced in a sheet metal operation is often subjected to a process called stamping (Poli, 2001). Sheet metal operations are always performed as cold working processes and are usually accomplished using a set of tools called a punch and die. The punch is the positive portion and the die is the negative portion of the tool set.

Equipment for Sheet-Metal Forming

For most general press working operations, the basic equipment consists of mechanical, hydraulic, pneumatic, or pneumatic-hydraulic presses with a wide variety of designs, features, capacities, and computer controls. Typical designs for press frames are shown in Figure 1. The proper design, stiffness, and construction of such equipment are essential to the efficient operation of the system and to achieving a high production rate, good dimensional control, and high product quality. The traditional C-frame structure has been used widely for ease of tool and work piece accessibility.

Figure 1. Schematic illustrations of types of press frames for sheet metal forming

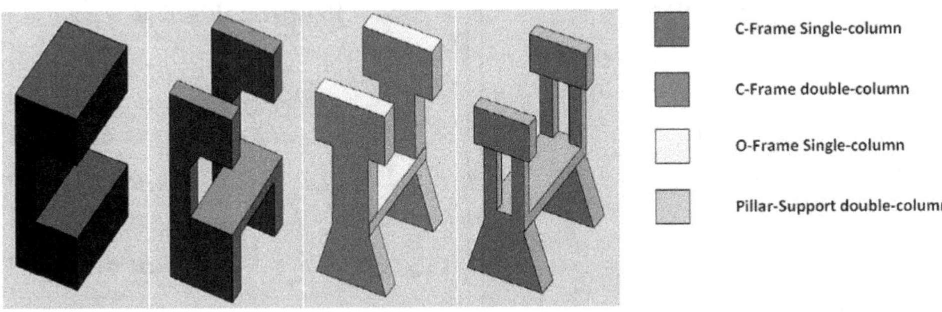

Tooling for Sheet-Metal Forming

Basic tools used with a metal working press are the punch and die. A punch (male) is a convex tool for making holes by shearing, or making a surface or displacing metal and a die can be defined as a concave tool, which is the female part. The tools for sheet metal work are made of heat-treated high alloy steels. Figure 2 illustrates various tool configurations for sheet metal working. There are many variations of progressive die designs. The design shown in Figure 2 illustrates some common operations and terminology associated with progressive dies.

State-of-the-Art in Sheet Metal Process Planning: CAPP

Computer Aided Process Planning (CAPP) involves the use of a computer to aid process planning. The tasks of set-up selection and bend sequencing, tolerance reasoning and tool selection are complicated ones, which can be very cumbersome when manually performed. Even where experienced personnel are involved in the manual sheet metal bending process planning, the resulting plan is usually not optimal.

Figure 2. Types of sheet metal dies

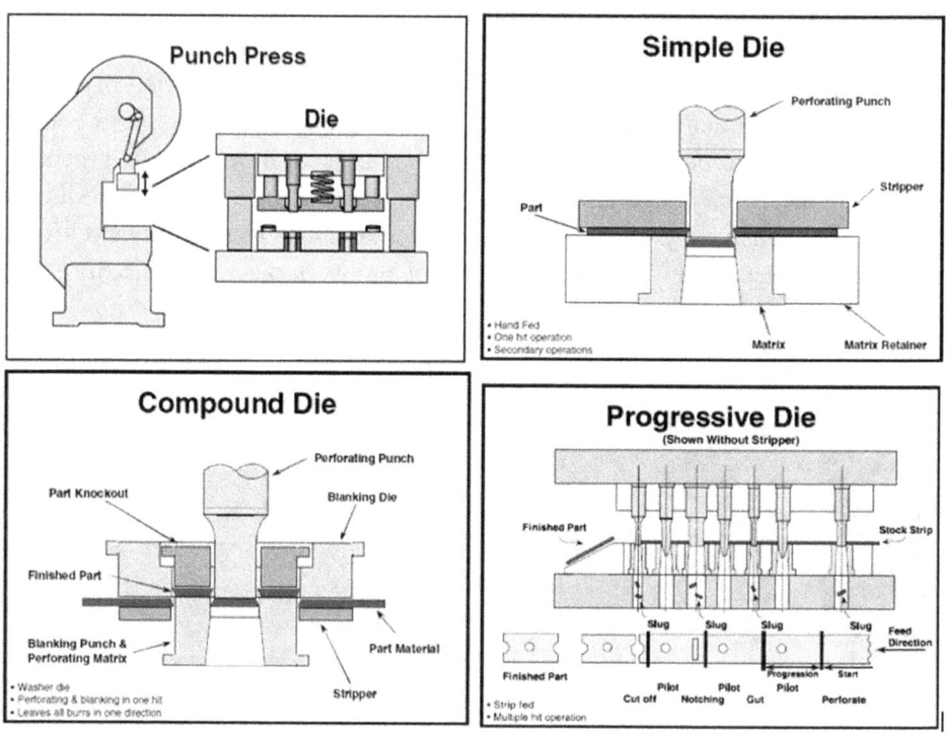

With technological development, powerful computers, with large memory capacities and processor speeds, are now available. They are also becoming more affordable, which is a very positive development to enable industries to use computer based planning systems. The major challenge remains the availability of technology to facilitate fully automated CAPP systems, which can generate sheet metal bending process plans without any human intervention.

Thorough research has been, and is still being, pursued in the field of sheet metal CAPP. De Vin et al. (2000) identified some modules which can be contained in a sheet metal CAPP. These are listed here:

1. **Geometry Related Modules:** These are for importing product models, identification of manufacturing features and geometrical manipulations;
2. **Technological Modules:** Such as selection of methods and equipment, determination of process conditions;
3. **Output Modules:** Such as the generation of NC code, operator instructions, tooling information;
4. Modules shared with other manufacturing functions.

De Vin et al. (1992) were the first to present a thorough sheet metal CAPP called PART-S. Pullant (2010) noted that automated feature recognition (AFR) is instrumental in developing a CAPP system, which is in turn essential for realising simultaneous design of product and process, also known as Concurrent Engineering (CE). Pullant's research was based on selective slicing of a given solid model at multiple levels and in multiple directions, and used contour based reasoning to identify form features. The methodology formulates rules to recognize manufacturing features by combining the information pertaining to the contours in successive slices. However, this methodology is only suitable in machining and other additive processes such as casting and moulding. Gupta et al. (1998) developed a CAPP system for robotic sheet metal bending press brakes. The research work also outlined the differences between sheet metal CAPP and other machining processes, where, for sheet metal bending, features may be easily identified as a collection of bend lines. However, selecting the most appropriate punch profile shape can be a very challenging problem whilst cutting tools for milling and drilling are selected based on their dimensions (diameters and lengths). The major outcome of the robotic bending cell CAPP was a planning system that employs a distributed planning architecture.

Various methods exist for solving process planning problems. Duflou et al. (2003) presented a procedure for automatic generation of process plans. Cattrysse et al. (2006) developed a process-planning module modelled both ways as a Travelling Purchaser Problem (TPP) and as a Generalized Travelling Salesperson Problem (GTSP). They used a hierarchical decomposition approach to solve the TPP, while the GTSP was solved using guided local search (GLS). Gupta and Rajagopal (2002) presented a methodology for creating sheet metal part families for generating shared press brake setups. Their motivation was to optimize sheet metal bending by producing more than one type of part without requiring a setup change. A greedy algorithm was presented first for forming a part family using a bottom-up approach that makes use of the mixed-integer linear programming formulation for generating shared setups for each part family. Second a mixed-integer linear programming formulation to generate a shared set up for a given set of parts was presented, if such a setup exists. The approach would improve machine tool utilisation and result in cost effective, small-batch manufacturing. Lichao et al. (2011) developed a CNC bending machine process planning software based on a robust method for sheet metal bend sequencing, in which hard reject criteria and a preclusion method of complex bending model are both involved. They developed a set of solution methods in bend sequencing based on an A-star algorithm. Kannan and Shunmugam (2009) proposed an integrated system of feature recognition and reasoning for 3D sheet metal components. Their approach did not use the usual modelling feature-based methodology; they claimed that such methodology lacks the information required for manufacturing and the entire feature information is lost when converted to a neutral format such as STEP AP-203. The proposed feature recognition system identifies manufacturing features in a generic manner, while feature reasoning furnishes the information required for manufacturing. 3D model data in STEP AP-203 format is the input into the feature recognition system, which first generates the central plane of the component. Further processing of faces is performed and various features with similar manufacturing attributes are identified, using a set of rules based on topology, geometry and Boolean logic. The system effectively identifies different types of manufacturing features as well as composite features, irrespective of their shape.

For the purpose of automating the process planning of sheet metal forming, Zhang et al. (2009) presented an approach for recognizing both isolated and intersecting geometric features of freeform surface models of parts. They developed a methodology with three major steps: subdivision of B-spline surfaces, detection of protrusions and depressions, and recognition of geometric features for sheet metal forming domain. The input geometry data format of the part is based on an IGES CAD surface model represented in the form of trimmed B-spline surfaces. The system classifies or subdivides each surface into different curvature regions with the aid of curvature property surfaces obtained by using symbolic computation of B-spline surfaces. Regions satisfying particular geometry and topology relations are recognized as protrusion and depression (DP) shapes. DP shapes are then classified into different geometric features using a rule-based approach.

Using Features to Aid Process Planning and Manufacturability

Almost all CAPP systems employ the concept of manufacturing features for part description. Manufacturing features can identify the portions of a part that have manufacturing significance. Because manufacturing features are domain dependent, they are significant only when the domain at hand and the task to be accomplished are clearly specified (Park, 2003). The sheet metal industry has focused on the automation of punching, shearing and nesting processes for sheet metal parts. Inui et al. (1998) researched on related product models with features and processes for bending operation. Zussman and Horsch (1994) proposed a motion planning approach for robot-assisted multiple bent parts based on configuration-space and potential filed. De Vin et al. (1992) developed a sheet metal CAPP system called PART-S which integrates cutting, nesting, bending and welding processes. Bourne and Wang (1995) developed an automatic bending process planner to generate plans with near minimum manufacturing costs.

Two approaches have been developed to identify features in a design:

1. Feature recognition and
2. Design-with-features

The feature recognition approach searches the geometric description of a part for known patterns and then labels the geometry subsets with a feature name. The design with- features approach allows the designer to directly label the geometry.

State-of-the-art sheet metal parts design systems represent the final part geometry. This geometry describes either the wire-frame or the boundary representation of a part. Unfolding software is then used to discover the topological relationships between the surfaces. With the unfolding result, the flat pattern with the bend deductions can be calculated.

Automated feature recognition (AFR) is instrumental in developing a CAPP system, which is in turn essential for realising simultaneous design of product and process, also known as Concurrent Engineering (CE). Lichao et al. (2011) developed a CNC bending machine process planning software based on a robust method for sheet metal bend sequencing, in which hard reject criteria and a preclusion method of complex bending model are both involved. They developed a set of solution methods in bend sequencing based on an A-star algorithm.

CAD and CAPP Integration

The literature survey above has reflects major efforts undertaken to develop fully automated CAPP systems. Current research efforts are dedicated towards feature reasoning, which is an essential element in full realization of a fully automatic CAPP system. Zhou et al. (2007) proposed a practical solution for a bi-directional integration of CAD and CAPP on the platform of commercial CAD systems. Their research work dealt with key techniques such as feature recognition and conversion, feature parameter and constraint extraction, feature tree reconstruction, technical information processing, process planning, automatic process drawing marking and 3D material stock CAD model generation. In this research work, the extracted features and their related technical information and knowledge are encapsulated together with the geometry-oriented CAD model, to form an integrated product information model to facilitate effective integration with the downstream activities. The proposed integrated CAD/CAPP system was implemented using a commercial CAD package, UGS/SolidEdge®. According to Khoshnevis et al. (1999), an integrated CAD and CAM system not only plays a key role in achieving digital manufacturing and computer integrated manufacturing, but is also vital to the competitiveness of mechanical manufacturing enterprises and their ability to respond rapidly to market changes. One of the most critical links for such integration is the link between design process planning activities. CAPP system exists along with other CAx systems used in all stages of the product lifecycle. CAD and CAM systems are directly linked to the CAPP system which enables the automated generation of technical documentation and the preparation of manufacturing operations which are performed on CNC machines (Milošević et al., 2017).

Web-Based CAPP

Social media is currently the biggest revolution taking place globally. This trend has its roots in the advent of the internet and the WWW two decades ago. Many social networks, such as Facebook for instance, have sprung up in recent years. Major efforts are currently underway to integrate social media applications into production planning and control, particularly in extracting the information from these applications and making it formal. Chin and Tang (2002) developed a methodology for enabling better concurrent stamping part and die development, using the Web technology. Their system development approach included knowledge-based part design and evaluation over the Web. Issues of cooperation and coordination among product designers and die designers were also involved. The research findings were that such a system allows part designers and die designers to share product data, communicate with each other, and maintain operational consistency in the distributed cooperative sites on different product design and development platforms. Alvares et al. (2008), developed an integrated web-based CAD/CAPP/CAM system for the remote design and manufacture of feature-based cylindrical parts. Their system is called WebMachining® and was developed in an e-manufacturing context, where the use of features allows integration among the activities of collaborative design (WebCADbyFeatures), generative process planning (WebCAPP) and manufacturing (WebTurning). The WebCADbyFeatures agent-based collaborative design module models cylindrical parts based on the synthesis of design features in a Concurrent Engineering context (CE). The developed system is based on a client-server architecture.

Web-Based Feature Modeller

In order to realize complete Web-based process planning, there have to be systems that can model and recognize part features on Web platforms without the need for generic CAD software. Thakur and Pande (2006) designed and implemented an Internet based system for feature modelling and process planning of sheet metal components commonly manufactured by blanking, shearing and bending processes. Their approach to modelling followed the design-by-features method, providing a very user-friendly environment for synthesis, validation and viewing of sheet metal components over the Web. The system is able to automatically generate flat pattern and efficient layout plans using a Genetic Algorithm.

Standardization for Enabling Smart Manufacturing and Interoperability

Multiple standards intended to work together are often defined as suites. Examples of this include the collection of standards from groups like WC3 and IETF that have aided the digital revolution. Common standards in manufacturing include the ISO 10303 suite of standards (known as STEP) for product data, the Quality Information Framework(QIF) for streamlining quality testing capabilities, and the standards emerging from consortia such as the Open Application Group, Inc. (OAGi) for enterprise level applications. These standards incorporate other, more fundamental, standards, e.g., the EXtensible Markup Language (XML) from W3C, and offer engineering and manufacturing content representation and interpretation (Lu et al., 2016). Standardization in the whole product life cycle (PLM), particularly in the areas of CAx (CAD, CAM, CAE, CAPP, etc.) have greatly improved engineering efficiency through enhancing modeling accuracy and reduce product innovation cycles. Table 1 shows some common standards in the PLM.

Product model and data exchange standards capture the representation of product and engineering information to enable data exchange between CAD software from different vendors. There are different file formats to exchange the product model data. The popular formats are DXF (Drawing Exchange Format), STL (Stereo Lithography), IGES (Initial Graphic Exchange Specifications) and STEP. DXF and STL files are

Table 1. Standards in PLM

Modeling Practice	Product Model and Data Exchange	Manufacturing Model Data	Product Catalog Data	Product Life Cycle Data Management
ASME Y14.41, Y14.36, Y4.5	ISO 10303-203/214/210/242	ISO 6983(G-Code)	ISO 13584	ISO 10303-239
ISO 16792, 7083	ISO 14306(JT)	ISO 14649	ISO 15926-4 (RDL)	LOTAR
ISO 1101, 5459	ISO 14739(PRC)	ISO 10303-207	ISO 22745 eOTD	PLMXML
ISO 14405, 2692	IGES, DXF	ISO 10303-224		
ISO 1302, 5458	ISO/ASTM 52915 (AMF),STL	ISO 10303-230		
ISO 8062, 17863		QIF Part1		
IEC 81724				

obsolete after the evolution of IGES and STEP. For seamless automation, CAD and CAM activities have to be linked by CAPP (Rameshbabu and Shunmugam, 2009). To enhance the interoperability and integration of CAx systems, a generic feature recognition system is needed in compliant with Standard for Exchange of Product model data (STEP) neutral file for seamless manufacturing (Venua et al., 2015). STEP is unique in that it addresses a broader range of information than is needed for CAD representations. The STEP architecture includes underlying methods for the definition and exchange of data and an architecture for integrating the data definitions into implementable units, called Application Protocols (AP), where CAD information is connected to product structure, as well as a wide array of other engineering and lifecycle information elements.

RESEARCH METHODOLOGY

An industrial survey based on a case study company that manufactures sheet metal accessories for the construction of roof trusses was done in order to acquire machine capabilities and specifications. A study on the existing process-sequencing documents was conducted as an instrument to gather data about the current production paths. Process plans for sample products from first stage to the last stage was acquired and data captured, including the machinery, their functional parameters and specific tasks performed at each stage. The data was examined by means of systematic recording, thus allowing the author to model the capabilities of machines. Process plans help to come up with capability descriptions for each machine at various work stations. Data acquired was populated in a cloud hosted database, on VB.net studio containing the capabilities, tools, parameters for the application utilization.

Analysis of the current operating environment for SMEs involved in sheet metal forming revealed the following basic problems:

1. Low level of automation and little to no integration on the few systems available.
2. Lack of tools which make it possible to cope in dynamic business environment characterized by rapid changing product design.
3. Non-collaborative process and business planning.
4. Inflexible ICT infrastructure.
5. Lack of tools, time, and space for alternative solutions, possibility of testing and optimization.
6. Lack of skilled human resources in one particular organization (there is need to share expertise across organizations).

A design approach is muted to address the challenges of the sheet metal forming industry. A cloud based CAPP system is conceptualized, with the following parameters; a module for feature recognition (FR): this module enables capturing of part characteristic features from a CAD model drawing of the new product, a module for process-sequencing: designed on VB.net studio, the module generates the part production path for the new product, proposes the required tools, and machinery with respect to the experience-based die-design guidelines. Four layers are perceived to be crucial to the system; knowledge, product, feature and parametric layers. A neutral file format based on the ISO 10303 (STEP) is adopted for system modelling.

CLOUD-BASED CAPP SYSTEM DEVELOPMENT

This section illustrates the development of the system and addresses the aspects involved in building up the sequencing system. The methods used and the guidelines that are followed in system development are discussed. The Unified Modeling Language (UML) was used to analyse and model the system.

Equipment and Process Planning at Case Study Company Before the New System

The capability assessment of machinery used in sheet metal forming at the case study company, particularly the machine type, supplier, capacity (general specifications e.g. tonnage, operating speed, etc.), nature of jobs suitable for the machine is shown in Figure 3 whilst Figure 4 shows a generic design procedure for a selected product (hurricane clips, BPH). The figure outlines specific parameters, machine, tools and geometry.

The process planning system that was being used at the case study company before the research intervention resulted in problems such as time delays, labour misallocation, human-error, scrap and rework. Figure 5 shows the functional flowchart for process flow in new product plans, with the process planning (activity 3) consuming the second longest interval, under the old system set-up.

Proposed System Model

The proposed solution is a STEP-compliant cloud based CAPP, which is based on a number of implementation methods for exchanging and manipulating information described by AP207. Figure 6 illustrates the framework of the cloud based CAPP system. The system has different interoperating modules, each providing essential functions to the system.

Feature Recognition (FR) module: Defines the part features and geometry required from the product model given in 3D format. The system input is provided and edited from a CAD/CAM application, Autodesk Inventor Professional®, which provides the sketch/draft model for the desired part geometry. The application is able to produce a developed flat pattern view of the 3D part for easy viewing of geometry and the part features. This developed flat pattern view provides the ease

Figure 3. Current machinery capabilities

	A	B	C	D	E	F	G	H
1	MACHINE TYPE	SUPPLIER/MNFTR	CAPACITY (tonnage)	Strokes/min	Ram Adj	Max Pa	TRANSMISSION	MORE INFO
2								
3	Eccentric/mechanical							
4	emc 1	Ter-Hart Holland	15 ton	180	0-75mm		vee-belt	fly-wheel
5	emc 2	Caxton Eng works	30 ton	120	2 1/8inch		vee-belt	G shaped
6	emc 3	Presse Cerini Italy	40 ton	107	3 1/2inch		vee-belt/gear	oil resevoirs
7	emc 4	Presse Cerini Italy	60 ton	95	3 1/2inch		vee-belt/gear	
8	emc 5	Caxton (SAMTMA)	600 kN	90	0-114mm		vee-belt	
9								
10	Hydraulic/mechanical							
11	hmc 1	Hytec (PVT) Ltd	100 ton					high grade oil
12	hmc 2	25 Coventry Rd	50 ton			175Bars		solenoid valves
13	hmc 3	P.O Box 621	20 ton			200Bars		limit switches
14	hmc 4	Harare	20 ton			200Bars		oil filters
15								
16	Pneumatic/Auto							
17	Minster P2-60-36	Minster MNF USA	60 ton	315	3inch		vee-belt/chain	solenoid valves
18	L & J	L & J Coperation,	50 ton	40	3inch		vee-belt	air filters
19								air resevoirs
20								
21	Guillotine	Gotenedes Sweden	max Fe gauge 3mm				vee-belt	
22								
23	Sheet bender		max Fe gauge 3mm					manual operation

Figure 4. Basic elements for a sample product: hurricane clip (BHP)

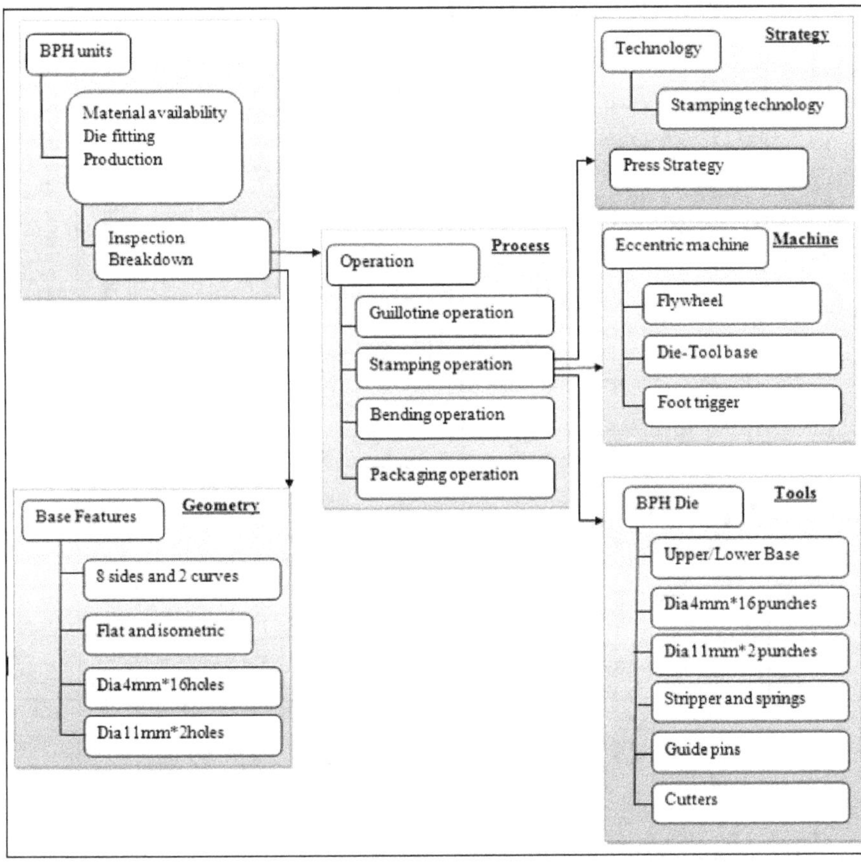

of viewing all the geometric features from the part at once, as compared to rotating the part so as to view the features from all different faces._

Capability taxonomy (Capability Tax.) module: The capability taxonomy is a method for storing the available tools in a hierarchical assortment of classes. The method allows searching for different devices that are able to perform the same function (e.g. making a hole) by different behaviour (e.g. blanking, notching, and perforating).

Database: The system database server stores the information extracted from the feature extraction module. The database concentrates on two information structures: the product features and the tools that are required. The features extracted are mapped to their corresponding tools from the source of storage. Apart from the tools and features, the database will be able to store and retrieve the information about parts processed at an earlier time.

Figure 5. Functional flowchart

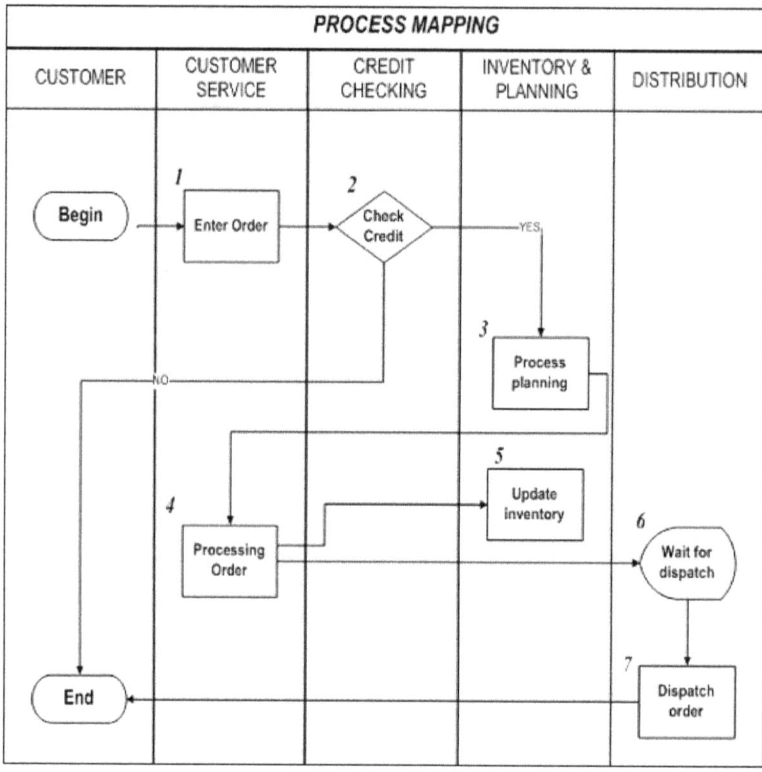

ACTIVITY	PROCESSING TIME/ hrs	CYCLE TIME / hrs
1	1/12	–
2	1/3	–
3	3/4	–
4	1+	–
5	1/6	–
6	1/2	–
7	1/4	–

Process-Planning (PP) module: The system module operates by resembling a central operation planner, being a governing module that keeps track of the potential alternatives and optimizes the operation sequence. It proposes alternative partial operation sequences to various specialized planners.

Database Design for the System

Unified Modelling Language (UML) was used to model the process sequencing system. The UML diagram is shown in Figure 7.

Feature Recognition Module

Feature recognition (FR) provides a communication medium between the system CAD model and the system process sequencing module, as illustrated in Figure 8.

The feature recognizing module breaks down a CAD model part and searches for known patterns that identify sheet metal features, such as faces, bends, holes,

Figure 6. Process sequencing system architecture

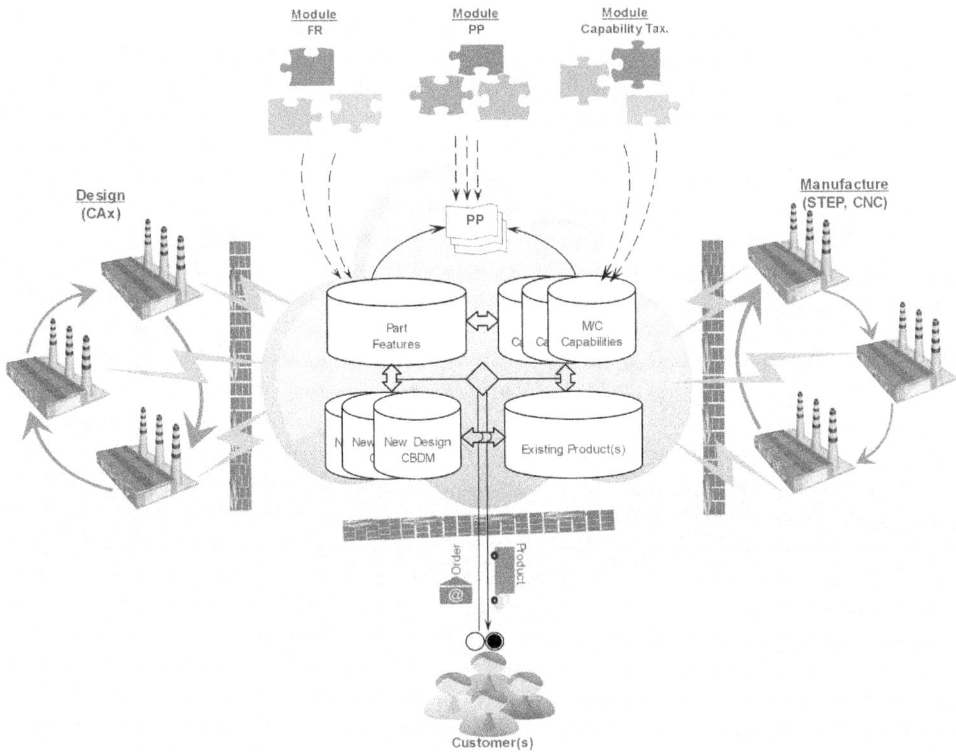

chamfers fillets, etc. The module contains a feature recognition (FR) unit and an unfolding unit shown in Figure 9. The unfolding unit is used to facilitate the design of sheet metal parts.

Process Sequencing Module

The module generates a step-by-step operation plan for the new product, based on the part features and the die-design guidelines; Figure 10 shows the units that make up this module.

SYSTEM IMPLEMENTATION

This section presents how the functional modules integrated into the prototype STEP-compliant cloud based CAPP system work. The prototype system is developed in VB.NET. During start-up of the system, a user validation authorizes the user to

Figure 7. Object-oriented approach for data modelling

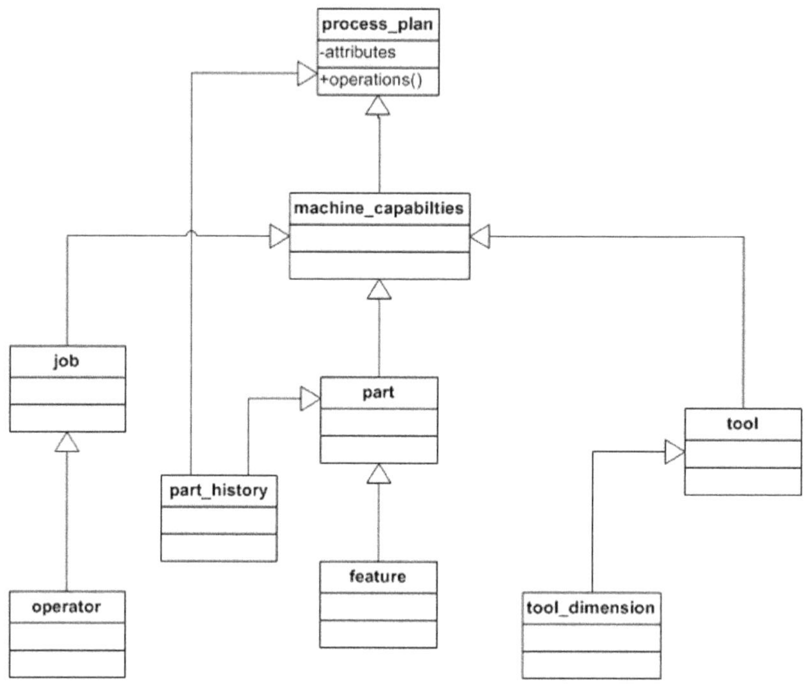

Figure 8. Feature recognition as the interface between CAD and process sequencing

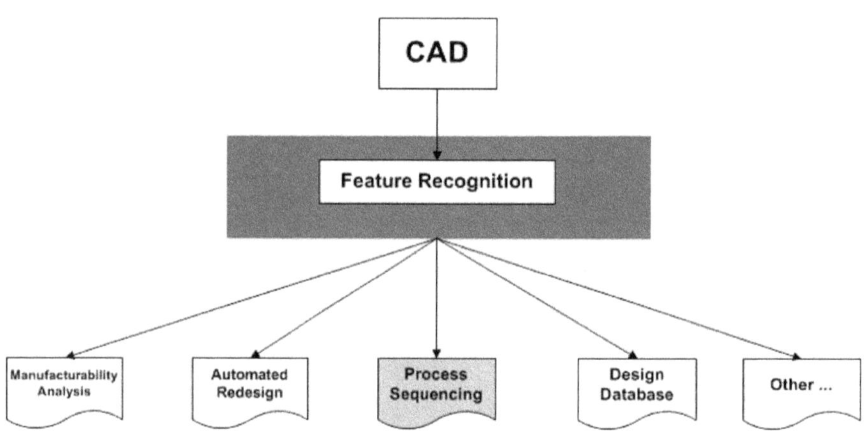

navigate the system. These rights are provided by a user validation platform provided by a username and password request as shown in Figure 11.

The login form prompts the user to the welcome page. After the welcome screen, the user views the main menu screen page; the system main menu page

Figure 9. Feature recognition module

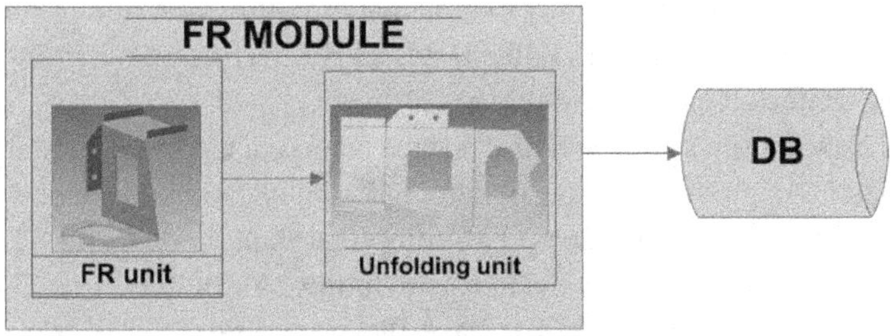

Figure 10. Process sequencing module

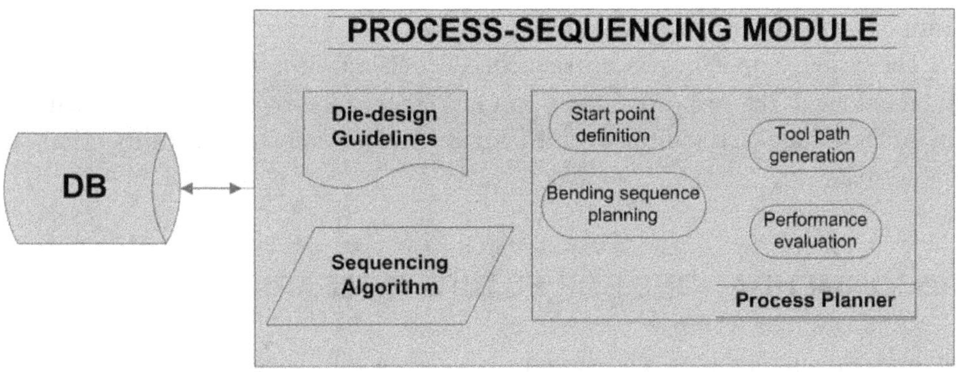

Figure 11. System login interface

is the graphical user interface (GUI) that users manipulate to navigate the system functionality and enter the basic information that is required into the system database before the subsequent interface is displayed. This interface requires input such as job, product, material, customer details.

Figure 12 shows the system main interface screen, a CAD/CAPP data interface for sheet metal design and process planning. The interface is designed to directly interpret the CAD model into the STEP–compliant data models and STEP part 21 files. Current versions of CAD systems, including AutoCAD, Pro/ENGINEER, SolidWorks, SolidEdge and CATIA, have the proprietary STEP translators that are incorporated into this module. The CAD/CAPP interface extracts geometric data from the part based on the defined STEP-compliant data models and displays them on the interface. When a part has been loaded, it is displayed together with its features on the interface as seen in the illustration on Figure 12.

The system output is a process sequence, which is an orderly list of the operations to be executed, the respective tools and machinery required. Figure 13 illustrates the process sequence interface whilst Figure 14 shows the report in its printable file format.

RECOMMENDATIONS FOR FUTURE RESEARCH

In order to build an integrated system, extensions will be needed to this research. There is always room for such extensions, and below are some of the recommendations

Figure 12. Part display

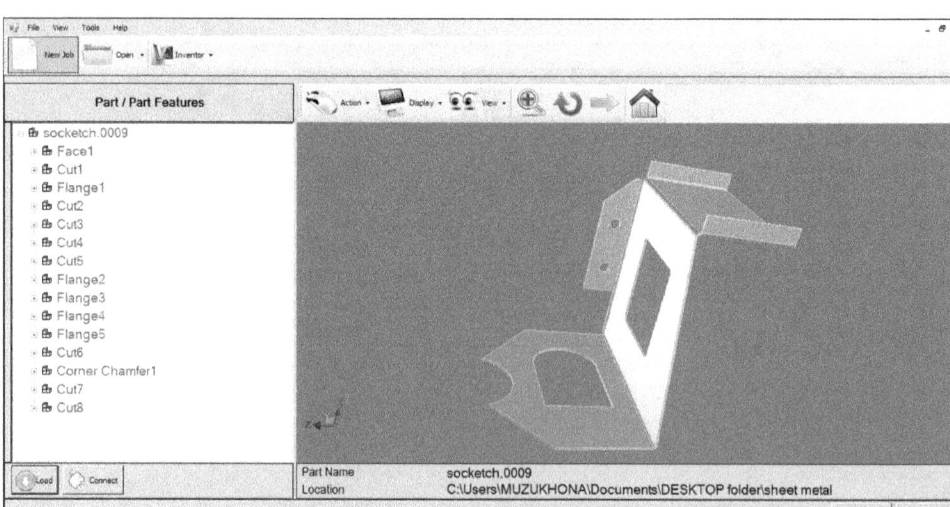

Figure 13. Process plan

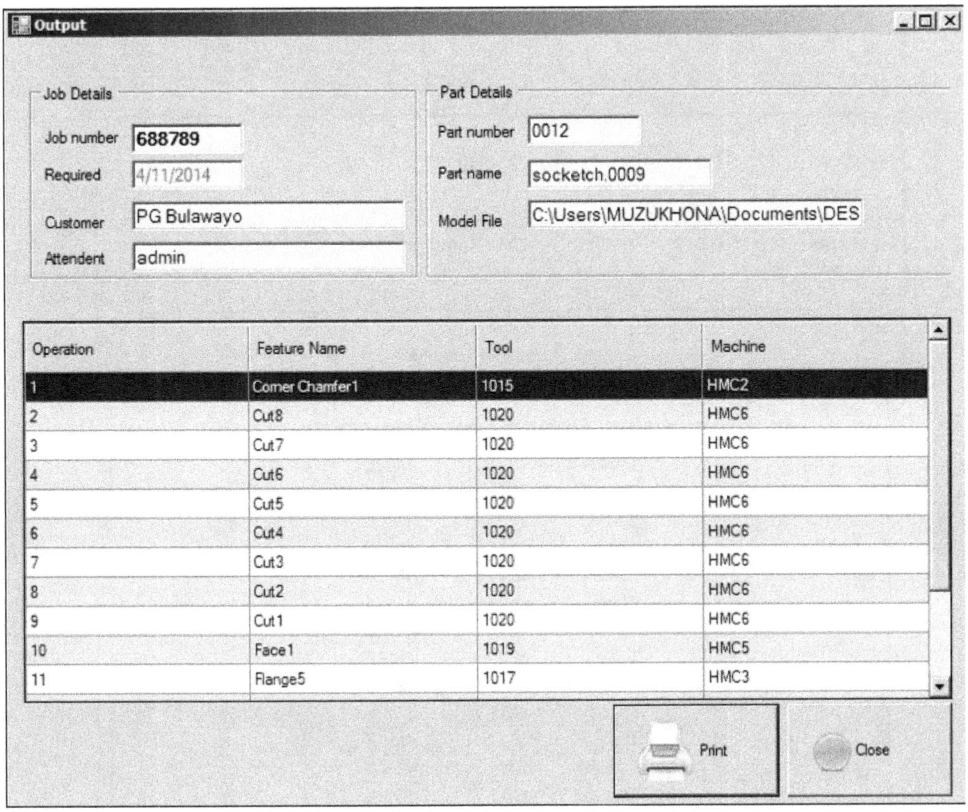

for future extensions. Feather research, including non-symmetrical features is highly recommended to extend this research. Developing the sheet metal part Extensible Mark-up Language (XML) file format as input is recommended because of XML's lightness and suitability for cloud based solutions. The design goals of XML emphasize simplicity, generality, and usability over the internet. Instead of using CAD interface functions for the system, a stand-alone feature modeller is required so that the feature-based systems become CAD software independent. In order to have a complete CAPP functionality, planning linked to computer numerically controlled (CNC) machine tools is needed, where the plan is sent to a machine controller unit (MCU) in code format. Automatically Programmed Tool (APT) or G-Code programming languages can be used to generate instructions for NC machinery from the output of the system. The recommended future extensions will enhance the usefulness of this research, and will result in the development of a fully integrated and intelligent cloud based CAPP system.

Figure 14. Printable process report

Process Plan Report

Run Date 4/11/2014 Run Time 8:00 PM

Job #: 688789 Customer: PG Bulawayo
Part name: socketch.0009 Date Required: 4/11/2014
Ran by: admin

Operation Sequence

Operation	Feature	Tool	Machine
1	Corner Chan	1015	HMC2
2	Cut8	1020	HMC6
3	Cut7	1020	HMC6
4	Cut6	1020	HMC6
5	Cut5	1020	HMC6
6	Cut4	1020	HMC6
7	Cut3	1020	HMC6
8	Cut2	1020	HMC6
9	Cut1	1020	HMC6
10	Face1	1019	HMC5
11	Flange5	1017	HMC3
12	Flange4	1017	HMC3
13	Flange3	1017	HMC3
14	Flange2	1017	HMC3

CONCLUSION

The chapter presented the development of a cloud based CAPP system that generates process sequences for new and existing products in sheet-metal forming. The system computerizes the planning phase for a CAD drawn model. This is achieved by the model's extracted features, which automatically retrieves the respective tools from the machinery capability database. These capabilities are in turn sequenced in order, based on the die-design guidelines to produce a step-by-step report of the operations. The interface of the system enables effective collaboration and knowledge sharing in the production environment. The application design was such that it can be easily implemented into an already existing ICT infrastructure using flexibly adjustable interfaces. A modern approach of CC was adopted, which brings many advantages to the SME sector: (1) flexible investment model, (2) agility, (3) modularity, (4) collaboration among enterprise experts and across industry partners, and (5) interoperability. The verification of the designed CAPP system has realised significant savings of 40% reduction in process planning effort, 5% in material, 15% in scrap and 10% in tooling. In order to design the cloud based CAPP system, data was gathered on the machinery capabilities for the sheet metal operations in a case study company that manufactures sheet metal accessories for the construction industry.

REFERENCES

Álvares, A. J., Ferreira, J. C. E., & Lorenzo, R. M. (2008). An integrated web-based CAD/CAPP/CAM system for the remote design and manufacture of feature-based cylindrical parts. *Journal of Intelligent Manufacturing*, *19*(6), 643–659. doi:10.1007/s10845-008-0117-1

Bourne, D. A., & Wang, C. H. (1995). Design and Manufacturing of Sheet Metal Parts. Boston: Academic Press.

Cattrysse, D., Beullens, P., Collin, P., Duflou, J., & Oudheusden, D. V. (2006). Automatic production planning of press brakes for sheet metal bending. *International Journal of Production Research*, *44*(20), 4311–4327. doi:10.1080/00207540600558031

Chin, K., & Tang, D. (2002). Web-Based Concurrent Stamping Part and Die Development. *Concurrent Engineering*, *10*(3), 213–228. doi:10.1177/1063293027616891 33

De Vin, L. J., De Vries, J., Streppel, A. H., & Kals, H. J. (1992). PART-S, a CAPP system for small batch manufacturing of sheet metal components. *Proceedings of the 24th CIRP International Seminar of Manufacturing Systems*, 171-182.

De Vin, L. J., De Vries, J., & Streppel, T. (2000). Process planning for small batch manufacturing of sheet metal parts. *International Journal of Production Research*, *38*(17), 4273–4283. doi:10.1080/00207540050205082

Duflou, J. R., Nguyen, T. H. M., Kruth, J. P., & Cattrysse, D. (2003). An optimization system for automated workpiece layout generation for bent sheet metal parts. *Proceedings of the SHE-MET Conference*, 235-244.

Faraz, Z., Ali, L., Waheedulhaq, S., Khan, M., & Jaffery, S. H. I. (2015). STEP-NC enabled manufacturing process planning of sheet metal bend parts. *Proceedings of the IRES 8th International Conference*. Retrieved from: http://www.worldresearchlibrary.org/up_proc/pdf/60-144101414113-17.pdf

Faulk, J., & Kotowski, F. J. (2003). *Dctionary of Engineering*. New York: McGraw-Hall.

Gernhardt, B., Miltner, F., Vogel, T., Brocks, H., Hemmje, M., & Wang, L. (2015). A Semantic representation for process-oriented knowledge management based on function block domain models supporting distributed and collaborative production planning. *ASME International Manufacturing Science and Engineering Conference:* Vol. 2. *Materials: Biomanufacturing; Properties, Applications and Systems; Sustainable Manufacturing*. Academic Press. doi:10.1115/MSEC2015-9485

Groover, M. P. (2010). *Fundamentals of Modern Manufacturing (Materials, Processes, and Systems)* (4th ed.). Danvers, MA: John Wiley & Sons, Inc.

Gupta, S. K., Bourne, D. A., Kim, K. H., & Khrishnan, S. S. (1998). Automated process planning for sheet metal bending operations. *Journal of Manufacturing Systems*, *17*(5), 338–360. doi:10.1016/S0278-6125(98)80002-2

Gupta, S. K., & Rajaqopal, D. (2002). Sheet metal bending: Forming part families for generating shared press-brake setups. *Journal of Manufacturing Systems*, *21*(5), 329–349. doi:10.1016/S0278-6125(02)80033-4

Inui, M., Kinosada, A., Suzuki, H., Kimura, F., & Sata, T. (1998). Automated process planning for sheet metal parts with bending simulation. *Intellingent and Integrated Manufacturing Analysis and Synthesis*, *25*, 245–258.

Kanna, T. R., & Shunmugam, M. S. (2009). Processing of 3D sheet metal components in STEP AP-203 format. Part I: Feature recognition system. *International Journal of Production Research*, *47*(4), 941–964. doi:10.1080/00207540701510055

Khoshnevis, B., Sormaz, D., & Park, J. Y. (1999). An integrated process planning system using feature reasoning and space search-based optimization. *IIE Transactions*, *31*(7), 597–616. doi:10.1080/07408179908969862

Lichao, Z., Yi, Z., Qiang, Z., & Fafu, H. (2011). Robust sheet metal bend sequencing method based on A-star algorithm. *IEEE International Conference on Computer Science and Automation Engineering*, 711-715. doi:10.1109/CSAE.2011.5952603

Lu, Y., Morris, K. C., & Frechette, S. (2016). Current standards landscape for smart manufacturing systems. *National Institute of Standards and Technology, 8107*. doi:10.6028/NIST.IR.8107

Lu, Y., & Xu, X. (2015). Process and production planning in a cloud manufacturing environment. *ASME International Manufacturing Science and Engineering Conference:* Vol. 2. *Materials: Biomanufacturing, Properties, Applications and Systems, Sustainable Manufacturing*. Academic Press. doi:10.1115/MSEC2015-9382

Lundgren, M., Hedlind, M., & Kjellberg, T. (2014). Model-based interactive learning of process planning. *6th Swedish Production Symposium*, Gothenburg, Sweden.

Milošević, M., Lukić, D., Borojević, S., Šimunović, G., & Antić, A. (2017). A model of collaborative process planning system (e-CAPP). *Tehnicki Vjesnik (Strojarski Fakultet)*, *24*(1), 97–103. Retrieved from https://hrcak.srce.hr/file/257838

Milošević, M., Todić, V., & Lukić, D. (2011). Web-based collaborative environment for process planning. *34th International Conference On Production Engineering, 28*.

Monkova, K., & Monka, P. (2014). Newly developed software application for multiple access process planning. Advances in Mechanical Engineering.

Mourtzis, D., Vlachoua, E., Xanthopoulos, N., Givehchi, M., & Wang, L. (2016). Cloud-based adaptive process planning considering availability and capabilities of machine tools. *Journal of Manufacturing Systems*, *39*, 1–8. doi:10.1016/j.jmsy.2016.01.003

Park, C. S. (2003). Knowledge capturing methodology in process planning. *Computer Aided Design*, *35*(12), 1109–1117. doi:10.1016/S0010-4485(02)00182-3

Poli, C. (2001). *Design For Manufacturing: A Structured Approach.* London: Elsevier Science and Technology Books.

Pullant, R. (2010). *Manufacturing feature recognition by 3D solid model slicing and contour based geometric reasoning* (Master of Science in Mechanical Engineering Thesis). University of Cincinnati, Department of Mechanical Engineering, College of Engineering and Applied Science.

Rameshbabu, V., & Shunmugam, M. S. (2009). Hybrid feature recognition method for setup planning from STEP AP-203. *Robotics and Computer-integrated Manufacturing*, *25*(2), 393–408. doi:10.1016/j.rcim.2007.09.014

Serope, K., & Schmid, S. R. (2009). *Manufacturing Engineering and Technology* (6th ed.). Singapore: Prentice Hall.

Siller, H. R., Estruch, A., Vila, C., Abellan, J. V., & Romero, F. (2008). Modeling workflow activities for collaborative process planning with product lifecycle management tools. *Journal of Intelligent Manufacturing*, *19*(6), 689–700. doi:10.1007/s10845-008-0120-6

Thakur, A. D., & Pande, S. S. (2006). Web-based feature modeling and process planning system for sheet metal components. *First International and 22nd All India Manufacturing Technology Design and Research Conference (22nd AIMTDR)*. IIT Roorkee.

Venua, B., Komma, V. R., & Srivastava, D. (2015). A methodology to recognize manufacturing features of prismatic parts from STEP AP 203 file. *Materials Today: Proceedings*, *2*(4-5), 3170–3177. doi:10.1016/j.matpr.2015.07.111

Wang, X. V., Givehchi, M., & Wang, L. (2017). Manufacturing system on the cloud: A case study on cloud-based process planning. *Procedia CIRP*, *63*, 39–45. doi:10.1016/j.procir.2017.03.103

Xie, S. Q., & Xu, X. (2006). A STEP-compliant process planning system for sheet metal parts. *International Journal of Computer Integrated Manufacturing*, *19*(6), 627–638. doi:10.1080/09511920600623708

Yusof, Y., Zakiah, N., Tan, Z., & Kasim, N. (2009). Exploring the ISO14649 (STEP-NC) for intelligent manufacturing system. *European Journal of Scientific Research*, *36*(3), 445–457.

Zhang, C., Zhou, X., & Li, C. (2009). Automatic recognition of intersecting features of freeform sheet metal parts. *Journal of Zhejiang University. Science A*, *10*(10), 1439–1449. doi:10.1631/jzus.A0820705

Zhou, X., Qui, Y., Hua, G., Wang, H., & Ruan, X. (2007). A feasible approach to the integration of CAD and CAPP. *Computer Aided Design*, *39*(4), 324–338. doi:10.1016/j.cad.2007.01.005

Zussman, E., & Horsch, T. (1994). A planning approach for robot-assisted multiple-bent profile handling. *Robotics and Computer-integrated Manufacturing*, *11*(1), 35–40. doi:10.1016/0736-5845(94)90005-1

ADDITIONAL READING

Gupta, S. K., & Bourne, D. A. (1999). Sheet Metal Bending: Generating Shared Setups. *ASME. Journal of Manufacturing Science and Engineering*, *121*(4), 689–694. doi:10.1115/1.2833107

Hussein, H. M. A., & Kumar, S. (2014). A framework for development of an intelligent system for design and manufacturing of stamping dies. *IOP Conference Series. Materials Science and Engineering*, *65*, 012029. doi:10.1088/1757-899X/65/1/012029

Kanna, T. R., & Shunmugam, M. S. (2009). Processing of 3D sheet metal components in STEP AP-203 format. Part II: Feature recognition system. *International Journal of Production Research*, *47*(5), 1287–1304. doi:10.1080/00207540701510063

Lin, A. C., & Chen, A. C. (2014). Sequence planning and tool selection for bending processes of 2.5d sheet metals. *Advances in Mechanical Engineering*, *6*, 204930. doi:10.1155/2014/204930

Salem, A. A., Abdelmaguid, T. F., Wifi, A. S., & Elmokadem, A. (2017). Towards an efficient process planning of the V-bending process: An enhanced automated feature recognition system. *International Journal of Advanced Manufacturing Technology*, *91*(9-12), 4163–4181. doi:10.1007/s00170-017-0104-9

KEY TERMS AND DEFINITIONS

Agile: The capability of surviving and prospering in a competitive environment of continuous and unpredictable change by reacting quickly and effectively to changing markets, driven by customer-designed products and services. Critical to the success of agile manufacturing are enabling technologies such as model-based engineering, supply chain integration, and flexible production systems with distributed intelligence.

Cloud Computing (CC): A model for enabling ubiquitous, convenient, on-demand network access to a shared pool of configurable computing resources (e.g., networks, servers, storage, applications, and services) that can be rapidly provisioned and released with minimal management effort or service provider interaction.

Computer-Aided Process Planning (CAPP): Involves using computer technology for generating processes for manufacture of physical products. The generated processes may involve a specified sequence of operations, particular uses of tools, and how to route products from one workstation to another. Two types of CAPP systems are "variant CAPP" and "generative CAPP."

Computer-Aided Technologies (CAx): The use of computer technologies to aid in the design, analysis, production planning, manufacture of products, etc.

Small and Medium-Sized Enterprises (SMEs): Are non-subsidiary, independent firms that employ less than a given number of employees. The number varies across countries, but the most frequent upper limit designating an SME is 250 employees.

XaaS: Is a collective term said to stand for a number of things including "X as a service," "anything as a service," or "everything as a service." The acronym refers to an increasing number of services that are delivered over the internet rather than provided locally or on-site. XaaS is the essence of cloud computing.

Chapter 11
Open Product Lifecycle Management (PLM) for Cloud Manufacturing and Cloud-Based Maintenance Integration Using XML

Norman Gwangwava
Botswana International University of Science and Technology, Botswana

ABSTRACT

Product lifecycle management (PLM) is concerned with managing all the processes of product transition from cradle to the grave. Phases of a product life are inception, engineering design, manufacture, service, and disposal. A product passes through different systems, organisations, and users as it completes the life cycle. Global markets of today have increased the complex nature of a product path. All the life cycle phases rely upon product data for efficient management. In order to ease the strain of managing products throughout the lifecycle, a common product data schema is needed. Currently many platforms for product design use different proprietary schemas that make it difficult to have smooth lifecycle management. The chapter illustrates applications of an open source, XML-based schema for product lifecycle management. The main focus is on the inclusion of the Cloud in order to have new generation cloud product life cycle management (CPLM). The main driver of CPLM is cloud-model-based systems engineering (CMBSE). Within the framework of CMBSE are cloud-based design (CBD), cloud manufacturing

DOI: 10.4018/978-1-5225-3628-4.ch011

(CM), and cloud-based maintenance (CBM). The three subsystems of CMBSE can be combined to form a single term, cloud-based design, manufacturing, and maintenance (CBDMM). Cloud computing, manufacturing, and maintenance are not new concepts, but many enterprises have not yet embraced them because of lack of complete seamless integration across various levels and processes in the product life. Many systems are still being run in silos of automation. CPLM is a service-oriented (SOA) model comprised of a pool of technologies such as cloud computing (CC), IoT, virtualization, and service-oriented technologies to support collaboration, sharing, and management across PLM phases.

INTRODUCTION

Large enterprises that are involved in the value chains for design, manufacture and servicing of complex engineered products or systems such as aircrafts, cars, trains and ships require strong collaborations. PLM systems have been in place for some time and they continue to be upgraded in order to match the demands for seamless integration and collaboration. Christophe et al. (2014) views PLM systems as key for managing collaboration between partners and supporting business process interoperability. Managing the life cycle of complex products require a large variety of tools to support different aspects during collaborative development projects as well as data and document management. Traditionally, collaborating partners would use computer aided applications (CAx) such as CAD/CAM/CAE/CAPP/CAQC within their organisation network. The process would facilitate automated workflows of editing or modifying the product designs. Sub-softwares that encompass the PLM systems are Product Portfolio Management (PPM), CAx applications, Product Data Management (PDM), Manufacturing Process Management (MPM), and Digital Manufacturing (DM). Cloud Computing (CC) came as a game changer which is revolutionizing the traditional approach to business. CC is considered as a field of multidisciplinary research as a result of the evolution and convergence of several computer trends such as Virtualization, Distributed Computting (DC), Storage, Content Outsourcing, Grid Computing (GC) (Zeballos and Quiroga, 2017). Xu (2012) assert that in CC everything is treated as a service, i.e. (XaaS.), e.g. SaaS (Software as a Service), PaaS (Platform as a Service), IaaS (Infrastructure as a Service), Design as a Service (DaaS), and Hardware as a Service (HaaS). Unlike the traditional approach where enterprises host silos of different software systems in-house, the CC era outsources computing power, data storage and services to third parties.

Enterprises are adopting the subscription-based cloud offerings that allow customers to fund IT investments as operational expenditures, without high up-front capital costs. The subscription-based business model is the major driving force behind the adoption of XaaS. Key characteristics of XaaS include:

1. High scalability,
2. Multi-tenancy,
3. Online and automated provisioning,
4. Try and buy,
5. Device independence, which enables users to access software regardless of what devices they are using,
6. Location independence,
7. Term-based billing or pay-as-you-go models.

Applications of CC in the manufacturing sector have led to the advent of Cloud-based Design (CBD), Cloud Manufacturing (CM), and Cloud-based Maintenance (CBM).

Wide ranges of methodologies, IT infrastructure and software relevant to Product Lifecycle Management (PLM) in global enterprises are heterogeneous. This calls for interoperability among CAx applications, Product Data Management (PDM) systems; which drive virtual product development, and Enterprise Resource Planning (ERP); which manage the real product. According to Thomas et al. (2011) a mediator approach is pertinent for the coordination of these systems.

Requirements for product representations include: platform/application independence, support for the product lifecycle, rapidly sharing information between geographically distributed applications and users, and protection of commercial security (Ding et al., 2007). In addition to interoperability, enterprises must implement model-based enterprise, advanced simulation and PLM initiatives supported by best-in-class engineering systems.

BACKGROUND

The manufacturing environment is ever becoming complex, as the market continues to have diversified demands. The world population continues to boom and the modern consumers have sophisticated demands that include shorter lead times, high quality and lower costs. These expectations mean that manufacturers cannot continue to adopt traditional means and systems of production and expect to satisfy the global markets. The manufacturing value chain now demands high visibility than ever before. Complete PLM approach gives the players involved in the whole manufacturing value chain a clear visibility and control throughout the process. Today's manufacturers need instant feedback, traceability, uniformity and transparency in order to enhance decision making. Traditionally many manufacturing organizations relied on systems that are disconnected from each other for planning and execution processes. This means manufacturers cannot fully explore market opportunities because market research, design, production planning, maintenance, and cetera cannot be adjusted to changes in each other quickly. The modern demands of globalisation have seen more specialization by individual companies in the whole manufacturing value chain. The specialized collaborating companies need common platforms of PLM systems in order to have seamless integration for visibility of all the phases of the processes involved. The need for open communication within and between all inter-company functions cannot be over emphasized.

A popular trend in many organizations nowadays is the use of Enterprise Resource Planning (ERP) systems, comprising of multiple modules for maintenance, production planning, customer relationship management and other business management functions such as finance and human resources management. In many cases the engineering functions are underutilized or their design in the particular ERP system cannot unlock high performance execution of the engineering functions in a typical manufacturing organization. The design aspect usually suffers the most. The advent of cloud computing (CC) has seen the adoption of modern approaches such as cloud manufacturing (CM). Cloud computing and the internet of things (IoT) are no longer just buzzy words but are practical business principles. The "cloud" has completely revolutionized the ways of doing business by many enterprises. CM is a multidisciplinary approach that comprise of state-of-the-art technologies such as globally networked manufacturing, virtual manufacturing, agile manufacturing, supply chain management and product service connection (Hao, 2016). CM uses IoT technologies such as RFID, wired and wireless sensor networks, and embedded systems to automatically manage, control and facilitate sharing of manufacturing resources and capabilities. There still remains a gape in facilitating all the perceived functions of PLM complete integration without leaving gaps across the physical world.

The chapter advocates for open PLM using XML standards in order to eliminate gaps between the physical world and the cloud for total implementation of all the PLM phases. The fundamental concept embedded in the proposed PLM structure is cloud model based systems engineering (CMBSE), that features aspects of cloud based design, manufacture and maintenance (CBDMM).

LITERATURE REVIEW

This section covers recent, historical and empirical reviews laying the foundation for the present study.

Product Lifecycle Management (PLM)

PLM can be taken as a strategic business approach that applies a consistent set of business solutions that support the collaborative creation, management, dissemination, and use of product definition information. PLM is sometimes interchangeable with product data management (PDM). PDM is the technology and associated software systems that support the management of both engineering data and process information during the product development phase and beyond (Chang, 2014). PDM is an essential subject in e-Design, which supports a cross-functional team for product design and development. PLM systems support the management of a portfolio of products, processes and services from initial concept, through design, launch, production and use to final disposal. The product's life cycle period usually consists of five major steps or phases: product development, product introduction, product growth, product maturity and finally product disposal. The phases can be split up into smaller ones depending on the product and must be considered when a new product is to be introduced into a market since they dictate the product's sales performance. Product Data Management (PDM) systems provide the tools to control access to and manage all product definition data. It does this by maintaining information (meta-data) about product information. In 3D CAD systems these meta-data files rely heavily on each other - they have relationships and contain dependencies that drive details like feature size and placement in other models. PDM systems allow engineers to better manage the complex interrelationships between the part, assembly and drawing files. They can share files with other team members and keep each other up to date on design modifications through a file check-in/check-out process.

PLM has not yet become a reality in many industries. A fair deployment is found in the automotive and aerospace industries. These two industries are associated with strong, broad networks of original equipment manufacturers (OEMs). There is great need for consistency in the standards across the network of suppliers hence the significant pressures to adopt PLM technologies. The majority of manufacturers of consumer goods, which mainly exist as single enterprises and or with limited supply chains, have been reluctant to apply PLM technologies. These types of industries do not quickly realize any significant gains of PLM. It is also difficult to organize and influence single enterprises to adopt common PLM standards. The cost factor of current PLM systems is also another deterrent factor. However, the new XaaS business model which is subscription based, is luring many enterprises regardless of their size. Automobiles and airplanes are complex systems compared to simple consumer products, hence adopting PLM in the automotive and aerospace industries bring attractive and significant benefits that help the entire lifecycle. Longer life span of products also influences the industries' early adoption of PLM systems. Theron and Hendry (2015) describes case studies of automotive and aerospace industries applying systems based approach to product lifecycle management. The approach helps manage complexities that emanate from technological advances, customer demand and the global supply chain. Modern automobiles, airlines and various other products are cross-domain engineered. Technology continues to advance at an alarming pace, with modern engineered systems featuring internet of things (IoT) capability, alongside the traditional mechanical and control technologies. Automobiles have become cyber physical systems (CPS), driven around online environments and serviced by online garages. Online manufacturers, in-turn use feedback data to optimize next generation products through Data Science (DS) platforms. Robust cross-domain engineering can only be realized through openness, integration and holistic improvement, rather than segregation and reinventing the wheel.

To develop products collaboratively in manufacturing industry, stable data exchange mechanisms are necessary. It is important to have standardized and compatible data formats and exchange methods (T. Systems, 2010). The ProSTEP consortium has addressed this issue for many years. ProSTEP iViP is a consortium based in Germany. The projects carried out by the ProSTEP iViP Association focus not only on the data but also on the processes and systems involved, and their efficient integration.

In the past years, there has been an advent of many software vendors supplying PLM and /or PDM systems. These vendors include among others; ENOVIA Smarteam of CATIA, Windchill by PTC, TeamCenter by Siemens UGS NX, SofTech ProductCenter® PLM, and Arena Cloud PLM (Chang, 2014). ENOVIA and Smarteam are two PLM solutions from Dassault Systemes. Windchill is offered by PTC and uses client-server technologies with the implementation, manageability, and usability benefits of the web. The web-centric solution is applicable for managing the entire product lifecycle. TeamCenter has end-to-end PLM solutions that empower engineering teams to capture, manage, synchronize product design data, facilitate engineering change, validation, and approval processes. TeamCenter supports NX, Pro/ENGINEER, CATIA, AutoCAD, Inventor, Solid Edge, and SolidWorks. SofTech ProductCenter is a commercial software with an integrated suite of PLM software. ProductCenter can be integrated with various CAD/CAM/CAE tools to ease product data management from design to manufacture across various engineering teams. The system's web-based services allow data sharing across enterprise networks. Arena Cloud PLM takes advantage of the modern software-as-a-service (SaaS) model to deploy its solutions in a collaborative environment.

Cloud Manufacturing (CM)

Cloud Manufacturing (CM) is a new service-oriented manufacturing model that integrates different technologies such as networked manufacturing, cloud computing

(CC), Internet of Things (IoT), virtualization and service oriented technologies to support collaboration, sharing and management of manufacturing resources (Tao et al., 2011). Some of the benefits of CM are listed below;

1. Closer partner relationships
2. Flexibility
3. Additional data analysis capabilities
4. Efficiency
5. Cost savings

CM has its foundation in Cloud Computing. Mell and Grance (2011) define Cloud Computing as a model for enabling ubiquitous, convenient, on-demand network access to a shared pool of configurable computing resources (e.g., networks, servers, storage, applications, and services) that can be rapidly provisioned and released with minimal management effort or service provider interaction. The Cloud is mainly categorized as private cloud, community cloud, public cloud and hybrid cloud (Ahmed and Hossain, 2014). Configurable manufacturing resources shared in CM include; manufacturing software tools, manufacturing equipment, and manufacturing capabilities, that can be rapidly provisioned and released with minimal management effort or service provider interaction (Liu and Li, 2012). The CC approach treats everything as a service (i.e. XaaS), e.g. SaaS (Software as a Service), PaaS (Platform as a Service), IaaS (Infrastructure as a Service), and Hardware as a Service (HaaS). These services define a layered system structure for CC. Implementing CC means a paradigm shift of business and IT infrastructure, where computing power, data storage and services are outsourced to third-parties and made available as commodities to enterprises and customers (Xu, 2012).

The cloud marketplace offers a wide range of cloud services that vary in complexity and value. The set of service categories are layered in a notional stack, with foundational offerings toward the bottom and more complex offerings toward the top. Many ICT enterprises have adopted CC and are offering various XaaS categories. Figure 1 illustrates Cloud Computing as a stack of layers as depicted by the SOA architecture (Raines, 2009).

Berry (2017) identified major CC vendors in the marketplace and the services they offer. Table 1 lists the major CC players and their products across the XaaS layers.

Amazon Web Services offer many types of cloud computing services whilst VMware is a provider of the technology underlying cloud computing and does not provide hosting services.

CM provides a collaborative work environment for all companies involved in the entire manufacturing value chain. IoT technologies such as RFID, wired and wireless sensor networks, and embedded systems enable the digitization and virtualization

Figure 1. Cloud Computing as a stack of layers according to SOA Architecture

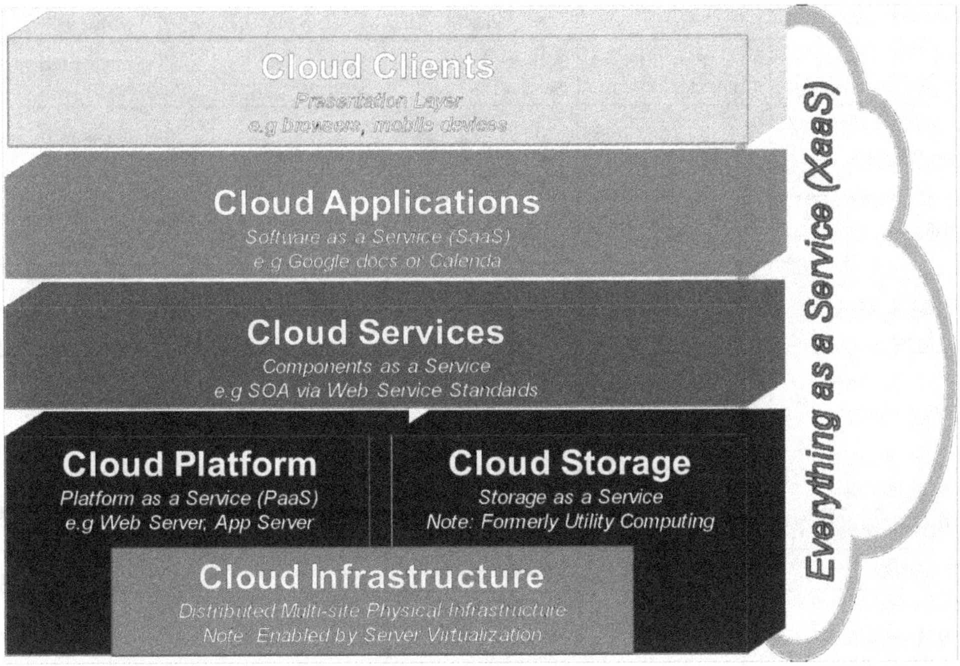

of shared manufacturing resources and capabilities for access through the cloud resource pool. IoT is an extension of internet into the real world, embracing everyday objects by utilizing the power of combining ubiquitous networking with embedded systems, RFID, sensors, and actuators. The objectives of IoT are the creation of smart environments/spaces and self-aware things (for example: smart transport, products, cities, buildings, rural areas, energy, health, living, etc.) for climate, food, energy, mobility, digital society and health applications (Vermesan et al., 2011). This allows physical entities to sense, communicate, act, interact, exchange data, information and knowledge. Each physical entity is accompanied by a rich, globally accessible virtual object, which contains both current and historical information on that object's physical properties, origin, ownership, and sensory context (Welbourne et al., 2009). The Internet acts as a storage and communication infrastructure that holds a virtual representation of things linking relevant information with the object (Uckelmann et al., 2011). Virtual objects serve as central hubs of object information, which combine and continuously update data from a wide range of sources (Verdouw et al., 2011).

Multi-layered architectures with modular approach are commonly used to build CM systems (He and Xu, 2015; Mourad et al., 2016). Table 2 and Table 3 show the key enabling technologies and core service management for CM respectively.

Table 1. Major players in the CC marketplace and their products

Vendor	IaaS	PaaS	SaaS	Storage
Amazon	EC2 (Elastic Cloud Compute)	Amazon Web Services*	Amazon Web Services*	S3 (Simple Storage Service)
Google	n/a	Google App Engine (Python, Java, Go)	Google Aps	Google Cloud Storage
HP	Enterprise Services Cloud – Compute	Cloud Application Delivery	HP Software as a Service	Enterprise Services Cloud – Compute
IBM	SmartCloud Enterprise	SmartCloud Application Services	SaaS products	SmartCloud Enterprise – object storage
Microsoft	Microsoft Private Cloud	Windows Azure (includes .NET, Node.js, Java, PHP)	MS Office 365	Microsoft Private Cloud
JoyentCloud	SmartMachines	Node.js	n/a	n/a
Rackspace	Cloud Servers	Cloud Sites	Email & Apps	Cloud Files
Salesforce.com	n/a	Force.com	Salesforce.com	n/a
VMware	VMware vSphere, vCloud	VMware vFabric (Java Spring), vCloud API	n/a	n/a
Dassault Systems			CATIA & 3D EXPERIENCE	
Autodesk			AutoCAD 360	

Tao et al. (2014) proposed a five (5) layer system architecture for intelligent perception and access of CM resources based on IoT consisting of resource layer, perception layer, network layer, service layer and application layer. The layers are briefly described as follows;

1. **Resource Layer:** Provides all kinds of manufacturing resources involved in the entire life cycle of manufacturing, and basic resources to the operation of manufacturing service platform. Manufacturing resources mainly include hardware manufacturing resources, computational resources, intellectual resources and other resources.
2. **Perception Layer:** Achieves intelligent perception and identification of all kinds of manufacturing resources through different sensing devices and adapters in product life cycle, thereby providing strong support for manufacturing service platform to intelligently identify and manage manufacturing resources.

Table 2. Key enabling technologies for CM (He and Xu, 2015)

Key Enabling Technology	Components or Characteristics of the Technology
manufacturing systems and technologies	CAD/CAM/CAE (computer aided design, computer aided manufacturing, computer aided engineering), MIDS (manufacturing information and database system), etc.
Cloud computing	Software as a Service (SaaS), Platform as a Service (PaaS), Infrastucture as a Service (IaaS); public cloud, private cloud, hybrid cloud, community cloud, etc
Internet of Things (IoT)	Radio frequency identification (RFID), wired and wireless sensor network, embedded system, intelligence in smart objects
Virtualisation	Virtual services, Virtual machine, data and information virtualisation; virtual description specification
Service-oriented technologies	Service-oriented architecture (SOA), web service, enterprise service bus, semantic web, ontology
High-performance computing (HPC)	Parallel processing, multiple tasks

Table 3. Core service management in CM (He and Xu, 2015).

Core Service	Related Content
Service publishing, request and discovery	Service registration; semantic publishing; cloud service searching and matching; information matching; hierarchy-searching; semantic discovery based on ontology;semantic web service matching
Resource allocation and service scheduling	Optimal and on-demand allocation of resources for jobs and tasks; scheduling algorithms (e.g., genetic algorithms, ant colony optimisation, immune algorithms)
Quality of service (QoS) management	Monitoring service execution; QoS assurance for multiple layers; QoS Modelling; trust-QoS; integrated evaluation
Service composition	Service combination and aggregation; research on service composition framework, service composition validation, service composition optimal-selection, and service composition methods; correlation-aware web services composition; service composition method based on business process and workflow, agents, artificial intelligence, etc.
Security, trust and reliability management	Identity check; authorisation; isolation; access control; certificate-based authentication; user trust management; data and information flow tracing
Evaluation of service execution results	Service execution result evaluation; transaction evaluation and billing

3. **Network Layer:** Provides all needed carrier network for accessing various resources in products' whole life cycle, including 2G networks, 3G networks, 4G networks, satellite networks, cable networks, corporate internal wireless networks, and so on. Construction of the network layer is conducive for numerous, security, and high-speed transmission of manufacturing resources, which provides network support for management service layer.

4. **Service Layer:** Primarily provides two categories of service, the CM service and the CM platform operational service. The former are the results of service encapsulation of manufacturing resources and capacities, which can be invoked by end users. The latter are the main services provided by the CM platform to realize the various operators to CM services, which includes knowledge management, transaction management, information assessment, billing management, resource calendar management, network management, virtual machine management, directory management, registration services, and so on.
5. **Application Layer:** Refers to the on-demand use of various CM services in the entire life cycle of manufacturing, including design, manufacturing, experimentation, simulation, management, maintain, recycling, and so on.

A typical representation of CM, according to SOA architecture, is as shown in Figure 2.

Manufacturing resource virtualization is an important factor for success in CM. Virtualization refers to abstraction of logical resources from their underlying physical resources (Xu, 2012). Different manufacturing resources are virtualized in different ways. Computational resources and manufacturing knowledge can be virtualized in similar ways as are the general CC resources. Manufacturing hardware is usually mapped to become virtual machines that are system-independent. The virtualized manufacturing resources are then packaged to become CM services. This requires the use of resource description protocols and service description languages (i.e.

Figure 2. Cloud Manufacturing representation according to SOA Architecture

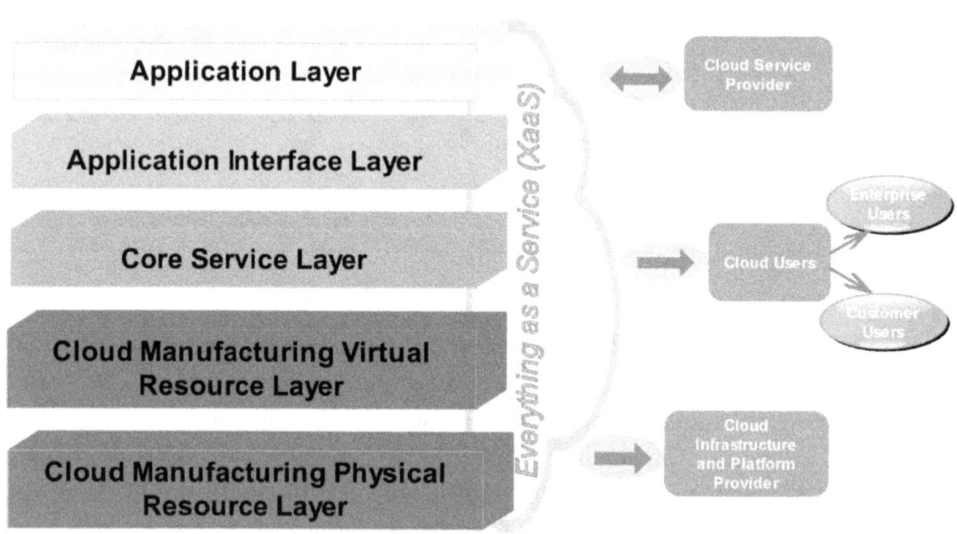

ontology languages e.g. Simple HTML Ontology Extension (SHOE), DARPA Agent Markup Language (DAML), and Web Ontology Language (OWL)).

The process of virtualizing manufacturing resource(s) can also be viewed as an encapsulating process, which can be carried out using three different mapping methods, one-to-one, many-to-one, and one-to-many. One-to-one mapping is the simplest situation, which applies to manufacturing resources that can only provide a single function and can therefore directly be encapsulated into one service. The CAD and CAE data format exchange service is one of the common types of such resource. In a many-to-one mapping, multiple resources (each providing a specific function) may be combined to create a more powerful or functional resource form. In cloud manufacturing, when multiple manufacturing resources are combined, more comprehensive manufacturing resource services called resource service composition can be provided to users to enable value-added services.

Virtualization removes fundamental constraints concerning: Place: virtual representations do not require geographic presence, i.e. physical proximity, to be observed, controlled or processed; Time: besides the representation of actual objects, virtualization can reproduce historical states, simulate future states or imagine a non-existing world; Human observation: virtual representations can visualize information about object properties (such as temperature information or X-ray images) that cannot be observed by the human senses (Verdouw et al., 2015). Virtual objects represent the current and historic state of objects that exist physically in the real-world. The physical objects are equipped with tags for identification, usually barcodes or RFID transponders, and with sensors that measure dynamic properties of physical things. The virtual object uses these sensor data to generate a representation of the object based on a meta model, which might be implicit. Usually, the sensor data are combined with external data to enrich the virtual representation. Four virtualization mechanisms can be identified: 1. Virtual representation of real objects; 2. Virtual projection of real objects; 3. Virtual representation of reference objects; 4. Virtual projection of reference objects.

SOA Architecture

SOA software architecture aims to implement an information system comprising independent but interconnected services. In that sense, the objective of SOA is to decompose functionalities in a set of services and describe their interactions. From a technical point of view, SOA defines software in terms of discrete services, implemented using components that can be called upon to perform a specified operation for a specific business task. The SOA concept changes the existing software concept of a function—a specific piece of code that performs one particular task—to include the notion of a contract, a technology-neutral but business-specific representation of the function (Ferreiro et al., 2012). The functionality of the SOA model is the exposure of one or more interfaces (services) through a given program (network access) (Avila et al., 2017). Those who consume these services are called service consumers, and those who provide them are called service providers. Each one of these services is independent, and includes the business logic and data associated with the service offered. W3C's working definition of a Web Service is, "a software application identified by a URI, whose interfaces and bindings are capable of being defined, described, and discovered as XML artefacts". A Web Service supports direct interaction with other software agents using XML-based messages exchanged via internet-based protocols (Oluwaseyi, 2008). Benefits of using the SOA architecture are:

1. Reusability promotion,
2. Interoperability integration,
3. Modular programming,
4. Better flexibility,
5. Agility and,
6. Easy to maintain services.

SOA promotes the goal of separating users (consumers) from the service implementations. Services can, therefore, be run on various distributed platforms and accessed across networks. Web Services for implementing SOA are based on a core set of technologies as stated below;

1. SOAP (Simple Object Access Protocol),
2. WSDL (Web Services Description Language),
3. BPEL (Business Process Execution Language),
4. WS-CDL (Web Service Choreography Description Language),
5. RESTful and,
6. UDDI (Universal Description, Discovery and Integration)

These set of technologies define the transport, interface description and discovery mechanism respectively. Each of these technologies is defined and implemented in XML (Zhao et al., 2005). Any application written in any language and running on any platform can interpret Web Services messages, interface description and discovery mechanisms that are implemented in XML using standard XML processing tools without the need for any specific middleware to be used. Web Services and SOA brought feasibility in realizing integration, accelerating responsiveness to customers, collaboration and optimizing the business processes among enterprises.

Unnamalai and Thresphine (2014) explain the four deployment models of SOA as follows;

1. **Public:** Services and resources are reachable to the public by using the internet. This environment emphasizes the advantages of rationalization (as a user has the ability to utilize only the needed services and pay only for their use), operational simplicity (as the system is organized and hosted by a third party) and scalability. The main concern in this type of cloud environment is the security; since this environment is accessible to the public and user data in one stage is hosted by a third party.
2. **Private:** Services and resources are reachable within a private institute. This environment emphasizes the advantages of integration, optimization of hardware deals and scalability. The main concern is the complexity, as this environment is organized and hosted by internal resources. Security is not a main issue compared to the public cloud as the services are reachable only through private and internal networks.
3. **Community:** Services and resources of this type are shared by various institutes with a common aim. It may be organized by one of the institutes or a third party.
4. **Hybrid:** This type combines the methods from the private and public clouds, where resources can be used either in a public or a private cloud environment. The advantages and the concerns are a mixture of the earlier type.

Maintenance Management

Maintenance management practices have evolved from the traditional approach of using papers to fill in plant details and service history to computerized systems. A Computerized Maintenance Management System (CMMS) is a software package that contains a computer database of information about an organization's maintenance operations. CMMS software can serve as the central hub of plant maintenance, holding a vast amount of real-time data that can give feedback on what's really going on at any given moment (MPULSE, 2015).

CMMS helps maintenance professionals in many ways as given below;

1. Monitor multiple meters (e.g., revolutions, psi, hours) on a single asset,
2. Specify multiple conditions, triggers, and responses,
3. Automate work order generation based on specified triggers,
4. Identify potential problems with data analysis tools.

CMMS forms the core of a PLM system, responsible for the use and service phase. IoT and Cloud Computing (CC) have also impacted on CMMS. The two trends mean more data in real time to maintenance professionals. This in turn gives better insights into assets, how they're working, and what they're actually costing the company. Better plant insight leads to improved efficiency, better customer service, and enhanced safety. Table 4 compares the on-premise and on-demand models of CMMS deployment. The latter is the modern approach, influenced by advances in CC research. The on-demand model follows the SaaS business model.

E-Maintenance

The scope of maintenance activities has been limited to the production and operation phases. The evolution of manufacturing towards achieving sustainability resulted in the changing focus of maintenance to take into account a life-cycle management oriented approach. The need for enhancing the eco-efficiency of the product life led to the advent of e-maintenance. E-Maintenance is maintenance management

Table 4. CMMS Deployment (WHO, 2011)

Solution	Description	Features	Networking Options
On-premise	CMMS installed and runs on premises of health facility	• Customer responsible for the technology infrastructure • Customer pays licensing fee to use and customize software • Customer may customize features and functions to meet requirements • Customer has full control of infrastructure and data	• Standalone workstation base system, which is useful only in small single workshops at small hospitals • Local area network system within clinical engineering department • Deployment over Internet at customer's site (self-hosted online solution) • Open architecture with integration to other applications on similar or different platforms • Built using standard Web Technologies
On-demand	Software As A Service (SAAS	• Vendor provides application licence to multiple customers • Infrastructure and application managed by vendor • System delivered over Internet • User does not maintain hardware or software • Customer pays for access to application	• 100% Internet-based application requiring no installation on client's machines • Deployment over Internet (hosted online solution) • Open architecture with integration to other applications on similar or different platforms • Built using standard Web Technologies
	Customized SAAS	Customizable version of SAAS	

through internet and connected computers to make better maintenance decisions. Globalization has led to a focus upon a more responsible use of resources and a more effective and sustainable production. E-maintenance can be considered as a maintenance management concept whereby assets are monitored and managed over the Internet (Levrat and Lung, 2007). Some e-maintenance platforms that were developed are ICAS-AME, PROTEUS, TELMA, CASIP and its up- graded version KASEM or DYNAMITE (Campos, 2009; Simon et al., 2015). Formal data descriptions, such as XML schemas, can be extracted from the data models and used in content based information system integration, while activity models help in the analysis of business processes. Maintenance data can be derived from wireless sensor networks, running log documents, surveillance image files, audio and video clips, complicated simulation and GPS-enabled spatiotemporal data, and much more.

NoSQL databases can be used to manage semi-structured data in a more scalable and agile manner. Maintenance applications and tools need to exchange maintenance-related data but these are often provided by heterogeneous sources and hence data integration in maintenance is an issue of concern. Research efforts to provide integrated open system architecture for e-maintenance can be illustrated through OSA-CBM (Emmanouilidis et al., 2011). An ISO-13374 standard regarding condition monitoring and diagnostics of machine was used to develop an Open System Architecture for Condition-Based Maintenance (OSA-CBM) (Mathew et al., 2006; Swearingen et al., 2007; Zhang, 2015). OSA-CBM architecture consists of seven layers: data acquisition, data manipulation, state detection, health assessment, prognostics, decision support and presentation. E-Maintenance requires semantic data modelling by means of domain-specific maintenance ontologies, which offer maintenance services as semantic mediators to ensure robust data exchange between subsystems (Jantunen et al., 2011).

Cloud Based Maintenance

Cloud based maintenance is a modern approach that uses CC technologies to perform equipment maintenance and facilities management. Cloud-based equipment maintenance and facility management service platforms perform continuous and remote collection of operating data (Yoshikawa et al., 2015). The platforms provide services that include monitoring, preventive maintenance, predictive failure diagnosis, energy efficiency analysis and management, and equipment maintenance management. Cloud-based maintenance solutions promise smoothen information logistics for maintenance decision support (Karim et al., 2015). Wang et al. (2015) defines cloud based maintenance as a new generation service-oriented technology to support multiple enterprises to deploy and manage predictive maintenance services including machinery condition monitoring, data analysis and diagnosis, prognosis, and maintenance planning over the internet. The adoption of the cloud helps facilitate sharing data, knowledge, processes, applications and locations via seamless collaboration.

Wang et al. (2015) proposed a mobile agent based cloud maintenance framework that features open and non-proprietary communication standard based on XML to support machine-to-machine (M2M) communications and possible interoperability. The cloud based predictive maintenance paradigm is service-oriented, accessible and robust, resource-aware, and collaborative and distributive. This shows that maintenance service providers are not limited to static solutions of wired infrastructures controlled by complex software suites. Employing centralized software solutions for maintenance support no longer constitutes the most efficient approach. Maintenance data are produced, stored, retrieved, read and acted upon all the time, and a centralized solution has limited capacity to provide the needed responses in such dynamic environments.

Wireless Sensing and Identification

Smart wireless sensor networks are being deployed as flexible alternatives to wired instrumentation systems. Advanced smart wireless sensor implementations incorporate system-on-a-chip (SOC) and has moved well beyond the data acquisition functionality. Their main advantages over wired solutions are ease of installation and operation, scalability and topological flexibility. Modern sensor boards contain powerful CPU architectures, huge flash memory, and diverse RF connectivity, e.g. WirelessHART or industry-grade Bluetooth, WiFi or ZigBee. (Jantunen et al., 2011; Zhang, 2015; Zhang et al., 2015). These sensing infrastructures are no longer only exploited by back-office staff and systems, but are also available to shop floor staff, who can be empowered to deliver better, more reliable and efficient maintenance services. The convergence of smart sensors and CC is pushing the frontiers of cloud based maintenance to a business reality rather than a theoretical perspective.

Interoperability: Open Source Schema

PLM is a collaborative approach that saves as a business strategy applying a consistent set of business solutions that supports lifecycle processes; research and pre-development, series development, production, market, and disposal. The ultimate benefits of PLM are improved product quality, time-to-market, costs benefits and cetera. Interoperability is a key enabler for effective collaboration at different levels across enterprises. Diana et al. (2014) and Dhuieb et al. (2014) identified three scopes of interoperability, which are;

1. **Conceptual Interoperability:** Concerns the definition of concepts and semantic supporting the communication between data and knowledge models;
2. **Organizational Interoperability:** Focuses on the connection between processes and,
3. **Technical Interoperability:** Deals with technological issues to support data exchange between software applications.

These approaches are currently used to support the interoperability and data exchange between CAx applications (x such as Design, simulation, etc.) and between CAx and other data management systems: 1) ontology and semantic web technologies, 2) standards, 3) web services. Many standards continue to be developed but the XML format is the most used standard language to support data mapping and communication between heterogeneous data servers in the web. The development of service-oriented architecture (SOA) allows enterprises to be able to access each other's product data at any time and from anywhere (Campos et al, 2014). A well-engineered SOA platform use a native XML database-powered metadata repository and data orchestration engine at its core.

Product life cycle data includes technical data of the product, such as design and manufacturing data, but also other data related to the product process and the business. During the whole product life cycle, a great deal of life cycle information such as CAD drawings, technical documents and structured and unstructured data is created, changed, transferred, stored and converted by and between different people and application systems. This results in complex information flows over the whole life cycle. There is no one complete data model for collecting and storing the product life cycle data, but a set of concepts and technologies that enable creating and modifying the data model based on the requirements for the specific product process and the users.

Companies are moving away from drawings and embedding all of the product manufacturing information inside the 3D model itself. The 3D models can come from a number of different proprietary software systems. Often those systems will produce various visualization formats that will include a 3D model or assembly with product manufacturing information embedded directly within the package. A large number of "neutral 3D formats" are available for the transfer of 3D models. Each of these formats has different attributes, such as a high level of precision regarding the images being displayed, small file sizes, versatility and many others. Lightweight representations support users at different stages of the product lifecycle in rapidly browsing, retrieving and manipulating product information (Ding et al., 2007). When parts are developed in 3D for the engineering domain, the data is initially stored in the original format of the software used to design the part (Fröhlich, 2011). Global engineering practice, which is promoted by the PLM approach, often brings together enterprises using different platforms for various activities of the product life cycle. Neutral 3D formats come into play when 3D CAD data is to be made available to people who do not have similar software. Some of the neutral formats to promote interoperability are discussed in the next paragraphs.

STEP is a standard for describing product data and is formally defined in the International Organization for Standardization (ISO) standard 10303. The most important parts of STEP relevant to discrete engineering are the application protocols (AP) 214 and 203. The APs are used in many CAx (CAD, CAM, CAE and cetera) systems to import and export product data. STEP is an ASCII format, which helps to significantly reduce file sizes by means of external compression, such as in ZIP format. A combination of AP 203 and AP 214 is being used to create AP 242, a convergent modular STEP Application Protocol for the aerospace and automotive industries. The business object model will be defined in UML (Unified Modeling Language) and XML, thus making it consistent with the current state of the art (Fröhlich, 2011).

JT is a format for describing 3D data that also supports object data and metadata. JT is currently being integrated as a format for the simplified representation of the geometry. JT files can be created by translating data from all major MCAD applications, such as UGS NX, SolidEdge®, Catia® and Autodesk Inventor®. The JT format is a mature lightweight data format, and can be used as a predominant, lightweight visualisation format for PLM, a CAD neutral exchanging format for collaborative product development, or a consistent 3D visualization up and down the supply chain. At present, the JT format is being widely used in the automobile, aerospace and various manufacturing industries (Ding, 2007).

The U3D ECMA standard defines the syntax and semantics of the Universal 3D file format for downstream 3D CAD repurposing and visualization, useful for many mainstream business applications. It enables embedding and presentation of 3D models within PDF documents (ECMA, 2017). Other content exchange formats which enable content visualization in Web browsers, include COLLADA and 3D XML. The U3D encoding method in 3D PDF is used as an interoperable file format for Computer-Aided Manufacturing (CAM), Product Life-Cycle Management (PLM) for supply-chain, training, marketing, archiving and engineering reports (PDF3D, 2017). Examples of applications of the U3D file format include:

1. Development of training tools based on interactive simulations created from 3D CAD models;

2. Electronic owner's manuals providing interactive guides for maintaining and repairing products;
3. Online catalogues equipped with interactive 3D models created during product development.

COLLAborative Design Activity (COLLADA) is an intermediate format with an open standard XML schema whose primary goal is to represent rich data in multiple forms, to exchange digital assets among various graphics software applications (Remi and Tony, 2007). COLLADA targets mainly the game industry, focusing on exchanging content and assets between diverse authoring tools, including DCC (Digital Content Creation) tools and Conditioning tools, and for archiving rich content. COLLADA enables free exchange of asset data, and therefore enables combining multiple tools into an integrated and flexible authoring pipeline. COLLADA documents are XML files, usually with a *.dae (digital asset exchange) filename extension.

3D XML is a proprietary XML-based 3D content representation format developed by Dassault Systèmes (Dassault Systemes, 2011). 3D XML format has been primarily developed to support PLM tools, and consequently it is limited to the description of 3D designs, without support for interaction and behaviour.

PLM XML is a protocol created by UGS to boost PLM interoperability by defining a set of XML schemas and associated services that facilitate open, high-content product lifecycle data sharing. Therefore the protocol is able to transport externally attached data while complying with the core of PLM XML. The schema definitions include structures which allow access to top-level metadata as well as detailed model data (Siemens PLM Software, 2008). Table 5 summarises the light weight representations that make interoperability possible.

X3D, 3D XML and PLM XML are all XML-based formats. XML is more preferable because:

1. It is related to HTML (HyperText Markup Language), which is conventionally used to create Web pages, thus it is a good compromise between being easy for humans to read and write, and being easy for computers to interpret;
2. It is extensible, which allows for tag definition reflecting the tailored structure of information for various applications.

RESEARCH METHODOLOGY

The research was organized into two stages; qualitative and design oriented research. Qualitative research was done in manufacturing enterprises that use PLM systems (CAx for product modeling, design and manufacture; CMMS for managing product use

Table 5. Summary of light weight representations (Ding, 2007).

	Compression Method	Developer	Supporting Tools	Applications	Characteristics
U3D	Domain specific Node/Resource mechanism	Intel and the 3D Industry Forum (3DIF)	Adobe Acrobat Version 7.0 Open source	-Sales & marketing -Customer support -Online promotions -Maintenance -Training	-Level-of-detail Progressive streaming -Rigid-body & skeletonbased animation File format and runtime extensibility
X3D	Domain-specific Type-specific field Fast InfoSe	Web3D Consortium	X3D Tools and Applications	- Technical publishing -Maintenance manuals, -Websites, -Database applications, -Visual simulations -Navigation systems	-No heavy browser XML-based open profile/ components - based architecture -Integration of advanced 3D techniques
3D XML	3D graphics compaction algorithm Reference/instance mechanism	Dassault Systemes/ IBM	V5R15, CATIA, Delmia, Enovia, Novia, Spatial, SmartTeam, SolidWorks, Virtools Dev 3.5 3D XML Player	-Technical documentation -Maintenance manuals -Marketing brochures -Websites -Email	-Level-of-Detail -Multi-file architecture -Easy to adopt -Extensibility
JT Format	Lossless compression algorithm Domain-specific	Engineering Animation, Inc./ UGS Inc	JT Open Toolkit JT2Go	-Lightweight visualisation format for PLM CAD-neural exchanging format -Consistent 3D visualisation	-Neutral exchange format Support of multiple files
PLM XML	References to external files	UGS Inc (Siemens)	UGS applications Open XML schemas	-Connection -UGS PLM Solutions products and third party adopter applications	-Incorporation of product, part and process information -Open source Support of multiple representations for shape definition -Extensibility

and service phase). The purpose of the study was to elicit business, user and system requirements for new generation cloud-based PLM (CPLM). Qualitative research is employed to understand people and their social and cultural context (Myers, 1997). The researcher found it imperative to seek the perspective of enterprises who are routinely exposed to PLM platforms as part of their business practice. These enterprises have firsthand experiences which best help to deduce current challenges, and map the intended characteristics of future technologies. Design-oriented research was used to develop the conceptual framework for future generation CPLM. The framework consists of an information system architecture for the CPLM which emphasises on the model based approach. Design-oriented research is typically involved with 'how' questions, i.e. how to solve a certain problem by the construction of a new artefact (March and Storey, 2008). A case study strategy usually fits best for this type of questions, because artefacts intended for real-life problems are influenced by many factors (Aken, 2004). Case studies can deal with such complex phenomena, which cannot be studied outside their rich, real-world context.

Interviews were conducted with several stakeholders in the case study companies and the results were recorded in the form of requirements specification. Companies used for the requirements survey consisted of different categories as follows;

1. Kitchen appliance manufacturers (*2),
2. HVAC equipment manufacturers (*2),
3. Liquid Petroleum Gas (LPG) accessory manufacturers (*1),
4. Solar Energy & Electrical appliances (*2),
5. Agricultural equipment producers (*1).

Although the companies chosen specialize in different products, the processes followed in the value chain are the same. All the PLM phases are exhibited across the different product lines. The enterprises rely on CAx systems for modeling, design, manufacture and servicing the products during the product use phase, until end-of-life (EoL). The product lines chosen are not as complex as in the aerospace, automotive and ship building industries. The research was based on enterprises in the developing world where there is limited manufacturing activity. However the researcher(s) tried to choose case study companies that offer products which have a broad scope life span. For example, LPG accessories such as the cylinders last for a long time, and are subject to routine checkups for pressure leaks among other issues. The same applies to agriculture equipment, kitchen appliances, HVAC equipment and electrical appliances. Customer or/ end user involvement is very crucial, right from the voice of customer gathering at the design stage and feedback during the product use phase. Integration is therefore seen as a necessity across all the PLM phases, and the CC era offers great possibilities. One major discovery across all the case study enterprises was that legacy CAx systems and ERPs are being used to facilitate the necessary business processes. There are silos of systems, which in the researcher(s)' views hamper smooth flow of activities and efficient business decisions.

Findings of the requirements survey were used to deduce patterns in the surveyed enterprises. PLM systems should generally not be customized to specific product lines, hence the different industries in the survey list are sufficient to provide a broad perspective of life cycle characteristics and key requirements for future CPLM. Benchmarking with complex engineered product manufacturers such as the automotive, aerospace, and ship building industries is necessary in order to deduce the appropriate requirements for broad applicable CPLM. These enterprises involved in the value chain for complex products have complex networks and supply chains that comprise of various suppliers, OE manufacturers, service or/ maintenance companies. The products' life is also much longer compared to that of products being handled by all of the case study companies.

Requirements Analysis

The case study companies were given pseudo names according to the alphabetical order as; company A & B (for kitchen appliances), C & D (HVAC equipment), E (LPG accessories), F & G (Solar Energy and Electrical appliances), and H (Agricultural equipment). Results of the requirements collection exercise from eight (8) companies were presented as common requirements specification document. The collected data set was recorded first, and then consolidated into a requirement document for further refinement.

Functional Requirements

Currently the surveyed companies have heterogeneous applications (CAD/CAM/ CAE, ERPs, and some cases standalone CMMS). There is no homogeneity in applications that perform same functions across different companies. The ICT architecture is therefore complicated. The approach is associated with high initial investment, difficulty in tying ICT costs to specific operations, lack of openness or interoperability. To sum it up, there is no integration across the systems. The summary of the functional requirements is outlined below;

1. Expectations for lower ICT infrastructure as well as systems costs. Currently there is high upfront cost in form of licensing of different systems in use. The ICT hardware requirements are sometimes different as per different systems.

The licensing model for self-hosted system is rigid and usually results in the companies paying for more than they are using.

2. Human resource investment costs are currently high since much of the expert work needs to be done in-house; installation, maintenance and new staff training.
3. Companies expect more openness across all the phases of the PLM. The proprietary applications currently in use are close ended, making it difficult to share ideas during the ideation processes and all the other phases. There are concerns for harmony and convergence even though there is a plethora of heterogeneous systems across the enterprises. At least let the difference be mainly a matter of style but not the fundamental functions and expected outputs.
4. Thoroughness in terms of capturing engineering detail throughout the PLM phases. The systems being used do not go beyond what the physical eye can see. This is rather a betrayal to the personnel and resources that go into the whole process of modeling, manufacture and use of the product. There is much detail, which unfortunately cannot be carried along with any presentation format of the product (CAD model, simulation, HMI and during use maintenance). There is consensus that MBSE approach should be cascaded throughout the PLM phases. Although companies alluded to existence of MBSE platforms, the current applications are limited to early life cycle phases and the integration of the cloud is not yet popular.
5. Customer involvement and access to the PLM phases. It appeared to be very difficult to engage customers throughout the PLM phases, besides through a simple questionnaire (whether on paper or online). Customers are undoubtedly becoming more knowledgeable and technology savvy than ever before. Especially with the ubiquitous growth and availability of the mobile technologies, almost everyone across all generations, is getting acquainted with various techniques. There is general feeling that the burden of assuming what the customer wants should be eliminated through total engagement in the PLM, with provision to key data about products.
6. Companies expressed concern about long turnaround times due to sluggish systems they are using. Agility is the pressing issue on their continuous improvement agenda.

The findings were analyzed and used to produce a requirements checklist that was in turn used to generate a framework for the open source CPLM architecture.

SOLUTIONS AND RECOMMENDATIONS

A new open source CPLM architecture is proposed based on the research and analysis of findings. The summary of findings is presented as a requirements checklist in Table 6.

Framework for Open Source Cloud-Based Product Life Cycle Management (CPLM)

This section presents the concept for the CPLM, based on the requirements specification checklist from case study enterprises. The framework is based on currently existing and new cloud-based service offerings. The SOA architecture is used as the base for modelling the CPLM. We propose the system architecture for CPLM to illustrate implementation of the layered service models (i.e., IaaS, PaaS, HaaS, SaaS, etc.) across the product life cycle phases. General life cycle phases are illustrated in Figure 3. PLM phases are iterative and therefore concentric model is widely acceptable rather than a linear model.

Table 6. A requirements checklist for open source CPLM

Requirement	Requirement Description
R1	Should provide everything as service – XaaS (IaaS, PaaS, HaaS, SaaS, and cetera) applications to users, based on the SOA architecture
R2	Should provide cloud-based distributed file systems that allow users to have ubiquitous access to PLM phases
R3	Should be based on open-source technologies that facilitate interoperability, integration and collaboration.
R4	Should have ability to process and analyse big data stored in the cloud
R5	Should provide a multi-tenancy environment that enables a single software instance can serve multiple tenants
R6	Should have intelligence and capability to collect and store in the cloud, real-time data from products, plant equipment and other enterprise resources (e.g., manufactured products, machines, robots, and assembly lines), remotely monitor and control these products, equipment and resources
R7	Should provide a framework for integration into mobile platforms (user applications should also have mobile versions)
R8	Should integrate social media to facilitate communication, information and knowledge sharing throughout the PLM phases
R9	Should be based on web services technologies and offer intelligent search engines
R10	Should support virtualization of all the entities involved in the PLM phases
R11	Should be based on MBSE approach for the cloud
R12	Should be agile and support simulation and optimization across all phases of PLM

Figure 3. PLM Phases

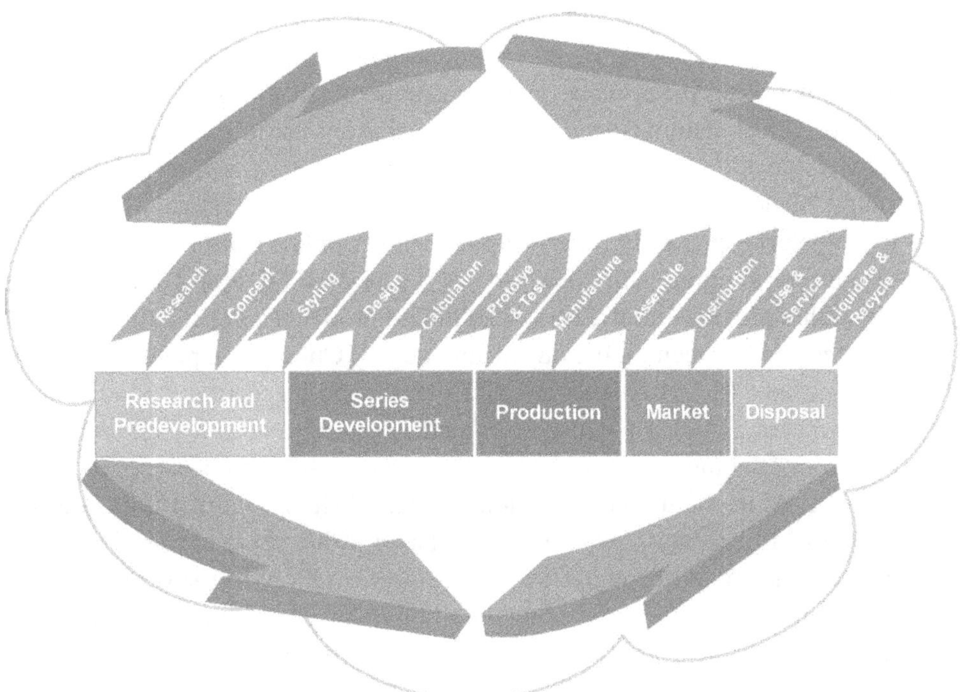

Cloud platforms such as IaaS and PaaS provided by Amazon Elastic Compute Cloud, Google BigQuery, and Salesforce allow the research and development (R&D) team to collaboratively participate during the Research and Predevelopment phase. Amazon provides storage of large datasets pertaining to customer preference research. The dataset, depending on the magnitude, can be regarded as Big Data. The data can be processed through Google BigQuery and Salesforce.com. Series Development can be done through online CAD platforms such as AutoCAD 360, SolidWorks and 3DEXPERIENCE®. Parts sourced off-the-shelf can be easily integrated into the design models as e-drawings from the OEM repository in order to allow all subsequent analyses and visualization. The SaaS CAD platforms are based on 3D design, analysis and simulation software supporting collaborative interaction. The solutions can be availed on premise, public or private cloud. But for the purposes of the CPLM proposal, the cloud solution is most preferred.

Prototyping and Manufacturing becomes less complex in the CPLM arena. Off-the-shelf parts and modules can be sourced from OEMs or various suppliers' online platforms. These are called Hardware as a Service (HaaS) in CC terminology. New designed parts can be outsourced to manufacturers' cloud platforms who will manufacture the parts according to design data. Examples of HaaS providers are

3D Systems On Demand Manufacturing (formerly Quickparts), MFG.COM® and Made-in-China.com™. 3D Systems On Demand Manufacturing provide online 3D printing and manufacturing services that empower designers and engineers with the tools to design, iterate, and produce quality parts directly from digital 3D files. MFG.COM® is a contract manufacturing marketplace that lets procurement professionals and engineers quickly and easily source CNC Machining Services, Injection Molding Companies, Metal Stamping Companies, Metal Fabrication Companies and many other manufacturing processes through an easy-to-use online platform. The platform offers a seamless integration of the world's best suppliers. Made-in-China.com™ is B2B portal that provides information on Chinese products and Chinese suppliers. It bridges the gap between global buyers and quality Chinese suppliers.

The proposed CPLM framework is SOA based and incorporates Model based System Engineering (MBSE). It integrates modern concepts of CM, CBD and CBM. The researcher(s) combine together the three cloud based approaches into Cloud Based Design, Manufacture and Maintenance (CBDMM). Figure 4 illustrates the proposed CPLM framework, where all the phases of life cycle are represented in stacked layers and offered through cloud platforms. The CPLM framework integrates the Internet into PLM systems by using the SOA architecture. These features allow enterprises to be able to access other enterprises' product data at any time and from anywhere. CPLM should use a native XML database-powered metadata repository and data orchestration engine at its core.

Whilst there is no one complete data model for collecting and storing the product life cycle data, a set of concepts and technologies are available to enable creating and modifying the data model based on the requirements for the specific product process and the users. MBSE relies on 3D models since they carry more information. Semantic data management systems can be used to model MBSE domain ontologies. The core of the product data model is the generic product ontology, which defines the upper-level concepts of the product data model. The product ontology is defined in more detail in different generic domain ontologies, such as generic design domain ontology. The generic domain ontologies are further defined in more detail in the generic sub-domain ontologies, such as generic CAD domain ontology for the design domain. Sub-domain ontologies define the common concepts and components of the particular sub-domain. Examples of sub-domain applications are REQM, SysML, Simulink, ECAD and MCAD. Many modelling techniques and approaches are applicable to MBSE. These include:

1. Structured Analysis and Design,
2. Data Flow Diagramming,
3. State Transition Diagramming,
4. Behavioural Modelling,

Figure 4. CPLM Architecture

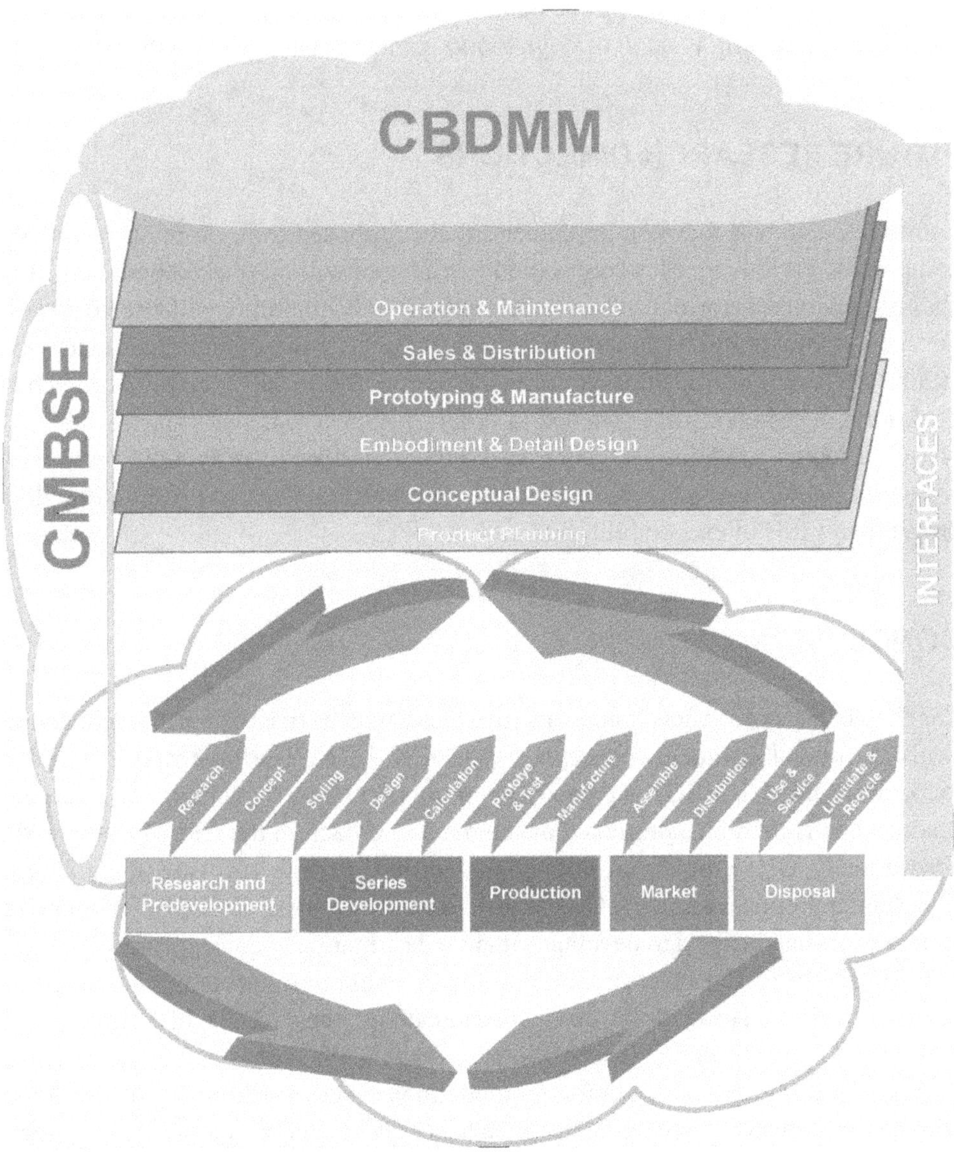

5. Entity Relationship Modelling,
6. Finite Element Modelling,
7. Environment Virtualisation,
8. Computer Aided Design (CAD),
9. Analytical Modelling and,
10. Process Modelling.

Data model structures can be defined for all the domain ontologies, such as the design, manufacture, marketing and maintenance domain ontology. CMBSE is when the MBSE modelling is done through cloud platforms using the XaaS concept.

FUTURE RESEARCH DIRECTIONS

Future research will focus on implementing the proposed framework on case study companies. Various metrics across the PLM phases will be tracked and compared against previous performance for the companies. Benchmarking will also be carried out against companies that are involved in the value chain for complex products such as aircrafts, automotive, ships and trains. Interoperability still remains a farfetched goal, more research effort is needed in order to achieve openness across the PLM phases. Product modeling is generally complex and more resources; personnel, computing power and programming skills, are needed. This has resulted in slow progress towards interoperability in PLM.

CONCLUSION

The chapter presented a new framework for CPLM, which is based on SOA architecture whilst incorporating modern concepts of CBD, CBM, CM. The three aspects combine to form CBDMM. MBSE has gained a lot of traction in the modern day engineering practice and is therefore promoted as the main approach in the CPLM framework. Traditionally MBSE has been offered through on premise software applications. The chapter proposes adopting CC to move towards the era of CMBSE. Web service technologies can be used to develop semantic data management systems, which are in turn used to model product data ontologies. Although ontology modeling has been of late limited towards the product design phase, the overall CPLM framework advocates for improved scope of MBSE across all the PLM phases. The chapter also discussed the applications of ontology modeling as a key method for virtualization of manufacturing assets, manufacturing services and the whole supply chain system. Other technologies such as RFID and smart sensors have been discussed in detail since they are the major drivers of virtualizing the physical entities in the PLM phases. They also help to gather Big Data which can later on be analysed through XaaS platforms offered by many vendors as presented in the chapter. Interoperability has been seen as the pivot for stakeholder collaboration, which leads to agility and process optimization.

REFERENCES

Ahmed, M., & Hossain, M. A. (2014). Cloud Computing and security issues in the cloud. *International Journal of Network Security & Its Applications*, *6*(1), 25–36. doi:10.5121/ijnsa.2014.6103

Avila, K., Sanmartin, P., Jabba, D., & Jimeno, M. (2017). Applications based on service-oriented architecture (SOA) in the field of home healthcare. *Sensors (Basel)*, *17*(18), 1703. doi:10.3390/s17081703 PMID:28757557

Berry, M. (2017). *Major cloud computing vendors*. Retrieved from: http://www.itmanagerdaily.com/cloud-computing-vendors/

Campos, J. (2009). Development in the application of ICT in condition monitoring and maintenance. *Computers in Industry*, *60*(1), 1–20. doi:10.1016/j.compind.2008.09.007

Campos, J., Kortelainen, J., & Jantunen, E. (2014). Industrial open source solutions for product life cycle management. *Cogent Engineering*, *1*(1), 939737. doi:10.1080/23311916.2014.939737

Chang, K. (2014). *Product Design Modeling using CAD/CAE*. The Computer Aided Engineering Design Series. doi:10.1016/B978-0-12-398513-2.00006-3

Dassault Systemes. (2011). *3D XML—a universal, lightweight XML-based format*. Retrieved from: http://www.3ds.com/products/3dvia/3d-xml/overview

Dhuieb, M. A., Belkadi, F., Laroche, F., & Bernard, A. (n.d.). *Interoperability framework for supporting information-based assistance in the factory*. The IFIP Working Group WG 5.1 11th International Conference on Product Lifecycle Management, Yokohama, France. Retrieved from: https://hal.archives-ouvertes.fr/hal-01066441/document

Ding, L., Ball, A., Matthews, J., McMahon, C. A., & Patel, M. (2007). Product representation in lightweight formats for product lifecycle management (PLM). *4th International Conference on Digital Enterprise Technology*. Retrieved from: http://www.ukoln.ac.uk/projects/grand-challenge/papers/DET2007_Ding_Paper.pdf

ECMA International. (2017). *Standard ECMA-363 Universal 3D File Format*. Retrieved from: http://www.ecma-international.org/publications/standards/Ecma-363.htm

Emmanouilidis, C., Jantunen, E. G., & Starr, A. (2011). e-Maintenance update: The road to success for modern industry. *Proceedings of the 24th International Congress on Condition Monitoring and Diagnostics Engineering Management, COMADEM*. Retrieved from: http://www.ipet.gr/~chrisem/Files/E-maintenance_update.pdf

Ferreiro, S., Arnaiz, A., Sierra, B., & Irigoien, I. (2012). Application of Bayesian networks in prognostics for a new integrated vehicle health management concept. *Expert Systems with Applications*, *39*(7), 6402–6418. doi:10.1016/j.eswa.2011.12.027

Friess, P., Guillemin, P., Gusmeroli, S., Sundmaeker, H., Bassi, A., Jubert, I. S., . . . Doody, P. (2011). *Internet of Things strategic research roadmap*. Retrieved from: http://www.internet-of-things-research.eu/pdf/IoT_Cluster_Strategic_Research_Agenda_2011.pdf

Fröhlich, A. (2011). 3D Formats in the field of engineering: a comparison. *PROSTEP AG*. Retrieved from: http://isicad.ru/uploads/http___www.prostep.pdf

Hao, Y. (2016). *Cloud Computing* (Doctoral thesis). University of Vaasa. Retrieved from: http://www.uva.fi/materiaali/pdf/isbn_978-952-476-715-1.pdf

He, W., & Xu, L. (2015). A state-of-the-art survey of cloud manufacturing. *International Journal of Computer Integrated Manufacturing*, *28*(3), 239–250. doi:10.1080/0951192X.2013.874595

Jantunen, E., Emmanouilidis, C., Arnaiz, A., & Gilabert, E. (2011). *e-Maintenance: trends, challenges and opportunities for modern industry.* 18th International Federation of Automatic Control (IFAC) World Congress, Milano, Italy. Retrieved from: http://www.ipet.gr/~chrisem/Files/e-Maintenance-Trends.pdf

Karim, R., Birk, W., & Larsson-Kråik, P. (2015). *Cloud-based e-maintenance solutions for condition-based maintenance of wheels in heavy haul operation. In 11th International Heavy Haul Conference*. Perth, Australia: International Heavy Haul Association. Retrieved from http://ltu.diva-portal.org/smash/get/diva2:1005837/FULLTEXT01.pdf

Komninos, I. (2002). *Product life cycle management.* Urban and regional Innovation research unit., Faculty of Engineering: Aristotle University of Thessaloniki. Retrieved from: https://www.urenio.org/tools/en/Product_Life_Cycle_Management.pdf

Levrat, E., & Lung, B. (2007). *TELMA: A full e-maintenance platform.* Second World Congress on Engineering Asset Management and the Fourth International Conference on Condition Monitoring, WCEAM/CM, Harrogate, UK. Retrieved from: http://dynamite.vtt.fi/conference_pres/telma_wceam_el_bi.pdf

Liu, N., & Li, X. (2012). A resource virtualization mechanism for cloud manufacturing systems. *Proceedings of the 4th International IFIP Working Conference, 122,* 46-59. Retrieved from: https://link.springer.com/content/pdf/10.1007%2F978-3-642-33068-1_6.pdf

March, S. T., & Storey, V. C. (2008). Design science in the information systems discipline: An introduction to the special issue on design science research. *MIS Quartely*, *32*(4), 725–730.

Mathew, A., Zhang, L., Zhang, S., & Ma, L. (2006). A Review of the MIMOSA OSA-EAI database for condition monitoring systems. In J. Mathew, J. Kennedy, L. Ma, A. Tan, & D. Anderson (Eds.), *Engineering Asset Management*. London: Springer. doi:10.1007/978-1-84628-814-2_88

Mell, P., & Grance, T. (2011). *The NIST definition of Cloud Computing*. National Institute of Standards and Technology, NIST Special Publication 800-145. Retrieved from: http://nvlpubs.nist.gov/nistpubs/Legacy/SP/nistspecialpublication800-145.pdf

Merlo, C., Vicien, G., & Ducq, Y. (2014). Interoperability modelling methodology for product design organisations. *International Journal of Production Research*, *52*(15), 4379–4395. doi:10.1080/00207543.2013.774484

Mourad, M., Nassehi, A., & Schaefer, D. (2016). Interoperability as a key enabler for manufacturing in the cloud. *Procedia CIRP*, *52*, 30-34. Retrieved from: https://www.sciencedirect.com/science/article/pii/S2212827116308010

MPULSE. (2015). *The future of maintenance management*. Retrieved from: https://www.mpulsesoftware.com/sites/default/files/MPulse_Future_of_Maintenance_Management_eBook.pdf

Myers, M. D. (1997). Qualitative research in information systems. *Management Information Systems Quarterly*, *21*(2), 241–242. doi:10.2307/249422

Oluwaseyi, A. (2008). *Service oriented architecture and web services: guidelines for migrating from legacy systems and financial consideration* (Master's thesis). Blekinge Institute of Technology. Retrieved from: https://www.diva-portal.org/smash/get/diva2:828526/FULLTEXT01.pdf

Paviot, T., Cheutet, V., & Lamouri, S. (2011). A PLCS framework for PDM / ERP interoperability. *International Journal of Product Lifecycle Management, 5*(2/3/4), 295-313. Retrieved from: https://hal.archives-ouvertes.fr/hal-00799829/document

PDF3D. (2017). Retrieved from: https://www.pdf3d.com/u3d/

Penciuc, D., Durupt, A., Belkadi, F., Eynard, B., & Rowson, H. (2014). Towards a PLM interoperability for a collaborative design support system. *Procedia CIRP*, *25*, 369-376. Retrieved from: http://www.sciencedirect.com/science/article/pii/S2212827114010828

Raines, G. (2009). *Cloud Computing and Service Oriented Architecture (SOA).* The MITRE Corporation. Retrieved from: https://www.mitre.org/sites/default/files/pdf/09_0743.pdf

Remi, A., & Tony, P. (2007). *Developing web application with COLLADA and X3D.* Retrieved from: https://www.khronos.org/collada/presentations/Developing_Web_Applications_with_COLLADA_and_X3D.pdf

Simón, V., Galar, D., & Baglee, D. (2015). eMaintenance platform for performing data Fusion mutation on machine tools. *Proceeding of Big Data Mining in real world applications,* 24-30. Retrieved from: http://www.worldcomp-proceedings.com/proc/proc2015/DMIN15_Final_Edition/DMIN15_Papers.pdf

Swearingen, K., Majkowski, W., Bruggeman, B., & Gilbertson, D. (2007). *An open system architecture for condition based maintenance overview* (pp. 1–8). IEEE Aerospace Conference. doi:10.1109/AERO.2007.352921

Tao, F., Zhang, L., Venkatesh, V. C., Luo, Y., & Cheng, Y. (2011). Cloud manufacturing: A computing and service-oriented manufacturing model. *Proceedings of the Institution of Mechanical Engineers. Part B, Journal of Engineering Manufacture, 225*(10), 1969–1976. doi:10.1177/0954405411405575

Tao, F., Zuo, Y., Xu, L. D., & Zhang, L. (2014). IoT-based intelligent perception and access of manufacturing resource toward cloud manufacturing. *IEEE Transactions on Industrial Informatics, 10*(2), 1547–1557. doi:10.1109/TII.2014.2306397

Theron, W., & Hendry, G. (2015). *The value of systems-driven product development: Case studies from the automotive & aerospace industries. Proceedings of the Eleventh (11th) SA INCOSE Conference*. Retrieved from https://www.incose.org.za/pubs/2015/Papers/SA_INCOSE_2015_submission_5.pdf

Uckelmann, D., Harrison, M., & Michahelles, F. (2011). An architectural approach towards the future Internet of Things. In D. Uckelmann, M. Harrison, & F. Michahelles (Eds.), *Architecting the Internet of Things* (pp. 1–24). Berlin: Springer-Verlag. doi:10.1007/978-3-642-19157-2_1

Unnamalai, V. E., & Thresphine, J. R. (2014). Service-Oriented Architecture for Cloud Computing. *International Journal of Computer Science and Information Technologies, 5*(1), 251–255. Retrieved from http://ijcsit.com/docs/Volume%205/vol5issue01/ijcsit2014050153.pdf

Verdouw, C. N., Beulens, A. J. M., Reijers, H. A., & van der Vorst, J. G. A. J. (2015). A control model for object virtualization in supply chain management. *Computers in Industry, 68*, 116–131. doi:10.1016/j.compind.2014.12.011

Vermesan, O., Friess, P., Guillemin, P., Gusmeroli, S., Sundmaeker, H., Bassi, A., . . . Doody, P. (2011). *Internet of Things – Global technological and societal trends*. River Publishers. Retrieved from: https://www.riverpublishers.com/pdf/ebook/chapter/RP_9788793609105C3.pdf

Wang, J., Zhang, L., Duan, L., & Robert, X. G. (2017). A new paradigm of cloud-based predictive maintenance for intelligent manufacturing. *Journal of Intelligent Manufacturing*, *28*(5), 1125–1137. doi:10.1007/s10845-015-1066-0

Welbourne, E., Battle, L., Cole, G., Gould, K., Rector, K., Raymer, S., & Borriello, G. et al. (2009). Building the Internet of Things using RFID: The RFID ecosystem experience. *IEEE Internet Computing*, *13*(3), 1089–7801. doi:10.1109/MIC.2009.52

World Health Organization (WHO). (2011). *Introduction to medical equipment inventory management*. Retrieved from http://apps.who.int/iris/bitstream/10665/44561/1/9789241501392_eng.pdf

Xu, X. (2012). From cloud computing to cloud manufacturing. *Robotics and Computer-integrated Manufacturing*, *28*(1), 75–86. doi:10.1016/j.rcim.2011.07.002

Yoshikawa, H., Tsubokura, T., Toida, S., Kawai, Y., & Hatori, F. (2015). Cloud-based equipment maintenance and facility management service platform. *Hitachi Review*, *64*(4). Retrieved from: http://www.hitachi.com/rev/pdf/2015/r2015_04_108.pdf

Zeballos, L., & Quiroga, O. (2017). From computer integrated manufacturing to cloud manufacturing. *SII, Simposio Argentino de Informatica Industrial*. Retrieved from: http://www.clei2017-46jaiio.sadio.org.ar/sites/default/files/Mem/SII/sii-16.pdf

Zhang, D., Liu, Y., Sheng, X., Li, J., Wu, T., Xue, C. J., & Yang, H. (2015). Deadline-aware task scheduling for solar-powered nonvolatile sensor nodes with global energy migration. In *Proceedings of the 52nd Annual Design Automation Conference*. ACM. Retrieved from: doi:10.1145/2744769.2744815

Zhang, L. (2015). *Big Data Analytics for e-Maintenance: Modeling of high-dimensional data streams* (Licentiate Thesis). Operation and Maintenance Engineering Luleå University of Technology. Retrieved from: https://www.diva-portal.org/smash/get/diva2:990005/FULLTEXT01.pdf

Zhao, Y. Z., Zhang, J. B., Zhuang, L., & Zhang, D. H. (2005). *Service-Oriented Architecture and technologies for automating integration of manufacturing systems and services*. 10th International Conference on Emerging Technologies and Factory Automation (ETFA), Catania, Italy. Retrieved from: http://fileadmin.cs.lth.se/ai/Proceedings/etfa2005/html/articles/CF-000361.pdf

ADDITIONAL READING

Cellary, W., & Walczak, K. (2012). Interactive 3D multimedia content: Models for creation, management, search and presentation. Retrieved from: http://www.springer.com/978-1-4471-2496-2

Claver, C. F., Debois-Felsmann, G. P., Delgado, F., Hascall, P., Marshall, S., Nordby, M., . . . Sebag, J. (2011). The large synoptic survey telescope (LSST): A system of systems. Retrieved from: https://www.lsst.org/sites/default/files/docs/aas/2011/217-RC-1779-AAS_Claver.ppt.pdf

Delarama, J., & Valilai, O. F. (2017). A novel solution for manufacturing interoperability fulfillment using interoperability service providers. *Prodecia CIRP*, *63*, 774–779. doi:10.1016/j.procir.2017.03.141

Foradis, T., & Thramboulidis, K. (2017). From mechatronic components to industrial automation things: An IoT model for cyber-physical manufacturing systems. *Journal of Software Engineering and Applications*, *10*(08), 734–753. doi:10.4236/jsea.2017.108040

Google Cloud Platform. (2017). Retrieved from: https://cloud.google.com/bigquery/docs/

Graf, N. A. (2012). 3DPDF: Open source solutions for incorporating 3D iformation in PDF files. *Presented at IEEE 2012 Nuclear Science Symposium, Medical Imaging Conference Anaheim, California*. Retrieved from: http://www.slac.stanford.edu/cgi-wrap/getdoc/slac-pub-15295.pdf

Hartman, N. W. (2009). Evaluating lightweight 3D graphics formats for product visualization and data exchange. *Journal of applied science & engineering technology*. Retrieved from: http://citeseerx.ist.psu.edu/viewdoc/download?doi=10.1.1.466.1631&rep=rep1&type=pdf

Hartman, N. W., & Lim, E. D. A. (2008). A comparison of functionality between STEP AP 203 E2, 3DXML, JT, and U3D Technologies. Retrieved from: www.catiacommunity.com/pdf/hartman.pdf

Hartman, N. W., & Lim, E. D. A. (2008). Examining Neutral Formats for Visualization and Data Exchange. *Proceedings of the 2008 IAJC-IJME International Conference ISBN 978-1-60643-379-9*. Retrieved from: http://www.ijme.us/cd_08/PDF/221%20IT302.pdf

Kapos, G. D., Dalakas, V., Tsadimas, A., Nikolaidou, M., & Anagnostopoulos, D. (2014). Model-based System Engineering using SysML: Deriving Executable Simulation Models with QVT. Retrieved from: http://galaxy.hua.gr/~mara/publications/syscon2014.pdf

Khronos Group™. (2017). COLLADA specification and homepage. Retrieved from: http://www.khronos.org/collada/

Leavitt, N., & Group, Y. (2009). Is cloud computing really ready for prime time? *IEEE Computer Society*. Retrieved from: http://www.leavcom.com/pdf/Cloudcomputing.pdf

Liu, N., & Li, X. (2012). A resource virtualization mechanism for cloud manufacturing systems. *IFIP International Federation for Information Processing*. Retrieved from: https://link.springer.com/content/pdf/10.1007%2F978-3-642-33068-1_6.pdf

Made-in-China.com™. (2017). Retrieved from: http://www.made-in-china.com

MFG. COM®. (2017). Retrieved from: https://www.mfg.com/

Salesforce.com. (2017). Retrieved from: https://www.salesforce.com/

Schallehn. Data Management for Engineering Applications. Lecture Notes. Retrieved from: http://www.dbse.ovgu.de/dbse/en/Lectures/Data+Management+for+Engineering+Applications/_/dmea-06-plm.pdf

Stark, J. (2015). *Product life cycle management, Vol 1*. Retrieved from: http://www.springer.com/978-3-319-17439-6

3D. Systems On Demand Manufacturing. (2017). Retrieved from: https://www.3dsystems.com/on-demand-manufacturing

Wu, D., Thames, J., Rosen, D. W., & Schaefer, D. (2013). Enhancing the product realization process with cloud-based design and manufacturing systems. *ASME. Journal of Computing and Information Science in Engineering*, *13*(4), 041004, 041004–041014. doi:10.1115/1.4025257

Yadekar, Y., Shehab, E., & Mehnen, J. (2016). Taxonomy and uncertainties of cloud manufacturing. *International Journal of Agile Systems and Management*, *9*(1), 48–66. ISSN 1741-9174. Retrieved from: 10.1504/IJASM.2016.076577

KEY TERMS AND DEFINITIONS

Cloud Computing (CC): A model for enabling ubiquitous, convenient, on-demand network access to a shared pool of configurable computing resources (e.g., networks, servers, storage, applications, and services) that can be rapidly provisioned and released with minimal management effort or service provider interaction.

Cloud Manufacturing (CM): Is a new service-oriented manufacturing model that integrates different technologies such as networked manufacturing, cloud computing, IoT, virtualization, and service-oriented technologies to support collaboration, sharing, and management of manufacturing resources.

Cyber Physical Systems (CPSs): Systems that link the physical world (e.g., through sensors or actuators) with the virtual world of information processing. CPSs bring advances in personalized healthcare, emergency response, traffic flow management, supply chain management, electric power generation and delivery, as well as in many other areas not yet envisioned.

Enterprise Resource Planning (ERP): Automation and integration of a company's core business to help them focus on effectiveness and simplified success. ERP software applications can be used to manage product planning, parts purchasing, inventories, interacting with suppliers, providing customer service, and tracking orders. ERP can also include application modules for the finance and human resources aspects of a business.

Internet of Things (IoT): Is a system of interrelated computing devices, mechanical and digital machines, objects, animals, or people that are provided with unique identifiers and the ability to transfer data over a network without requiring human-to-human or human-to-computer interaction.

Machine-to-Machine (M2M): A term used to describe any technology that enables networked devices to exchange information and perform actions without the manual assistance of humans. M2M is considered an integral part of the internet of things (IoT) and brings several benefits to industry and business in general as it has a wide range of applications such as industrial automation, logistics, smart grid, smart cities, health, defence, etc. mostly for monitoring but also for control purposes.

Manufacturing Execution Systems (MES): A control system for managing and monitoring work-in-process on a factory floor. An MES keeps track of all manufacturing information in real time, receiving up-to-the-minute data from robots, machine monitors, and employees.

Manufacturing Operations Management (MOM): The subsequent system to manufacturing execution system (MES) software which expands focus from a single facility to the entire supply network and monitors a variety of aspects of the manufacturing process, including production capacity analysis, Work-in-Process (WIP), inventory turns and standard lead times.

Model-Based Systems Engineering (MBSE): Is a systems engineering paradigm that emphasizes the application of rigorous visual modeling principles and best practices to systems engineering activities throughout the system development life cycle (SDLC).

OEM: Original equipment manufacturer.

Product Data Management (PDM)/Product Life-Cycle Management (PLM): PDM is a category of computer software used to control data related to products. PDM creates and manages relations between sets of data that define a product, and store those relationships in a database. It is an important tool in product lifecycle management. PLM is a strategic business approach that applies a consistent set of business solutions that support the collaborative creation, management, dissemination, and use of product definition information throughout the lifecycle of the product.

Service-Oriented Architecture (SOA): Is a style of software design where services are provided to the other components by application components, through a communication protocol over a network. The basic principles of service-oriented architecture are independent of vendors, products, and technologies. SOA promotes the goal of separating users (consumers) from the service implementations. Services can, therefore, be run on various distributed platforms and accessed across networks. SOA is implemented using web services technologies.

STEP/STEP-NC: Standard for the exchange of product model data, a comprehensive ISO standard (ISO 10303) that describes how to represent and exchange digital product information. STEP is a means by which graphical information is shared among unlike computer systems around the world. It is designed so that virtually all essential information about a product, not just CAD files, can be passed back and forth among users.

Supply Chain Management (SCM): An integrated approach to planning, implementing, and controlling the flow of information, materials, and services from raw material and component suppliers through the manufacturing of the finished product for ultimate distribution to the end customer. It includes the systematic integration of processes for demand planning, customer relationship collaboration, order fulfillment/delivery, product/service launch, manufacturing/operations planning and control, supplier relationship collaboration, life cycle support, and reverse logistics and their associated risks.

Compilation of References

3gpp.org. (2016). *LTE*. Available at: http://www.3gpp.org/technologies/keywords-acronyms/98-lte

Adams, P. (2005). *Update to the IT asset management maturity model*. Retrieved 11 February 2017, from https://www.gartner.com/doc/484622/update-it-asset-management-maturity

Adams, D. A., Nelson, R. R., & Todd, P. A. (1992). Perceived usefulness, ease of use, and usage of information technology: A replication. *Management Information Systems Quarterly*, *16*(2), 227–247. doi:10.2307/249577

Adibi, H. (2015). Social perspectives on health: mHealth implications. In mHealth Multidisciplinary Verticals (pp. 219 – 236). Taylor & Francis Group, LLC.

Agarwal, R., Rastogi, S., & Mehrotra, A. (2009). Customers' perspectives regarding e-banking in an emerging economy. *Journal of Retailing and Consumer Services*, *16*(5), 340–351. doi:10.1016/j.jretconser.2009.03.002

Agostinho, C., Ducq, Y., Zacharewicz, G., Sarraipa, J., Lampathaki, F., Poler, R., & Jardim-Goncalves, R. (2016). Towards a sustainable interoperability in networked enterprise information systems: Trends of knowledge and model-driven technology. *Computers in Industry*, *79*, 64–76. doi:10.1016/j.compind.2015.07.001

Ahire, S. L. (1997). Management Science: Total Quality Management interfaces: An integrative framework. *Interfaces*, *27*(6), 91–105. doi:10.1287/inte.27.6.91

Ahmed, M., & Hossain, M. A. (2014). Cloud Computing and security issues in the cloud. *International Journal of Network Security & Its Applications*, *6*(1), 25–36. doi:10.5121/ijnsa.2014.6103

AIIM. (2010). *What is Enterprise Content Management?* Retrieved 11 February 2017, from http://www.aiim.org/What-is-ECM-Enterprise-Content-Management.aspx

Ajzen, I. (2006). *Theory of planned behavior*. Available at: http://people.umass.edu/aizen/tpb.diag.html#null-link

Akinci, S., Atilgan-Inan, E., & Aksoy, S. (2010). Re-assessment of E-S-Qual and E-RecS-Qual in a pure service setting. *Journal of Business Research*, *63*(3), 232–240. doi:10.1016/j.jbusres.2009.02.018

Akkermans, H., & van Helden, K. (2002). Vicious and virtuous cycles in ERP implementation: A case study of interrelations between critical success factors. *European Journal of Information Systems*, *11*(1), 35–46. doi:10.1057/palgrave.ejis.3000418

Al-Ajam, A. S., & Md Nor, K. (2015). Challenges of adoption of internet banking service in Yemen. *International Journal of Bank Marketing*, *33*(2), 178–194. doi:10.1108/IJBM-01-2013-0001

Al-alak, B. A. (2014). Impact of marketing activities on relationship quality in the Malaysian banking sector. *Journal of Retailing and Consumer Services*, *21*(3), 347–356. doi:10.1016/j.jretconser.2013.07.001

Alalwan, A.A., Dwivedi, Y.K., Rana, N.P., Lal, B., & Williams, M.D. (2015). Consumer adoption of internet banking in Jordan: examining the role of hedonic motivation, habit, self-efficacy and trust. *Journal of Financial Services Marketing, 20*(2), 145-157.

Aldás-Manzano, J., Lassala-Navarré, C., Ruiz-Mafé, C., & Sanz-Blas, S. (2009). The role of consumer innovativeness and perceived risk in online banking usage. *International Journal of Bank Marketing*, *27*(1), 53–75. doi:10.1108/02652320910928245

Aldas-Manzano, J., Ruiz-Mafe, C., Sanz-Blas, S., & Lassala-Navarré, C. (2011). Internet banking loyalty: Evaluating the role of trust, satisfaction, perceived risk and frequency of use. *Service Industries Journal*, *31*(7), 1165–1190. doi:10.1080/02642060903433997

Al-hawari, M. A. (2015). How the personality of retail bank customers interferes with the relationship between service quality and loyalty. *International Journal of Bank Marketing*, *33*(1), 41–57. doi:10.1108/IJBM-09-2013-0096

Al-Hawari, M. A. A. (2014). Does customer sociability matter? Differences in e-quality, e-satisfaction, and e-loyalty between introvert and extravert online banking users. *Journal of Services Marketing*, *28*(7), 538–546. doi:10.1108/JSM-02-2013-0036

Al-Jabri, I. M., & Sohail, M. S. (2012). Mobile banking adoption: Application of diffusion of innovation theory. *Journal of Electronic Commerce Research*, *13*(4), 379–391.

Almeida, L. H. (2009). WebPrisma: An Interactive Web-Based Tool for Exploratory Visualization Using Multiple Coordinated Views. *Information Visualisation, 2009 13th International Conference*. doi:10.1109/IV.2009.17

Alsajjan, B., & Dennis, C. (2010). Internet banking acceptance model: Cross-market examination. *Journal of Business Research*, *63*(9), 957–963. doi:10.1016/j.jbusres.2008.12.014

Al-smadi, M. O. (2012). Factors affecting adoption of electronic banking: An analysis of the perspectives of banks' customers. *International Journal of Business and Social Science*, *3*(17).

Álvares, A. J., Ferreira, J. C. E., & Lorenzo, R. M. (2008). An integrated web-based CAD/CAPP/CAM system for the remote design and manufacture of feature-based cylindrical parts. *Journal of Intelligent Manufacturing*, *19*(6), 643–659. doi:10.1007/s10845-008-0117-1

Amin, M., Isa, Z., & Fontaine, R. (2011). The role of customer satisfaction in enhancing customer loyalty inMalaysian Islamic banks. *Service Industries Journal*, *31*(9), 1519–1532. doi:10.1080/02642060903576076

Amin, M., Isa, Z., & Fontaine, R. (2013). Islamic banks: Contrasting the drivers of customer satisfaction on image, trust, and loyalty of Muslim and non-Muslim customers in Malaysia. *International Journal of Bank Marketing*, *31*(2), 79–97. doi:10.1108/02652321311298627

Amin, M., & Nasharuddin, S. Z. (2013). Hospital service quality and its effects on patient satisfaction and behavioural intention. *Clinical Governance: An International Journal*, *18*(3), 238–254. doi:10.1108/CGIJ-05-2012-0016

Anderson, J. C., & Gerbing, D. (1988). Structural equation modeling in practice: A review and recommended two-step approach. *Psychological Bulletin*, *103*(3), 411–423. doi:10.1037/0033-2909.103.3.411

Angelakopoulos, G., & Mihiotis, A. (2011). E-banking: Challenges and opportunities in the Greek banking sector. *Electronic Commerce Research*, *11*(3), 297–319. doi:10.1007/s10660-011-9076-2

ARC White Paper. (2015). *The Industrial Revolution Continues*. Available at: http://www.treasurebox.com.sg/meap/nl_201605/pdf/ARC_White_Paper_e-Factory_final_2015-12.pdf

ARMA International. (2013). *General Accepted Recordkeeping Principles®: Information Governance Maturity Model*. Retrieved 3 March 2016, from http://www.arma.org/docs/bookstore/theprinciplesmaturitymodel.pdf?sfvrsn=2

ARMA International. (2014, February 20). *General Accepted Recordkeeping Principles®*. Retrieved 3 March 2016, from http://www.arma.org/docs/sharepoint-roadshow/the-principles_executive-summaries_final.doc

Arnett, J. J. (2004). *Emerging adulthood: The winding road from the late teens through the twenties*. Oxford, UK: Oxford University Press.

Ashton, K. (2009). That 'Internet of Things' thing. *RFID Journal*, *22*(7), 97–114.

Atzori, L., Iera, A., & Morabito, G. (2010). The internet of things: A survey. *Computer Networks*, *54*(15), 2787–2805. doi:10.1016/j.comnet.2010.05.010

Avila, K., Sanmartin, P., Jabba, D., & Jimeno, M. (2017). Applications based on service-oriented architecture (SOA) in the field of home healthcare. *Sensors (Basel)*, *17*(18), 1703. doi:10.3390/s17081703 PMID:28757557

Babakus, E., & Boller, G. W. (1992). An empirical assessment of the SERVQUAL scale. *Journal of Business Research*, *24*(3), 253–268. doi:10.1016/0148-2963(92)90022-4

Baccarani, C., Ugolini, M., & Bonfanti, A. (2010). A conceptual service quality map: The value of a wide opened perspective. *Proceedings of 13th Toulon-Verona Conference "Excellence in Services"*.

Bagozzi, R. P., & Yi, Y. (1988). On the Evaluation of Structural Equation Models. *Journal of the Academy of Marketing Science*, *16*(1), 74–94. doi:10.1007/BF02723327

Bajaj, A., & Nidumolu, S. R. (1998). A feedback model to understand information system usage. *Information & Management*, *33*(4), 213–224. doi:10.1016/S0378-7206(98)00026-3

Barclay, D., Higgins, C., & Thompson, R. (1995). The partial least squares (PLS) approach to causal modeling: Personal computer adoption and use as an illustration. *Technology Studies*, *2*(2), 285–309.

Barett, J. (2016). *The Disadvantages of Bluetooth Technology*. Available at: https://www.techwalla.com/articles/the-disadvantages-of-bluetooth-technology

Bartikowski, B., Chandon, J. L., & Muller, B. (2008). *Mesurer la confiance des internautes dans les sites web marchands: adaptation de l'échelle de Mcknight, Kacmar et Choudhury*. Communication au 7ème congrès sur les tendances du marketing en Europe, Venise.

Baskarada, S. (2009). *Information Quality Management Capability Maturity Model*. Wiesbaden, Germany: Vieweg and Teubner. doi:10.1007/978-3-8348-9634-6

Bateson, J. E. G. (2000). Perceived Control and the Service Experience. In Handbook of Service Marketing and Management. Thousand Oaks, CA: SAGE. doi:10.4135/9781452231327.n11

Baumann, C., Burton, S., Elliott, G., & Kehr, H. M. (2007). Prediction of attitude and behavioural intentions in retail banking. *International Journal of Bank Marketing*, *25*(2), 102–116. doi:10.1108/02652320710728438

Bennett, B. (2012). *The power of Bluetooth 4.0: It'll change your life*. Available at: http://www.cnet.com/news/the-power-of-bluetooth-4-0-itll-change-your-life

Bennington, L., & Cummane. (1998). Measuring service quality: A hybrid Methodology. *Total Quality Management, 9*(6), 395-405.

Berry, M. (2017). *Major cloud computing vendors*. Retrieved from: http://www.itmanagerdaily.com/cloud-computing-vendors/

Bertot & Choi. (n.d.). Big data and e-government: Issues, policies, and recommendations. *University of Maryland College Park.*

BETaaS Consortium. (2012). *Building the environment for the things as a service*. Retrieved from http://www.betaas.eu/

BETaaS Consortium. (2013). *D1.4.2 – TaaS Reference Model*. Retrieved from http://www.betaas.eu/docs/deliverables/BETaaS%20-%20D1.4.2%20%20TaaS%20Reference%20Model%20 v1.0.pdf

Bevan, N., & Macleod, M. (1993). Usability assessment and measurement. In The Management and Measurement of Software Quality. Ashgate Technical/Gower Press.

Bevan, N., Petrie, H., & Claridge, N. (2007). Strategies for providing guidance on usability and accessibility. *Proceedings HCI International*. doi:10.1007/978-3-540-73283-9_3

Bittman, T. J. (2004). *Gartner introduces the infrastructure maturity model*. Retrieved 26 March 2017, from https://www.gartner.com/doc/461820/gartner-introduces-infrastructure-maturity-model

Black, N. J., Lockett, A., Ennew, C., Winklhofer, H., & McKechnie, S. (2002). Modelling consumer choice of distribution channels: An illustration from financial services. *International Journal of Bank Marketing*, *20*(4), 161–173. doi:10.1108/02652320210432945

Bobbitt, L. M., & Dabholkar, P. A. (2001). Integrating attitudinal theories to understand and predict use of technology-based self-service. *International Journal of Service Industry Management*, *12*(5), 423–450. doi:10.1108/EUM0000000006092

Bodet, G. (2008). Customer satisfaction and loyalty in service: Two concepts, four constructs, several relationships. *Journal of Retailing and Consumer Services*, *15*(3), 156–162. doi:10.1016/j.jretconser.2007.11.004

Bohu, L., Lin, Z., & Xudong, C. (2010). *Introduction to cloud manufacturing*. Retrieved from: http://wwwen.zte.com.cn/endata/magazine/ztecommunications/2010Year/no4/articles/201012/t20101219_196752.html

Bourne, D. A., & Wang, C. H. (1995). Design and Manufacturing of Sheet Metal Parts. Boston: Academic Press.

Bowen, G., & Ozuem, W. (2015). *Computer-mediated marketing strategies: Social media and online brand communities*. Hershey, PA: IGI Global. doi:10.4018/978-1-4666-6595-8

Bower, A. (2005). *The diffusion and value of healthcare information technology*. RAND Corporation.

Boyer, K. K., Hallowell, R., & Roth, A. V. (2002). E-services: Operating strategy—a case study and a method for analyzing operational benefits. *Journal of Operations Management*, *20*(2), 175–188. doi:10.1016/S0272-6963(01)00093-6

Brady, M. K., & Cronin, J. J. Jr. (2001). Some new thoughts on conceptualizing perceived service quality: A hierarchical approach. *Journal of Marketing*, *65*(3), 34–49. doi:10.1509/jmkg.65.3.34.18334

Brender, J. (2006). *Handbook of Evaluation Methods for Health Informatics*. Burlington, MA: Elsevier Academic Press Publications.

Bressolles, G., Durrieu, F., & Senecal, S. (2014). A consumer typology based on e-service quality and e-satisfaction. *Journal of Retailing and Consumer Services*, *21*(6), 889–896. doi:10.1016/j.jretconser.2014.07.004

Brown, A., Katuu, S., Sebina, P., & Seles, A. (2009). Module 4: Preserving electronic records. *Training in Electronic Records Management* Retrieved 26 March 2017, from http://www.irmt.org/documents/educ_training/term%20modules/IRMT%20TERM%20Module%204.pdf

Burgess, L. (2004). A conceptual framework for understanding and measuring perceived service quality in net-based customer support systems. *CollECTeR LatAm Conference.*

Cai, S., & Jun, M. (2003). Internet user's perceptions of online service quality: A comparison of online buyers and information searchers. *Managing Service Quality*, *13*(6), 504–519. doi:10.1108/09604520310506568

Calabrese, A., & Scoglio, F. (2012). Reframing the past: A new approach in service quality assessment. *Total Quality Management & Business Excellence*, *23*(11), 1329–1343. doi:10.1080/14783363.2012.733259

Caldwell, B., Cooper, M., Reid, L., & Vanderheiden, G. (2007). *Web Content Accessibility Guidelines 2.0, W3C Working Draft*. Biot.

Calisir, F., & Gumussoy, C. A. (2008). Internet banking versus other banking channels: Young consumers' view. *International Journal of Information Management*, *28*(3), 215–221. doi:10.1016/j.ijinfomgt.2008.02.009

Cameron, S. A. (2011). *Enterprise content management - A business and technical guide*. Swindon, UK: The Chartered Institute for IT.

Campos, J. (2009). Development in the application of ICT in condition monitoring and maintenance. *Computers in Industry*, *60*(1), 1–20. doi:10.1016/j.compind.2008.09.007

Campos, J., Kortelainen, J., & Jantunen, E. (2014). Industrial open source solutions for product life cycle management. *Cogent Engineering*, *1*(1), 939737. doi:10.1080/23311916.2014.939737

Candido, C. J. F., & Morris, D. S. (2000). Charting service quality gaps. *Total Quality Management*, *11*(4-6), 463–472. doi:10.1080/09544120050007779

Candra, S. (2013). Revisit technology acceptance model for internet banking (Case Study: Public Banking in Indonesia). *International Conference on Computer, Networks and Communication Engineering (ICCNCE 2013).*

Cao, X., & Mokhtarian, P. (2005). *The intended and actual adoption of online purchasing: a brief review of recent literature*. University of California Transportation.

Capatina, A., & Rancati, E. (2017). *Key challenges and opportunities in web entrepreneurship*. Hershey, PA: IGI Global. doi:10.4018/978-1-5225-2466-3

Carlson, J., & O'Cass, A. (2011). Developing a framework for understanding e-service quality, its antecedents, consequences, and mediators. *Managing Service Quality*, *21*(3), 264–286. doi:10.1108/09604521111127965

Carman, J. M. (1990). Consumer perceptions of service quality: An assessment of the SERVQUAL dimensions. *Journal of Retailing*, *66*(l), 33–55.

Carr, D. (1999). *Guidelines for designing information visualization applications*. Paper presented at the ECUE'99.

Casaló, L. V., Flavián, C., & Guinalíu, M. (2008). The role of satisfaction and website usability in developing customer loyalty and positive word-of-mouth in the e-banking services. *International Journal of Bank Marketing*, *26*(6), 399–417. doi:10.1108/02652320810902433

Cattrysse, D., Beullens, P., Collin, P., Duflou, J., & Oudheusden, D. V. (2006). Automatic production planning of press brakes for sheet metal bending. *International Journal of Production Research*, *44*(20), 4311–4327. doi:10.1080/00207540600558031

Celik, H. (2008). What determines Turkish consumers' acceptance of internet banking? *International Journal of Bank Marketing*, *26*(5), 353–370. doi:10.1108/02652320810894406

Chang, K. (2014). *Product Design Modeling using CAD/CAE*. The Computer Aided Engineering Design Series. doi:10.1016/B978-0-12-398513-2.00006-3

Chang, S., & Lu, M. (2004). Understanding internet adoption and use behavior: A Hong Kong perspective. *Journal of Global Information Management*, *12*(3), 21–43. doi:10.4018/jgim.2004070102

Charfeddine, L., & Nasri, W. (2013). The behavior intention of Tunisian banks' customers on using internet banking. *International Journal of Innovation in the Digital Economy*, *4*(1), 16–30. doi:10.4018/jide.2013010102

Chatterjee, S., & Chatterjee, A. (2005). Prioritization of service quality parameters based on ordinal responses. *Total Quality Management & Business Excellence*, *16*(4), 477–489. doi:10.1080/14783360500078482

Chau, P., & Hu, P. (2001). Information technology acceptance by individual professionals: A model comparison approach. *Decision Sciences*, *32*(4), 699–719. doi:10.1111/j.1540-5915.2001.tb00978.x

Chemingui, H., & Ben lallouna, H. (2013). Resistance, motivations, trust and intention to use mobile financial services. *International Journal of Bank Marketing*, *31*(7), 574–592. doi:10.1108/IJBM-12-2012-0124

Chen, C. (2013). Perceived risk, usage frequency of mobile banking services. *Managing Service Quality: An International Journal*, *23*(5), 410–436. doi:10.1108/MSQ-10-2012-0137

Chen, C. P., & Zhang, C. (2014). Data-intensive applications, challenges, techniques and technologies: A survey on big data. *Inf. Sci.*, *275*, 314–347. doi:10.1016/j.ins.2014.01.015

Cheng, K., & Bateman, R. J. (2008). e-Manufacturing: Characteristics, applications and potentials. *Progress in Natural Science*, *18*(11), 1323–1328. doi:10.1016/j.pnsc.2008.03.027

Cheng, T. C., Lam, D. Y. C., & Yeung, A. C. L. (2006). Adoption of internet banking: An empirical study in Hong Kong. *Decision Support Systems*, *42*(3), 1558–1572. doi:10.1016/j.dss.2006.01.002

Chen, M.-Y., & Teng, C.-I. (2013). A comprehensive model of the effects of online store image on purchase intention in an e-commerce environment. *Electronic Commerce Research*, *13*(1), 1–23. doi:10.1007/s10660-013-9104-5

Chen, Q., & Wells, W. D. (2002). Attitude toward the site II: New information. *Journal of Advertising Research*, *42*(2), 33–45. doi:10.2501/JAR-42-2-33-45

Chen, R.-F., Hsiao, J.-L., & Hwang, H.-G. (2012). Measuring customer satisfaction of Internet banking in Taiwan: Scale development and validation. *Total Quality Management & Business Excellence*, *23*(7/8), 749–767. doi:10.1080/14783363.2012.704284

Chen, Y. H., & Barnes, S. (2007). Initial trust and online buyer behaviour. *Industrial Management & Data Systems*, *107*(1), 21–36. doi:10.1108/02635570710719034

Cheong, J. H., & Park, M. C. (2005). Mobile internet acceptance in Korea. *Internet Research*, *15*(2), 125–140. doi:10.1108/10662240510590324

Chin, K., & Tang, D. (2002). Web-Based Concurrent Stamping Part and Die Development. *Concurrent Engineering*, *10*(3), 213–228. doi:10.1177/106329302761689133

Chin, W. W. (1998). The Partial Least Squares Approach to Structural Equation Modeling. In G. A. Marcoulides (Ed.), *Modern Methods for Business Research* (pp. 295–358). London: Lawrence Erlbaum Associates.

Chiou, J.-S., & Shen, C.-C. (2012). The antecedents of online financial service adoption: The impact of physical banking services on Internet banking acceptance. *Behaviour & Information Technology*, *31*(9), 859–871. doi:10.1080/0144929X.2010.549509

Chiu, H. C., & Lin, N. P. (2004). A service quality measurement derived from the theory of needs. *Service Industries Journal*, *24*(1), 187–204. doi:10.1080/02642060412331301202

Chouk, I., & Perrien, J. (2005). La confiance du consommateur vis-a'-vis d'un marchand internet: Proposition d'une échelle de mesure. *Revue Française du Marketing*, *205*(1), 5–20.

Chowdhary, N., & Prakash, M. (2007). Prioritizing service quality dimensions. *Managing Service Quality: An International Journal*, *17*(5), 493–509. doi:10.1108/09604520710817325

Clark, B. Y., Brudney, J. L., & Jang, S.-G. (2013). Coproduction of government services and the new information technology: Investigating the distributional biases. *Public Administration Review*, *73*(5), 687–701. doi:10.1111/puar.12092

CMTC's Guide to Smart Manufacturing. (2015). Available at: https://cdn2.hubspot.net/hubfs/103829/Ebooks_Whitepapers/CMTC-Smart-Manufacturing_Ebook_Final.pdf?t=1509064707903

Cohen, J. (1988). *Statistical Power Analysis for the Behavioral Sciences*. Hillsdale, NJ: Lawrence Erlbaum.

Collier, J. E., & Bienstock, C. C. (2006). Measuring service quality in e-retailing. *Journal of Service Research*, *8*(3), 260–275. doi:10.1177/1094670505278867

Connell, N., & Young, T. (2007). Evaluating healthcare information systems through an 'enterprise' perspective. *Information & Management*, *44*(4), 433–440. doi:10.1016/j.im.2007.04.002

Cook, C., & Thompson, B. (2000). Reliability e-service quality in e-retailing. *Journal of Service Research*, *8*(3), 260–275.

Cooley, R., Mobasher, B., & Srivastava, J. (1997). *Web mining: Information and pattern discovery on the world wide web.* Paper presented at the Tools with Artificial Intelligence. doi:10.1109/TAI.1997.632303

Coronel, C., Morris, S., & Rob, P. (2016). Database Systems: Design, Implementation, and Management (11th ed.). Cengage Learning.

Costanza, F. (2017). Social media marketing and value co-creation: A system dynamics approach. In S. Rozenes & Y. Cohen (Eds.), *Handbook of research on strategic alliances and value co-creation in the service industry* (pp. 205–230). Hershey, PA: IGI Global. doi:10.4018/978-1-5225-2084-9.ch011

Coye, C., & Loftesness, B. S. (2012). *Interoperability in electronic payments: Lessons and opportunities.* Retrieved from: http://www.cgap.org/sites/default/files/Interoperability_in_Electronic_Payments.pdf

Cronin, J. J. Jr. (2003). Looking back to see forward in services marketing: Some ideas to consider. *Managing Service Quality*, *13*(5), 332–337. doi:10.1108/09604520310495813

Cronin, J. J. Jr, & Taylor, S. A. (1992). Measuring service quality: A reexamination and extension. *Journal of Marketing*, *56*(3), 55–68. doi:10.2307/1252296

Cruz-Cunha, M. M. (2010). *Enterprise information systems for business integration in SMEs: Technological, organizational, and social dimensions*. Hershey, PA: IGI Global. doi:10.4018/978-1-60566-892-5

Cruz-Cunha, M. M., & Varajao, J. (2011). *Enterprise information systems design, implementation and management: Organizational applications*. Hershey, PA: IGI Global. doi:10.4018/978-1-61692-020-3

Curtis, B., Hefley, W. E., & Miller, S. (1995). *Overview of the people capability maturity model.* Retrieved 26 March 2017, from http://www.sei.cmu.edu/reports/95mm001.pdf

Dabholkar, P. A. (1996). Consumer evaluations of new technology-based self-service options: An investigation of alternative models of service quality. *International Journal of Research in Marketing*, *13*(1), 29–51. doi:10.1016/0167-8116(95)00027-5

Dabholkar, P. A., Shepherd, C. D., & Thorpe, D. I. (2000). A comprehensive framework for service quality: An investigation of critical conceptual and measurement issues through a longitudinal study. *Journal of Retailing*, *76*(2), 139–173. doi:10.1016/S0022-4359(00)00029-4

Dahlstrom, R., Nygaard, A., Kimasheva, M., & Ulvnes, A. M. (2014). How to recover trust in the banking industry? A game theory approach to empirical analyses of bank and corporate customer relationships. *International Journal of Bank Marketing*, *32*(4), 268–278. doi:10.1108/IJBM-03-2014-0042

Dassault Systemes. (2011). *3D XML—a universal, lightweight XML-based format*. Retrieved from: http://www.3ds.com/products/3dvia/3d-xml/overview

Davis, F. D. (1989). Perceived usefulness, perceived ease of use and user acceptance of computer technology. *Management Information Systems Quarterly*, *13*(3), 319–337. doi:10.2307/249008

Davis, F. D. (1993). User acceptance of information technology: System characteristics, user perceptions and behavioral impacts. *International Journal of Man-Machine Studies*, *38*(3), 475–487. doi:10.1006/imms.1993.1022

Davis, F. D., Bagozzi, R. P., & Warshaw, P. R. (1989). User acceptance of computer technology: A comparison of two theoretical models. *Management Science*, *35*(8), 982–1003. doi:10.1287/mnsc.35.8.982

Davis, F. D., Bagozzi, R., & Warshaw, P. R. (1992). Extrinsic and intrinsic motivation to use computers in the workplace. *Journal of Applied Social Psychology*, *22*(14), 1109–1130. doi:10.1111/j.1559-1816.1992.tb00945.x

De Vin, L. J., De Vries, J., Streppel, A. H., & Kals, H. J. (1992). PART-S, a CAPP system for small batch manufacturing of sheet metal components. *Proceedings of the 24th CIRP International Seminar of Manufacturing Systems*, 171-182.

De Vin, L. J., De Vries, J., & Streppel, T. (2000). Process planning for small batch manufacturing of sheet metal parts. *International Journal of Production Research*, *38*(17), 4273–4283. doi:10.1080/00207540050205082

Delena, D., & Demirkanb, H. (2017). Data, information and analytics as services. *Decision Support Systems*, *55*(1), 359–363. doi:10.1016/j.dss.2012.05.044

Demb, S. R. (2008). A case study of the use of the Records Management Capacity Assessment System (RMCAS) Tool across the London Museum Hub. *Records Management Journal*, *18*(2), 130–139. doi:10.1108/09565690810882986

Department of Health [South Africa] & Council for Scientific and Industrial Research [South Africa]. (2014). *National Health Normative Standards Framework for Interoperability in eHealth in South Africa*. Retrieved 26 March 2017, from http://www.samed.org.za/Filemanager/userfiles/hnsf-complete-version.pdf

Desouza, K. C., & Vanapalli, G. K. (2005). Securing knowledge in organizations: Lessons from the defense and intelligence sectors. *International Journal of Information Management*, *25*(1), 85–98. doi:10.1016/j.ijinfomgt.2004.10.007

DeYoung, R., Lang, W. W., & Nolle, D. L. (2007). How the internet affects output and performance at community banks. *Journal of Banking & Finance*, *31*(4), 1033–1060. doi:10.1016/j.jbankfin.2006.10.003

Dhuieb, M. A., Belkadi, F., Laroche, F., & Bernard, A. (n.d.). *Interoperability framework for supporting information-based assistance in the factory.* The IFIP Working Group WG 5.1 11th International Conference on Product Lifecycle Management, Yokohama, France. Retrieved from: https://hal.archives-ouvertes.fr/hal-01066441/document

Dick, A. S., & Basu, K. (1994). Customer loyalty, toward an integrated conceptual framework. *Journal of the Academy of Marketing Science*, *22*(2), 99–113. doi:10.1177/0092070394222001

Dickey, I. J., & Lewis, W. F. (2012). An overview of digital media and advertising. In I. Management Association (Ed.), E-marketing: Concepts, methodologies, tools, and applications (pp. 31–61). Hershey, PA: IGI Global. doi:10.4018/978-1-4666-1598-4.ch003

Ding, L., Ball, A., Matthews, J., McMahon, C. A., & Patel, M. (2007). Product representation in lightweight formats for product lifecycle management (PLM). *4th International Conference on Digital Enterprise Technology*. Retrieved from: http://www.ukoln.ac.uk/projects/grand-challenge/papers/DET2007_Ding_Paper.pdf

Dishawa, M. T., & Strong, D. M. (1999). Extending the technology acceptance model with task-technology fit constructs. *Information & Management*, *36*(1), 9–21. doi:10.1016/S0378-7206(98)00101-3

Doan, T., & Kennedy, M. L. (2009). Innovation, creativity, and meaning: Leading in the information age. *Journal of Business & Finance Librarianship*, *14*(4), 348–358. doi:10.1080/08963560802424011

Dollar, C., & Ashley, L. J. (2014). *Assessing digital preservation capability using a maturity model process improvement approach.* Retrieved 3 March 2016, from http://static1.squarespace.com/static/52ebbb45e4b06f07f8bb62bd/t/53559340e4b058b6b2212d98/1398117184845/DPCMM+White+Paper_Revised+April+2014.pdf

Dollar, C. M., & Ashley, L. J. (2015). Ensuring Long-term availability, usability and trustworthiness of digital information. *The Sedona Conference Journal*, *16*, 178–199.

Dowell, J., & Long, J. B. (1998). Conception of the cognitive engineering design problem. *Ergonomics*, *41*(2), 126–139. doi:10.1080/001401398187125

Driving New Business Economies with the New Triple Play, IoT, RTC, and OTT. (2016). Available at: https://www.kaaproject.org/wp-content/uploads/2016/07/telecom-whitepaper.pdf

Duflou, J. R., Nguyen, T. H. M., Kruth, J. P., & Cattrysse, D. (2003). An optimization system for automated workpiece layout generation for bent sheet metal parts. *Proceedings of the SHEMET Conference*, 235-244.

Durga, A. (2015). *State of ECM technology: Customer survey results.* Retrieved 26 March 2017, from https://www.youtube.com/watch?v=obn6qFFqeN4

Durga, A. (2016). *Benchmarking with Excel and paper? There's a better way...* Retrieved 26 March 2017, from https://www.realstorygroup.com/Blog/2990-Benchmarking-with-excel-and-paper-Theres-a-better-way

Durga, A. (2012). A maturity model-based approach for DAM implementations. *Journal of Digital Media Management*, *1*(2), 183–194.

Durvasula, S., Lysonshi, S., & Madhavi, A. D. (2011). Beyond service attributes: Do personal values matter? *Journal of Services Marketing*, *25*(1), 33–45. doi:10.1108/08876041111107041

Eastin, M. S., Daugherty, T., & Burns, N. M. (2011). *Handbook of research on digital media and advertising: User generated content consumption*. Hershey, PA: IGI Global. doi:10.4018/978-1-60566-792-8

ECM Maturity Model. (2012). *About*. Retrieved 26 March 2017, from https://web.archive.org/web/20130825194705/http://ecm3.org/about/

ECMA International. (2017). *Standard ECMA-363 Universal 3D File Format*. Retrieved from: http://www.ecma-international.org/publications/standards/Ecma-363.htm

Eick, S. G. (2000). Visualizing multi-dimensional data. *Computer Graphics*, *34*(1), 61–67. doi:10.1145/563788.604454

Einasto, O. (2014). E-service quality criteria in university library: A focus group study. *Procedia: Social and Behavioral Sciences*, *147*, 561–566. doi:10.1016/j.sbspro.2014.07.163

El Kadiri, S., Grabot, B., Thoben, K. D., Hribernik, K., Emmanouilidis, C., von Cieminski, G., & Kiritsis, D. (2016). Current trends on ICT technologies for enterprise information systems. *Computers in Industry*, *79*, 14–33. doi:10.1016/j.compind.2015.06.008

El-Gohary, H., & Eid, R. (2013). *E-marketing in developed and developing countries: Emerging practices*. Hershey, PA: IGI Global. doi:10.4018/978-1-4666-3954-6

Emmanouilidis, C., Jantunen, E. G., & Starr, A. (2011). e-Maintenance update: The road to success for modern industry. *Proceedings of the 24th International Congress on Condition Monitoring and Diagnostics Engineering Management, COMADEM*. Retrieved from: http://www.ipet.gr/~chrisem/Files/E-maintenance_update.pdf

Eng, T. R. (2002). E-health research and evaluation: Challenges and opportunities. *Journal of Health Communication*, *7*(4), 267–272. doi:10.1080/10810730290001747 PMID:12356287

Ericsson Mobility Report. (2016). *Ericsson Mobility Report: On the Pulse of the Network Society*. Available at https://www.ericsson.com/res/docs/2016/ericsson-mobility-report-2016.pdf

Eriksson, K., Kerem, K., & Nilsson, D. (2005). Customer acceptance of internet banking in commerce: Using and extension of the technology acceptance model. *Journal of Internet Banking and Commerce*, *12*(3), 1–26.

Eroğlu, S., & Çakmak, T. (2016). Enterprise information systems within the context of information security: A risk assessment for a health organization in Turkey. *Procedia Computer Science*, *100*, 979–986. doi:10.1016/j.procs.2016.09.262

European Commission. (2003). *E-health.* Retrieved from http://ec.europea.eu/information_society/eeurope/ehealth/conference/2003/doc/min_dec_22_may_03.pdf

European Union Report. (2013). *IoT Privacy, Data Protection, Information Security*. Available at: http://ec.europa.eu/information_society/newsroom/cf/dae/document.cfm?doc_id=1753

Eysenbach, G. (2001). What is e-health? *Journal of Medical Internet Research, 3*(2), e20. doi:10.2196/jmir.3.2.e20 PMID:11720962

Ezzi, S. Z. (2014). A theoretical model for internet banking: Beyond perceived usefulness and ease of use. *Archives of Business Research, 2*(2), 31–46. doi:10.14738/abr.22.184

Fang, Y.-H., Chiu, C.-M., & Wang, E. T. (2011). Understanding customers' satisfaction and repurchase intentions: An integration of IS success model, trust, and justice. *Internet Research, 21*(4), 479–503. doi:10.1108/10662241111158335

Fan, L., Tiejun, W., & Liuyi, W. (2012). Research and implementation on model for high availability of enterprise information system. *IERI Procedia, 3*, 181–185. doi:10.1016/j.ieri.2012.09.030

Fan, S., Lau, R. Y., & Zhao, J. L. (2015). Demystifying big data analytics for business intelligence through the lens of marketing mix. *Big Data Res., 2*(1), 28–32. doi:10.1016/j.bdr.2015.02.006

Faraz, Z., Ali, L., Waheedulhaq, S., Khan, M., & Jaffery, S. H. I. (2015). STEP-NC enabled manufacturing process planning of sheet metal bend parts. *Proceedings of the IRES 8th International Conference.* Retrieved from: http://www.worldresearchlibrary.org/up_proc/pdf/60-144101414113-17.pdf

Farwick, M., Schweda, C. M., Breu, R., & Hanschke, I. (2016). A situational method for semi-automated enterprise architecture documentation. *Software & Systems Modeling, 15*(2), 397–426. doi:10.1007/s10270-014-0407-3

Fassnacht, M., & Koese, I. (2006). Quality of electronic services – conceptualizing and testing a hierarchical model. *Journal of Service Research, 9*(1), 19–37. doi:10.1177/1094670506289531

Faulk, J., & Kotowski, F. J. (2003). *Dctionary of Engineering*. New York: McGraw-Hall.

Ferreiro, S., Arnaiz, A., Sierra, B., & Irigoien, I. (2012). Application of Bayesian networks in prognostics for a new integrated vehicle health management concept. *Expert Systems with Applications, 39*(7), 6402–6418. doi:10.1016/j.eswa.2011.12.027

Finkelstein, A. (1992). *A software process immaturity model*. Retrieved 26 March 2017, from http://www0.cs.ucl.ac.uk/staff/A.Finkelstein/papers/immaturity.pdf

Fishbein, M., & Ajzen, I. (1975). *Belief, Attitude, Intention and Behavior: An Introduction to Theory and Research*. Reading, MA: Addison-Wesley.

Fonseca, J. R. (2014). e-banking culture: A comparison of EU 27 countries and Portuguese case in the EU 27 retail banking context. *Journal of Retailing and Consumer Services, 21*(5), 708–716. doi:10.1016/j.jretconser.2014.05.006

Fornell, C., & Larcker, D. F. (1981). Evaluation Structural Equation Models with Unobservable Variables and Measurement Error. *JMR, Journal of Marketing Research*, *18*(1), 39–50. doi:10.2307/3151312

Forza, C. (2002). Survey research in operations management: A process-based perspective. *International Journal of Operations & Production Management*, *22*(2), 152–194. doi:10.1108/01443570210414310

Franks, P. (2016a). *Enterprise content management and digital preservation online education: A case study.* Paper presented at the European Conference on Knowledge Management, Kidmore End, UK.

Franks, P. (2016b). Integrated ECM solutions: Where records managers, knowledge workers converge. *Information & Management*, *50*(4), 18–22.

Friedman, C., & Wyatt, J. (2000). *Evaluation Methods in Medical Informatics*. New York, NY: Springer-Verlag.

Friess, P., Guillemin, P., Gusmeroli, S., Sundmaeker, H., Bassi, A., Jubert, I. S., . . . Doody, P. (2011). *Internet of Things strategic research roadmap*. Retrieved from: http://www.internet-of-things-research.eu/pdf/IoT_Cluster_Strategic_Research_Agenda_2011.pdf

Fröhlich, A. (2011). 3D Formats in the field of engineering: a comparison. *PROSTEP AG*. Retrieved from: http://isicad.ru/uploads/http___www.prostep.pdf

Fry, B. J. (2004). *Computational information design*. Massachusetts Institute of Technology.

Fusilier, M., & Durlabhji, S. (2005). An exploration of student Internet use in India. *Campus-Wide Information Systems*, *22*(4), 233–246. doi:10.1108/10650740510617539

Gandomi, A., & Haider, M. (2015). Beyond the hype: Big Data Concepts, Methods, and Analytics. *International Journal of Information Management*, *35*(2), 137–144. doi:10.1016/j.ijinfomgt.2014.10.007

Ganguli, S., & Roy, S. K. (2011). Generic technology-based service quality dimensions in banking: Impact on customer satisfaction and loyalty. *International Journal of Bank Marketing*, *29*(2), 168–189. doi:10.1108/02652321111107648

Gargeya, V. B., & Brady, C. (2005). Success and Failure Factor of Adopting SAP in ERP System Implementation. *Business Process Management Journal*, *11*(5), 501–516. doi:10.1108/14637150510619858

Gaur, A. S., & Gaur, S. S. (2006). Statistical methods for practice and research: A guide to data analysis using SPSS. *Sage (Atlanta, Ga.)*.

Gefen, D., & Straub, D. (2003). Managing user trust in b to c e- service. *e-Service Journal*, *2*(2), 7-24.

Gefen, D. (2000). E-commerce: The role of familiarity and trust. *Omega*, *28*(6), 725–737. doi:10.1016/S0305-0483(00)00021-9

Gefen, D., Karahanna, E., & Straub, D. W. (2003). Trust and TAM in online shopping: An integrated model. *Management Information Systems Quarterly*, *27*(1).

Gefen, D., Ridgon, E. E., & Straub, D. W. (2011). Editor's Comments: An Updated and Extension to SEM Guidelines for Administrative and Social Science Research. *Management Information Systems Quarterly*, *35*(2), iii–xiv.

Geisser, S. (1974). A Predictive Approach to the Random Effect Model. *Biometrika*, *61*(1), 101–107. doi:10.1093/biomet/61.1.101

Geisser, S. (1975). The predictive sample reuse method with applications. *Journal of the American Statistical Association*, *70*(350), 320–328. doi:10.1080/01621459.1975.10479865

Gernhardt, B., Miltner, F., Vogel, T., Brocks, H., Hemmje, M., & Wang, L. (2015). A Semantic representation for process-oriented knowledge management based on function block domain models supporting distributed and collaborative production planning. *ASME International Manufacturing Science and Engineering Conference:* Vol. 2. *Materials: Biomanufacturing; Properties, Applications and Systems; Sustainable Manufacturing*. Academic Press. doi:10.1115/MSEC2015-9485

Gerrard, P., & Cunningham, J. B. (2003). The diffusion of internet banking among Singapore consumers. *International Journal of Bank Marketing*, *21*(1), 16–28. doi:10.1108/02652320310457776

Getty, J. M., & Getty, R. L. (2003). Lodging quality index (LQI): Assessing customers' perceptions of quality delivery. *International Journal of Contemporary Hospitality Management*, *15*(2), 94–104. doi:10.1108/09596110310462940

Getty, J. M., & Thompson, K. N. (1994). A procedure for scaling perceptions of lodging quality. *Journal of Hospitality Research*, *18*(2), 75–96.

Gifford, C. (2013). *The MOM Chronicles: ISA-95 Best Practices Book 3.0*. International Society of Automation.

Gikandi, J., & Bloor, C. (2010). Adoption and effectiveness of electronic banking in Kenya. *Electronic Commerce Research and Applications*, *9*(1), 277–282. doi:10.1016/j.elerap.2009.12.003

Gillies, A. (2000). Information support for general practice in the new NHS. *Health Libraries Review*, *17*(2), 91–96. doi:10.1046/j.1365-2532.2000.00229.x

Giordani, G., Floros, C., & Judge, G. (2014). Econometric investigation of internet banking adoption in Greece. *Journal of Economic Studies (Glasgow, Scotland)*, *41*(4), 586–600. doi:10.1108/JES-04-2011-0042

Glasgow, R. E. (2007). E-health evaluation and dissemination research. *American Journal of Preventive Medicine*, *32*(5), S119–S26. doi:10.1016/j.amepre.2007.01.023 PMID:17466816

Godinho, P. I. A., Meiguins, B. S., Meiguins, A. S. G., do Carmo, R. M. C., de Brito Garcia, M., Almeida, L. H., & Lourenco, R. (2007). *PRISMA-A multidimensional information visualization tool using multiple coordinated views.* Paper presented at the Information Visualization, 2007. IV'07. 11th International Conference. doi:10.1109/IV.2007.90

Goh, E., Katuu, S., Miller, A., Nordland, L. P., & Sebina, P. (2009). Module 3: Managing the Creation, use and disposal of electronic records. *Training in Electronic Records Management.* Retrieved 26 March 2017, from http://www.irmt.org/documents/educ_training/term%20modules/IRMT%20TERM%20Module%203.pdf

Gonzalez, G. R., Organero, M. M., & Kloos, C. D. (2008, July). Early Infrastructure of an Internet of Things in Spaces for Learning. In ICALT (pp. 381-383). doi:10.1109/ICALT.2008.210

Gonzalez-Fuentes, M. (2017). The organization of the future and the marketing function: Marketers' competencies in the era of information technology. In R. Batko & A. Szopa (Eds.), *Strategic imperatives and core competencies in the era of robotics and artificial intelligence* (pp. 126–145). Hershey, PA: IGI Global. doi:10.4018/978-1-5225-1656-9.ch007

Goodwin, M. (2006). *The enterprise document management maturity model whitepaper.* London: McLaren Software.

Gosain, S. (2004). Enterprise Information Systems as Objects and Carriers of Institutional Forces: The New Iron Cage. *Journal of the Association for Information Systems*, *5*(4), 151–182.

Grabner–Krauter, S., & Faullant, R. (2008). Consumer acceptance of internet banking: The influence of internet trust. *International Journal of Bank Marketing*, *16*(7), 483–504. doi:10.1108/02652320810913855

Graham, R., & Caverlee, J. (2008). *Exploring feedback models in interactive tagging.* Paper presented at the Web Intelligence and Intelligent Agent Technology, 2008. WI-IAT'08. IEEE/WIC/ACM International Conference on. doi:10.1109/WIIAT.2008.419

Grewal, D., Iyer, G., & Levy, M. (2004). Internet retailing: Enablers, limiters and market consequences. *Journal of Business Research*, *57*(7), 703–7013. doi:10.1016/S0148-2963(02)00348-X

Griffin, A. (2004). Records Management Capacity Assessment System (RMCAS). *Archival Science*, *4*(1-2), 71–97. doi:10.1007/s10502-005-6991-9

Griffith, D. A. (1996). *Some guidelines for specifying the geographic weights matrix contained in spatial statistical models.* Practical Handbook of Spatial Statistics.

Grönroos, C. (1984). A service quality model and its marketing implications. *European Journal of Marketing*, *18*(4), 36–44. doi:10.1108/EUM0000000004784

Groover, M. P. (2010). *Fundamentals of Modern Manufacturing (Materials, Processes, and Systems)* (4th ed.). Danvers, MA: John Wiley & Sons, Inc.

Guédria, W., Naudet, Y., & Chen, D. (2015). Maturity model for enterprise interoperability. *Enterprise Information Systems*, *9*(1), 1–28. doi:10.1080/17517575.2013.805246

Guiry, M., Scott, J. J., & Vequist, D. G. IV. (2013). Experienced and potential medical tourists' service quality expectations. *International Journal of Health Care Quality Assurance*, *26*(5), 433–446. doi:10.1108/IJHCQA-05-2011-0034 PMID:23905303

Gu, J. C., Lee, S. C., & Suh, Y. H. (2009). Determinants of behavioral intention to mobile banking. *Expert Systems with Applications*, *36*(9), 11605–11616. doi:10.1016/j.eswa.2009.03.024

Gupta, A., McDaniel, J. C., & Herath, S. K. (2005). Quality management in service firms: Sustaining structures of total quality service. *Managing Service Quality*, *15*(4), 389–402. doi:10.1108/09604520510606853

Gupta, S. K., Bourne, D. A., Kim, K. H., & Khrishnan, S. S. (1998). Automated process planning for sheet metal bending operations. *Journal of Manufacturing Systems*, *17*(5), 338–360. doi:10.1016/S0278-6125(98)80002-2

Gupta, S. K., & Rajaqopal, D. (2002). Sheet metal bending: Forming part families for generating shared press-brake setups. *Journal of Manufacturing Systems*, *21*(5), 329–349. doi:10.1016/S0278-6125(02)80033-4

Guriting, P., & Ndubisi, N. O. (2006). Borneo online banking: Evaluating customer perceptions and behavioural intention. *Management Research News*, *29*(1/2), 6–15. doi:10.1108/01409170610645402

Gustafson, D., & Wyatt, J. (2004). Evaluation of e-health systems and services. *British Medical Journal*, *328*(7449), 1150. doi:10.1136/bmj.328.7449.1150 PMID:15142895

Gwangwava, N., Nyanga, L., Moyongewenyu, B., & Mpofu, K. (2013) A framework for a Technology Transfer Decision Support System (TTDSS). *SAIIE25 Proceedings*.

Hair, J. F., Hult, G. T. M., Ringle, C. M., & Sarstedt, M. (2013). *A Primer on Partial Least Squares Structural Equation Modeling (PLS-SEM)* (1st ed.). Thousand Oaks, CA: Sage.

Hair, J. F., Ringle, C. M., & Sarstedt, M. (2011). PLS-SEM: Indeed a Silver Bullet. *Journal of Marketing Theory and Practice*, *18*(2), 139–152. doi:10.2753/MTP1069-6679190202

Hair, J. F., Sarstedt, M., Ringle, C. M., & Mena, J. A. (2012). An Assessment of the Use of Partial Least Squares Structural Equation Modeling in Marketing Research. *Journal of the Academy of Marketing Science*, *40*(3), 414–433. doi:10.1007/s11747-011-0261-6

Hajli, N. (2015). *Handbook of research on integrating social media into strategic marketing*. Hershey, PA: IGI Global. doi:10.4018/978-1-4666-8353-2

Hamdaqa, M., & Tahvildari, L. (2012). Cloud computing uncovered: A research landscape. *Advances in Computers*, *86*, 41–84. doi:10.1016/B978-0-12-396535-6.00002-8

Hamed, T., Samshul, S., & Neda, J. (2015). *A Review paper on e-service technologies and concepts. 8th International Conference Interdisciplinary in Engineering*, Romania.

Hamid, A., & Sarmad, A. (2008). Evaluation of e-health services: user's perspective criteria. *Transforming Government: People. Process and Policy*, *2*(4), 243–255.

Hamlet, C., & Strube, M. (2000). Community banks go online. *ABA Banking Journal, 92*(3), 61-65.

Hanafizadeh, P., Keating, B. W., & Khedmatgozar, H. R. (2014). A systematic review of internet banking adoption. *Telematics and Informatics*, *31*(3), 492–510. doi:10.1016/j.tele.2013.04.003

Hao, Y. (2016). *Cloud Computing* (Doctoral thesis). University of Vaasa. Retrieved from: http://www.uva.fi/materiaali/pdf/isbn_978-952-476-715-1.pdf

Harris, K. (2006). *A knowledge management maturity model explains where you're going and how to get there*. Retrieved 26 March 2017, from https://www.gartner.com/doc/489448/knowledge-management-enables-highperformance-workplace

Harrison, T. S., Onyia, O. P., & Tagg, S. K. (2014). Towards a universal model of internet banking adoption: Initial conceptualization. *International Journal of Bank Marketing*, *32*(7), 647–687. doi:10.1108/IJBM-06-2013-0056

Ha, S., & Stoel, L. (2008). Promoting customer-retailer relationship building: Influence of customer trustworthiness of customer loyalty programme marketing. *Journal of Customer Behaviour*, *7*(3), 215–229. doi:10.1362/147539208X369652

Hasim, M. S., & Salman, A. (2010). Factors affecting sustainability of internet usage among youth. *The Electronic Library*, *28*(2), 300–313. doi:10.1108/02640471011033657

Hassan, Q. F. (2011, January). Demystifying Cloud Computing. *The Journal of Defense Software Engineering. Crosstalk*, *2011*, 16–21. Retrieved from http://www.crosstalkonline.org/storage/issue-archives/2011/201101/201101-Hassan.pdf

Hausman, A., & Siekpe, J. (2008). The effect of web interface features on consumer online purchase intentions. *Journal of Business Research, 8*(1), 1-9.

Headley, D. E., & Miller, S. J. (1993). Measuring service quality and its relationship to future consumer behavior. *Journal of Health Care Marketing*, *13*(4), 32–41. PMID:10131732

Henseler, J., Ringle, C. M., & Sinkovics, R. R. (2009). The Use of Partial Least Squares Path Modeling in International Marketing. In R. R. Sinkovics & P. N. Ghauri (Eds.), *Advances in International Marketing 20* (pp. 277–320). Bingley, UK: Emerald. doi:10.1108/S1474-7979(2009)0000020014

Herbst, A., Simons, A., vom Brocke, J., & Derungs, R. (2014). Critical success factors in enterprise content management: Toward a Framework for readiness assessment. In J. vom Brocke & A. Simons (Eds.), *Enterprise content management in information systems research* (pp. 109–124). Berlin: Springer. doi:10.1007/978-3-642-39715-8_7

Herington, C., & Weaven, S. (2009). E-retailing by banks: E-service quality and its importance to customer satisfaction. *European Journal of Marketing*, *43*(9/10), 1220–1231. doi:10.1108/03090560910976456

Hernandez, J. M., & Mazzon, J. A. (2007). Adoption of internet banking: Proposition and implementation of an integrated methodology approach. *International Journal of Bank Marketing*, *25*(2), 72–88. doi:10.1108/02652320710728410

He, W., & Xu, L. (2015). A state-of-the-art survey of cloud manufacturing. *International Journal of Computer Integrated Manufacturing*, *28*(3), 239–250. doi:10.1080/0951192X.2013.874595

Hickson, M. (2008). *Research handbook for health care professionals*. Chichester, UK: Wiley-Blackwell.

Ho, C.-T. B., & Lin, W.-C. (2010). Measuring the service quality of internet banking: Scale development and validation. *European Business Review*, *22*(1), 5–24. doi:10.1108/09555341011008981

Hoecht, A. (2004). Control in collaborative research and technology development: A case study in the chemical industry. *Journal of Managerial Psychology*, *19*(3), 218–234. doi:10.1108/02683940410527720

Ho, L.-A., Kuo, T.-H., & Lin, B. (2012). The mediating effect of website quality on Internet searching behavior. *Computers in Human Behavior*, *28*(3), 840–848. doi:10.1016/j.chb.2011.11.024

Holsapplea, C., Lee-Postb, A., & Pakath, R. (2014). A unified foundation for business analytics. *Decision Support Systems*, *64*, 130–141. doi:10.1016/j.dss.2014.05.013

Howcroft, B., Hamilton, R., & Hewer, P. (2002). Consumer attitude and the usage and adoption of home-based banking in the United Kingdom. *International Journal of Bank Marketing*, *20*(3), 111–121. doi:10.1108/02652320210424205

HR-Focus. (2000). Online recruiting: what works, what doesn't. *HR Focus*, *00*(3), 11-15.

Hug, M. (2008). *Will cloud-based multi-enterprise information systems replace extranets?* Retrieved from https://www.infoq.com/articles/will-meis-replace-extranets

Hullavarad, S., O'Hare, R., & Roy, A. K. (2015). Enterprise content management solutions—Roadmap strategy and implementation challenges. *International Journal of Information Management*, *35*(2), 260–265. doi:10.1016/j.ijinfomgt.2014.12.008

Hu, P. J., Chau, P. Y. K., Sheng, O. R. L., & Tam, K. Y. (1999). Examining the technology acceptance model using physician acceptance of telemedicine technology. *Journal of Management Information Systems*, *16*(2), 91–112. doi:10.1080/07421222.1999.11518247

Hurley, R., Tahir Jan, M., & Abdullah, K. (2014). The impact of technology CSFs on customer satisfaction and the role of trust. *International Journal of Bank Marketing*, *32*(5), 429–447. doi:10.1108/IJBM-11-2013-0139

IDC Manufacturing Insight. (2015). *Transforming Manufacturing with the Internet of Things*. Available at https://www.cognizant.com/InsightsWhitepapers/transforming-manufacturing-with-the-Internet-of-Things.pdf

IDC White Paper. (2013). *The 3rd Platform: Enable Digital Transformation*. Available at: http://achievabledigitaltransformation.com/TCS-DSS/tcs-white-paper_244515.pdf

IEC 62264-3. (2007). *Enterprise-Control System Integration – Part 3: Activity Models of Manufacturing Operations Management*. IEC.

Igbaria, M., Parasuraman, S., & Baroudi, J. J. (1996). A motivational model of microcomputer usage. *Journal of Management Information Systems*, *13*(1), 127–144. doi:10.1080/07421222.1996.11518115

Inder Jit Singh, M., Vinod, K., Hanuv, M., & Uma, K. (2015). *Scope of City E-Government Initiative*. Sprott School of Business, Carleton University.

Intel® Internet of Things Solutions Alliance for Premier Members (2016). Available at: https://www.intel.com/content/dam/m/enrollment/isalegalpdfs/us/en/IOTPremier_Lug_093015.pdf

Inui, M., Kinosada, A., Suzuki, H., Kimura, F., & Sata, T. (1998). Automated process planning for sheet metal parts with bending simulation. *Intellingent and Integrated Manufacturing Analysis and Synthesis*, *25*, 245–258.

IoT Semantic Interoperability: Research Challenges, Best Practices, Solutions and Next Steps (2015). Available at http://www.internet-of-things-research.eu/pdf/IERC_Position_Paper_IoT_Semantic_Interoperability_Final.pdf

IoT@Work Consortium. (2013). *D1.2 – Final framework architecture specification*. Retrieved from https://www.iot-at-work.eu/data/D1.3 IoT@Work Architecture final v1.0-submitted.pdf

ITU Internet Reports. (2005, Nov.). *The Internet of Things*. Available at https://www.itu.int/net/wsis/tunis/newsroom/stats/The-Internet-of-Things-2005.pdf

Jain, N., Ahuja, V., & Medury, Y. (2015). Digital marketing optimization. In M. Khosrow-Pour (Ed.), *Strategic e-commerce systems and tools for competing in the digital marketplace* (pp. 162–170). Hershey, PA: IGI Global. doi:10.4018/978-1-4666-8133-0.ch008

Janouri, F., & Gharbi, J. E. (2008). *Affrontement entre le risque perçu et le capital marque dans la formation de la confiance dans un site marchand*. Available at: www.esscpeap. net/conferences/marketing/2008-cp/Materiali/Paper/Fr/Janouri-Gharbi.pdf

Jantunen, E., Emmanouilidis, C., Arnaiz, A., & Gilabert, E. (2011). *e-Maintenance: trends, challenges and opportunities for modern industry*. 18th International Federation of Automatic Control (IFAC) World Congress, Milano, Italy. Retrieved from: http://www.ipet.gr/~chrisem/Files/e-Maintenance-Trends.pdf

Jaruwachirathanakul, B., & Fink, D. (2005). Internet banking adoption strategies for a developing country: The case of Thailand. *Internet Research*, *15*(3), 295–311. doi:10.1108/10662240510602708

Järvinen, R. A. (2014). Consumer trust in banking relationships in Europe. *International Journal of Bank Marketing*, *32*(6), 551–566. doi:10.1108/IJBM-08-2013-0086

Javalgi, R. G., Martin, C. L., & Todd, P. R. (2004). The export of e-services in the age of technology transformation: Challenges and implications for international service providers. *Journal of Services Marketing*, *18*(7), 560–573. doi:10.1108/08876040410561884

Jiang, J. J., Klein, G., & Crampton, S. M. (2000). A note on SERVQUAL, reliability and validity in information system service quality measurement. *Decision Sciences*, *31*(3), 725–734. doi:10.1111/j.1540-5915.2000.tb00940.x

Kahraman, C., Yazıcı, İ., & Karaşan, A. (2016). A literature survey on the usage of fuzzy MCDM methods for digital marketing. In A. Kumar & M. Dash (Eds.), *Fuzzy optimization and multi-criteria decision making in digital marketing* (pp. 1–19). Hershey, PA: IGI Global. doi:10.4018/978-1-4666-8808-7.ch001

Kaltenrieder, P., D'Onofrio, S., & Portmann, E. (2016). Applying the fuzzy analytical network process in digital marketing. In A. Kumar & M. Dash (Eds.), *Fuzzy optimization and multi-criteria decision making in digital marketing* (pp. 202–232). Hershey, PA: IGI Global. doi:10.4018/978-1-4666-8808-7.ch010

Kana, M., & Karni, R. (2004). A capability maturity model for knowledge-based decision-making. *Information, Knowledge, Systems Management*, *4*(4), 225–252.

Kanna, T. R., & Shunmugam, M. S. (2009). Processing of 3D sheet metal components in STEP AP-203 format. Part I: Feature recognition system. *International Journal of Production Research*, *47*(4), 941–964. doi:10.1080/00207540701510055

Karahanna, E., Straub, D. W., & Chervany, N. L. (1999). Information technology adoption across time: A cross-sectional comparison of pre-adoption and post-adoption beliefs. *Management Information Systems Quarterly*, *23*(2), 183–213. doi:10.2307/249751

Karim, R., Birk, W., & Larsson-Kråik, P. (2015). *Cloud-based e-maintenance solutions for condition-based maintenance of wheels in heavy haul operation. In 11th International Heavy Haul Conference*. Perth, Australia: International Heavy Haul Association. Retrieved from http://ltu.diva-portal.org/smash/get/diva2:1005837/FULLTEXT01.pdf

Karjaluoto, H., Mattila, M., & Pento, T. (2002). Factors underlying attitude formation towards online banking in Finland. *International Journal of Bank Marketing*, *20*(6), 261–272. doi:10.1108/02652320210446724

Kasemsap, K. (2018c). Customer lifetime value. In M. Khosrow-Pour (Ed.), Encyclopedia of information science and technology (4th ed.; pp. 1584–1593). Hershey, PA: IGI Global. doi:10.4018/978-1-5225-2255-3.ch137

Kasemsap, K. (2014a). The role of brand loyalty on CRM performance: An innovative framework for smart manufacturing. In Z. Luo (Ed.), *Smart manufacturing innovation and transformation: Interconnection and intelligence* (pp. 252–284). Hershey, PA: IGI Global. doi:10.4018/978-1-4666-5836-3.ch010

Kasemsap, K. (2014b). The role of social media in the knowledge-based organizations. In I. Lee (Ed.), *Integrating social media into business practice, applications, management, and models* (pp. 254–275). Hershey, PA: IGI Global. doi:10.4018/978-1-4666-6182-0.ch013

Kasemsap, K. (2015a). Implementing enterprise resource planning. In M. Khosrow-Pour (Ed.), *Encyclopedia of information science and technology* (3rd ed.; pp. 798–807). Hershey, PA: IGI Global. doi:10.4018/978-1-4666-5888-2.ch076

Kasemsap, K. (2015b). The role of information system within enterprise architecture and their impact on business performance. In M. Wadhwa & A. Harper (Eds.), *Technology, innovation, and enterprise transformation* (pp. 262–284). Hershey, PA: IGI Global. doi:10.4018/978-1-4666-6473-9.ch012

Kasemsap, K. (2015c). The role of electronic commerce in the global business environments. In F. Cipolla-Ficarra (Ed.), *Handbook of research on interactive information quality in expanding social network communications* (pp. 304–324). Hershey, PA: IGI Global. doi:10.4018/978-1-4666-7377-9.ch019

Kasemsap, K. (2015d). The role of customer relationship management in the global business environments. In T. Tsiakis (Ed.), *Trends and innovations in marketing information systems* (pp. 130–156). Hershey, PA: IGI Global. doi:10.4018/978-1-4666-8459-1.ch007

Kasemsap, K. (2015e). The role of social media in international advertising. In N. Taşkıran & R. Yılmaz (Eds.), *Handbook of research on effective advertising strategies in the social media age* (pp. 171–196). Hershey, PA: IGI Global. doi:10.4018/978-1-4666-8125-5.ch010

Kasemsap, K. (2015f). The role of business analytics in performance management. In M. Tavana & K. Puranam (Eds.), *Handbook of research on organizational transformations through big data analytics* (pp. 126–145). Hershey, PA: IGI Global. doi:10.4018/978-1-4666-7272-7.ch010

Kasemsap, K. (2016a). Implementing electronic commerce in global marketing. In I. Lee (Ed.), *Encyclopedia of e-commerce development, implementation, and management* (pp. 591–602). Hershey, PA: IGI Global. doi:10.4018/978-1-4666-9787-4.ch043

Kasemsap, K. (2016b). Advocating electronic business and electronic commerce in the global marketplace. In S. Dixit & A. Sinha (Eds.), *E-retailing challenges and opportunities in the global marketplace* (pp. 1–24). Hershey, PA: IGI Global. doi:10.4018/978-1-4666-9921-2.ch001

Kasemsap, K. (2016c). Role of social media in brand promotion: An international marketing perspective. In A. Singh & P. Duhan (Eds.), *Managing public relations and brand image through social media* (pp. 62–88). Hershey, PA: IGI Global. doi:10.4018/978-1-5225-0332-3.ch005

Kasemsap, K. (2016d). The roles of social media marketing and brand management in global marketing. In W. Ozuem & G. Bowen (Eds.), *Competitive social media marketing strategies* (pp. 173–200). Hershey, PA: IGI Global. doi:10.4018/978-1-4666-9776-8.ch009

Kasemsap, K. (2017a). Mastering intelligent decision support systems in enterprise information management. In G. Sreedhar (Ed.), *Web data mining and the development of knowledge-based decision support systems* (pp. 35–56). Hershey, PA: IGI Global. doi:10.4018/978-1-5225-1877-8.ch004

Kasemsap, K. (2017b). The importance of customer satisfaction and customer loyalty in the service sector. In S. Rozenes & Y. Cohen (Eds.), *Handbook of research on strategic alliances and value co-creation in the service industry* (pp. 59–82). Hershey, PA: IGI Global. doi:10.4018/978-1-5225-2084-9.ch004

Kasemsap, K. (2017c). Mastering social media in the modern business world. In N. Rao (Ed.), *Social media listening and monitoring for business applications* (pp. 18–44). Hershey, PA: IGI Global. doi:10.4018/978-1-5225-0846-5.ch002

Kasemsap, K. (2017d). Professional and business applications of social media platforms. In V. Benson, R. Tuninga, & G. Saridakis (Eds.), *Analyzing the strategic role of social networking in firm growth and productivity* (pp. 427–450). Hershey, PA: IGI Global. doi:10.4018/978-1-5225-0559-4.ch021

Kasemsap, K. (2017e). Digital storytelling and digital literacy: Advanced issues and prospects. In D. Loveless, P. Sullivan, K. Dredger, & J. Burns (Eds.), *Deconstructing the education-industrial complex in the digital age* (pp. 151–171). Hershey, PA: IGI Global. doi:10.4018/978-1-5225-2101-3.ch009

Kasemsap, K. (2017f). Mastering electronic procurement, green public procurement, and public procurement for innovation. In R. Shakya (Ed.), *Digital governance and e-government principles applied to public procurement* (pp. 29–55). Hershey, PA: IGI Global. doi:10.4018/978-1-5225-2203-4.ch002

Kasemsap, K. (2018a). Cloud computing, green computing, and green ICT. In K. Munir (Ed.), *Cloud computing technologies for green enterprises* (pp. 28–50). Hershey, PA: IGI Global. doi:10.4018/978-1-5225-3038-1.ch002

Kasemsap, K. (2018b). Facilitating customer relationship management in modern business. In M. Khosrow-Pour (Ed.), *Encyclopedia of information science and technology* (4th ed.; pp. 1594–1604). Hershey, PA: IGI Global. doi:10.4018/978-1-5225-2255-3.ch138

Kashif, M., Wan Shukran, S. S., Rehman, M. A., Sarifuddin, S., Estelami, H., & Heinonen, K. (2015). Customer satisfaction and loyalty in Malaysian Islamic banks: A PAKSERV investigation. *International Journal of Bank Marketing*, *33*(1), 23–40. doi:10.1108/IJBM-08-2013-0084

Kassim, N., & Abdullah, N. A. (2010). The effect of perceived service quality dimensions on customer satisfaction, trust, and loyalty in e-commerce settings: A cross cultural analysis. *Asia Pacific Journal of Marketing and Logistics*, *22*(3), 351–371. doi:10.1108/13555851011062269

Kastenhofer, J., & Katuu, S. (2016). Declassification: A clouded environment. *Archives and Records: The Journal of the Archives and Records Association*, *37*(2), 1–27. doi:10.1080/23257962.2016.1194814

Katuu, S. (2002b). Sustainable management of electronic records: The challenge for developing countries. *World Forum Proceedings of the International Research Foundation for Development*. Retrieved 26 March 2017, from https://www.researchgate.net/publication/315381902_Sustainable_management_of_electronic_records_-_the_challenge_for_the_public_sector_in_developing_countries

Katuu, S. (2012b). *Enterprise content management and digital curation applications – Maturity model connections*. Paper presented at the Memory of the World in the Digital Age: Digitization and Preservation, Vancouver. Retrieved from http://www.unesco.org/new/fileadmin/MULTIMEDIA/HQ/CI/CI/pdf/mow/VC_Katuu_28_D_1130.pdf

Katuu, S. (2013). The utility of maturity models – The ECM maturity model within a South African context. *iPRES 2013 Workshop: Digital Preservation Capabilities - How to Assess and Improve Capabilities in Digital Preservation.* Retrieved 3 March 2016, from http://benchmarkdigitalpreservation.files.wordpress.com/2013/09/caiw2013katuu.pdf

Katuu, S. (2002a). Are we information providers or the information police? The uneasy marriage between access and security. In G. E. Gorman (Ed.), *International yearbook of library and information management 2002/2003* (pp. 361–378). London, UK: Facet Publishing.

Katuu, S. (2012a). Enterprise content management – An overview of phases, standards and best practices guidelines. *Bilgi Dünyası*, *13*(2), 457–476.

Katuu, S. (2016a). Assessing the functionality of the enterprise content management maturity model. *Records Management Journal*, *26*(2), 218–238. doi:10.1108/RMJ-08-2015-0030

Katuu, S. (2016b). Managing digital records in a global environment – A review of the landscape of international standards and best practice guidelines. *The Electronic Library*, *34*(5), 869–894. doi:10.1108/EL-04-2015-0064

Katuu, S. (2016c). Transforming South Africa's health sector - The eHealth strategy, the implementation of electronic document and records management systems (EDRMS) and the utility of maturity models. *Journal of Science and Technology Policy Management*, *7*(3), 330–345. doi:10.1108/JSTPM-02-2016-0001

Katuu, S. (2017). Health information systems, eHealth strategy and the management of health records: The quest to transform South Africa's public health sector. In K. H. Moahi, K. J. Bwalya, & P. M. Sebina (Eds.), *Health information systems and the advancement of medical practice in developing countries* (pp. 237–261). Hershey, PA: IGI Global. doi:10.4018/978-1-5225-2262-1.ch014

Kaura, V., Prasad, C. S. D., & Sharma, S. (2015). Service quality, service convenience, price and fairness, customer loyalty, and the mediating role of customer satisfaction. *International Journal of Bank Marketing*, *33*(4), 404–422. doi:10.1108/IJBM-04-2014-0048

Keim, D. A., Hao, M. C., Ladisch, J., Hsu, M., & Dayal, U. (2001). *Pixel bar charts: A new technique for visualizing large multi-attribute data sets without aggregation.* Paper presented at the IEEE Symposium on Information Visualization, 2001. INFOVIS 2001. doi:10.1109/INFVIS.2001.963288

Keim, D. A. (2000). Designing pixel-oriented visualization techniques: Theory and applications. *IEEE Transactions on Visualization and Computer Graphics*, *6*(1), 59–78. doi:10.1109/2945.841121

Keim, D. A., Hao, M. C., Dayal, U., & Hsu, M. (2002). Pixel bar charts: A visualization technique for very large multi-attribute data sets. *Information Visualization*, *1*(1), 20–34. doi:10.1057/palgrave.ivs.9500003

Kerzner, H. R. (2011). *Using the project management maturity model: Strategic planning for project management*. Hoboken, NJ: Wiley.

Ketter, W., Banjanin, M., Guikers, R., & Kayser, A. (2009). *Introducing an agile method for enterprise mash-up component development.* Paper presented at the Commerce and Enterprise Computing, 2009. CEC'09. IEEE Conference on. doi:10.1109/CEC.2009.40

Khalil, N., Sutanonpaiboon, J., & Mastor, N. H. (2010). Malay, Chinese, and internet banking. *Chinese Management Studies*, *4*(2), 141–153. doi:10.1108/17506141011053069

Khoshnevis, B., Sormaz, D., & Park, J. Y. (1999). An integrated process planning system using feature reasoning and space search-based optimization. *IIE Transactions*, *31*(7), 597–616. doi:10.1080/07408179908969862

Khosrow-Pour, M. (2015). *Strategic e-commerce systems and tools for competing in the digital marketplace*. Hershey, PA: IGI Global. doi:10.4018/978-1-4666-8133-0

Kidd, P. T. (2011). Enterprise information systems: Aligning and integrating strategy, technology, organization and people. In I. Management Association (Ed.), Enterprise information systems: Concepts, methodologies, tools and applications (pp. 314–328). Hershey, PA: IGI Global. doi:10.4018/978-1-61692-852-0.ch201

Kilbourne, W. E., Duffy, J. A., Duffy, M., & Giarchi, G. (2004). The applicability of SERVQUAL in cross-national measurements of health-care quality. *Journal of Services Marketing*, *18*(6/7), 524–533. doi:10.1108/08876040410561857

Kim, M. S., & Ahn, J. H. (2006). Comparison of trust sources of an online market-maker in the e-marketplace: Buyer's and seller's perspectives. *Journal of Computer Information Systems*, *47*(1), 84–94.

Kim, M., Kim, J. H., & Lennon, Sh. J. (2006). Online service attributes available on apparel retail web sites: An E-S-Qual approach. *Managing Service Quality*, *16*(1), 51–77. doi:10.1108/09604520610639964

Knutson, B., Stevens, P., Wullaert, C., & Patton, M. (1991). Lodgserv: A service quality index for the lodging industry. *Hospitality Research Journal*, *14*(7), 277–284.

Koç, M., & Lee, J. (2003). E-Manufacturing-fundamentals, requirements and expected impacts. *International Journal of Advanced Manufacturing Systems*, *6*(1), 29–46.

Koehler-Kruener, H., Chin, K., & Hobert, K. A. (2015). *Magic quadrant for enterprise content management.* Retrieved 26 March 2017, from http://www.project-consult.de/files/Gartner_MQ_ECM_Enterprise_Content_Management_2015.pdf

Komninos, I. (2002). *Product life cycle management, Urban and regional Innovation research unit, faculty of engineering*. Aristotle University of Thessaloniki.

Komninos, I. (2002). *Product life cycle management.* Urban and regional Innovation research unit., Faculty of Engineering: Aristotle University of Thessaloniki. Retrieved from: https://www.urenio.org/tools/en/Product_Life_Cycle_Management.pdf

Kosala, R., & Blockeel, H. (2000). Web mining research: A survey. *ACM Sigkdd Explorations Newsletter*, *2*(1), 1–15. doi:10.1145/360402.360406

Kshetri, N. (2013). Cybercrime and cyber-security issues associated with China: Some economic and institutional considerations. *Electronic Commerce Research*, *13*(1), 41–69. doi:10.1007/s10660-013-9105-4

Ladhari, R. (2010). Developing e-service quality scales: A literature review. *Journal of Retailing and Consumer Services*, *17*(6), 464–477. doi:10.1016/j.jretconser.2010.06.003

Laeeq, K., & Shamsi, J. A. (2015). *A Study of Security Issues, Vulnerabilities and Challenges in Internet of Things*. Securing Cyber-Physical Systems. doi:10.1201/b19311-9

Laforet, S., & Li, X. (2005). Consumers: Attitudes towards online and mobile banking in China. *International Journal of Bank Marketing*, *23*(5), 362–380. doi:10.1108/02652320510629250

Lahuerta-Otero, E., & Cordero-Gutiérrez, R. (2016). How to promote teaching centers by using online advertising: A case study. In I. Lee (Ed.), *Encyclopedia of e-commerce development, implementation, and management* (pp. 262–271). Hershey, PA: IGI Global. doi:10.4018/978-1-4666-9787-4.ch020

Laksamana, P., Wong, D., Kingshott, R. P., & Muchtar, F. (2013). The role of interaction quality and switching costs in premium banking services. *Marketing Intelligence & Planning*, *31*(3), 229–249. doi:10.1108/02634501311324591

Lam, S. K. S. (1997). SERVQUAL: A tool for measuring patients' opinions of hospital service quality in Hong Kong. *Total Quality Management*, *8*(4), 145–152. doi:10.1080/0954412979587

Lam, T. K. P. (2002). Making sense of SERVQUAL's dimensions to the Chinese customers in Macau. *Journal of Market-Focused Management*, *5*(10), 43–58. doi:10.1023/A:1012575412058

Lapointe, L., Lamothe, L., & Fortin, J. P. (2002). The dynamics of IT adoption in a major change process in healthcare delivery. *Proceedings of the 35th HICSS Conference*. doi:10.1109/HICSS.2002.994117

Las Vegas. (2016). *Wi-Fi Alliance® introduces low power, long range Wi-Fi HaLow™ | Wi-Fi Alliance*. Available at: http://www.wi-fi.org/news-events/newsroom/wi-fi-alliance-introduces-low-power-long-range-wi-fi-halow

Lee, I. (2014). *Integrating social media into business practice, applications, management, and models*. Hershey, PA: IGI Global. doi:10.4018/978-1-4666-6182-0

Lee, I. (2016). *Encyclopedia of e-commerce development, implementation, and management*. Hershey, PA: IGI Global. doi:10.4018/978-1-4666-9787-4

Lemieux, V. (2004). *Managing risks for records and information*. Lenexa, KS: ARMA International.

Lenhart, A., Horrigan, J., Rainie, L., Allen, K., Boyce, A., Madden, M., & O'Grady, E. (2003). *The Ever-shifting Internet Population. A New Look at Internet Access and the Digital Divide*. Washington, DC: Pew Internet & American Life Project.

Levrat, E., & Lung, B. (2007). *TELMA: A full e-maintenance platform*. Second World Congress on Engineering Asset Management and the Fourth International Conference on Condition Monitoring, WCEAM/CM, Harrogate, UK. Retrieved from: http://dynamite.vtt.fi/conference_pres/telma_wceam_el_bi.pdf

Levy, S. (2014). Does usage level of online services matter to customers' bank loyalty? *Journal of Services Marketing*, *28*(4), 292–299. doi:10.1108/JSM-09-2012-0162

Lezoche, M., Yahia, E., Aubry, A., Panetto, H., & Zdravković, M. (2012). Conceptualising and structuring semantics in cooperative enterprise information systems models. *Computers in Industry*, *63*(8), 775–787. doi:10.1016/j.compind.2012.08.006

Liao, Z., & Cheung, M. T. (2002). Internet-based e-banking and consumer attitudes: An empirical study. *Information & Management*, *39*(4), 283–295. doi:10.1016/S0378-7206(01)00097-0

Lichao, Z., Yi, Z., Qiang, Z., & Fafu, H. (2011). Robust sheet metal bend sequencing method based on A-star algorithm. *IEEE International Conference on Computer Science and Automation Engineering*, 711-715. doi:10.1109/CSAE.2011.5952603

Li, E. Y., Loh, S., Evans, C., & Lorenzi, F. (2013). *Organizations and social networking: Utilizing social media to engage consumers*. Hershey, PA: IGI Global. doi:10.4018/978-1-4666-4026-9

Li, H., Liu, Y., & Suomi, R. (2009). Measurement of e-service quality: An empirical study on online travel service. *Proceedings of 17th European Conference on Information Systems*.

Li, H., & Suomi, R. (2009). A Proposed Scale for Measuring E-Service Quality. *International Journal of u- and e-service. Science and Technology*, *2*(1), 1–10.

Li-hua, Y. (2012). Customer satisfaction antecedents within service recovery context: Evidences from 'Big 4' banks in China. *Nankai Business Review International*, *3*(3), 284–301. doi:10.1108/20408741211264585

Lim, N. (2003). Consumers' perceived risk: Sources versus consequences. *Electronic Commerce Research and Applications, 2*(3), 216–228. doi:10.1016/S1567-4223(03)00025-5

Lin, C. H., Shih, H. Y., & Sher, P. J. (2007). Integrating technology readiness into technology acceptance: The TRAM model. *Psychology and Marketing, 24*(7), 641–657. doi:10.1002/mar.20177

Lin, H. F. (2011). An empirical investigation of mobile banking adoption: The effect of innovation attributes and knowledge-based trust. *International Journal of Information Management, 31*(3), 252–260. doi:10.1016/j.ijinfomgt.2010.07.006

Lin, J., & Lu, H. (2000). Towards an understanding of the behavioural intention to use a web site. *International Journal of Information Management, 20*(1), 197–208.

Link Labs. (2016). *The Complete List of Wireless IoT Network Protocols.* Available at: http://www.link-labs.com/complete-list-iot-network-protocols

Lipsey, D., & Regli, T. (2012). The DAM maturity model: An interview with David Lipsey and Theresa Regli. *Journal of Digital Media Management, 1*(2), 202–205.

Liu, N., & Li, X. (2012). A resource virtualization mechanism for cloud manufacturing systems. *Proceedings of the 4th International IFIP Working Conference, 122,* 46-59. Retrieved from: https://link.springer.com/content/pdf/10.1007%2F978-3-642-33068-1_6.pdf

Liu, R. Y. L. (2002). *Capability maturity model integration: Origins and applications* (Master's thesis). California State University, Long Beach, CA.

Loiacono, E. T., Watson, R. T., & Goodhue, D. L. (2002). Webqual: A web site quality instrument. *Proceedings of the AMA Winter Marketing Educators' Conference.*

LOPEZ Research. (2014). *Part 2. of "The IoT Series", "Building Smarter Manufacturing with The Internet of Things (IoT)".* Available at: http://cdn.iotwf.com/resources/6/iot_in_manufacturing_january.pdf

Louw, D. A., Van Ede, D. M., & Louw, A. E. (Eds.). (1998). *Human development* (2nd ed.). Cape Town, South Africa: Kagiso Tertiary.

Lovelock, C. H. (1984). *Services Marketing.* Englewood Cliffs, NJ: Prentice Hall, Inc.

Lu, Y., & Xu, X. (2015). Process and production planning in a cloud manufacturing environment. *ASME International Manufacturing Science and Engineering Conference:* Vol. 2. *Materials: Biomanufacturing, Properties, Applications and Systems, Sustainable Manufacturing.* Academic Press. doi:10.1115/MSEC2015-9382

Lu, Y., Morris, K. C., & Frechette, S. (2016). Current Standards Landscape for Smart Manufacturing Systems. *NISTIR 8017.* Available at: 10.6028/NIST.IR.8107

Luarn, P., & Lin, H. H. (2005). Toward an understanding of the behavioral intention to use mobile banking. *Computers in Human Behavior, 21*(6), 873–891. doi:10.1016/j.chb.2004.03.003

Luk, S. T. K., & Layton, R. (2002). Perception gaps in customer expectations: Managers versus service providers and customers. *Service Industries Journal*, *22*(2), 109–128. doi:10.1080/714005073

Lundgren, M., Hedlind, M., & Kjellberg, T. (2014). Model-based interactive learning of process planning. *6th Swedish Production Symposium*, Gothenburg, Sweden.

Maduku, D. K. (2013). Predicting retail banking consumers' attitude towards internet banking services in South Africa. *Southern African Business Review*, *17*(3).

Mahesh Kumar, S., & Kunwar Singh, V. (2016). E-health for Rural Areas of Uttarakhand under Governance Service Delivery Model. *IEEE 1st International Conference on Recent Advances in Information Technology*.

Mahmud & Sattar. (2016). Deployment of Contextual E-healthcare System: A prospective e-service based on context aware conceptual framework and ICTization framework model. *IEEE 11th Conference on Industrial Electronics and Applications (ICIEA)*.

Manna, A., Sengupta, A., & Mazumdar, C. (2016). A survey of trust models for enterprise information systems. *Procedia Computer Science*, *85*, 527–534. doi:10.1016/j.procs.2016.05.212

Mantzana, V., & Themistocleous, M. (2006). A method for the identification of actors involved in the adoption of innovations in healthcare organizations. *Proceedings of the 39th Hawaii International Conference on System Sciences*.

March, S. T., & Storey, V. C. (2008). Design science in the information systems discipline: An introduction to the special issue on design science research. *MIS Quartely*, *32*(4), 725–730.

Marques, R. P. (2017). Continuous assurance and business compliance in enterprise information systems. In M. Tavana (Ed.), *Enterprise information systems and the digitalization of business functions* (pp. 99–119). Hershey, PA: IGI Global. doi:10.4018/978-1-5225-2382-6.ch005

Marshall, S. (2005). Determination of New Zealand tertiary institution e-learning capability: An application of an e-learning maturity model. *Journal of Open, Flexible and Distance Learning*, *9*(1), 58–63.

Martinez, N. (2013). *Localisation workflows: The impact of process well-handledness on automation* (Master's thesis). University of Limerick, Limerick, Ireland.

Masatsugu, T. (2015). Analysis of the Long-run Effect of e-Health Intervention on Chronic Diseases: A DID-PSM Approach. *17th International Conference on E-health Networking, Applications& Services.*

Maseh, E., & Katuu, S. (2017). The Kenyan judiciary's open government initiative: Prospects and challenges. *Journal of Science and Technology Policy Management*, *8*(1), 76–94. doi:10.1108/JSTPM-04-2016-0007

Mashaqbeh. (2016). Computers and e-Health: Roles and New Applications. *IEEE Transactions*.

Mathew, A., Zhang, L., Zhang, S., & Ma, L. (2006). A Review of the MIMOSA OSA-EAI database for condition monitoring systems. In J. Mathew, J. Kennedy, L. Ma, A. Tan, & D. Anderson (Eds.), *Engineering Asset Management*. London: Springer. doi:10.1007/978-1-84628-814-2_88

Mauri, A. G., Minazzi, R., & Muccio, S. (2013). A review of literature on the gaps model on service quality: a 3-decades period: 1985–2013. *International Business Research*, *6*(12), 134. doi:10.5539/ibr.v6n12p134

McFarland, D., & Hamilton, D. (2004). Adding contextual specificity to the technology acceptance model. *Computers in Human Behavior*, *22*(3), 427–447. doi:10.1016/j.chb.2004.09.009

McGraw, G., & Fick, N. (2011). Separating threat from the hype: What Washington needs to know about cyber security. In K. Lord & T. Sharp (Eds.), American's cyber future: Security and prosperity in the information age volumes I and II (pp. 41-53). Washington, DC: Center for a New American Security.

McKechnie, S., Winklhofer, H., & Ennew, C. (2006). Applying the technology acceptance model to the online retailing of financial services. *International Journal of Retail & Distribution Management*, *34*(4/5), 388–410. doi:10.1108/09590550610660297

Mcknight, D. H., Choudhury, V., & Kacmar, C. (2002a). Developing and validating trust measures for ecommerce: An integrative typology. *Information Systems Research*, *13*(3), 334–359. doi:10.1287/isre.13.3.334.81

McKnight, D., Choudhury, V., & Kamcmar, C. (2002b). The impact of initial consumer trust on intentions to transact with a web site: A trust building model. *The Journal of Strategic Information Systems*, *11*(1), 297–323. doi:10.1016/S0963-8687(02)00020-3

McNeish, J. (2015). Consumer trust and distrust: Retaining paper bills in online banking. *International Journal of Bank Marketing*, *33*(1), 5–22. doi:10.1108/IJBM-08-2013-0088

McRoberts, H. A., & Sloan, B. C. (1998). Financial management capability model. *International Journal on Government Auditing*, *25*(3), 8–11.

Mekonnen, A. (2016). Digital marketing strategy for affinity marketing: Utilising the new marketing arena. In W. Ozuem & G. Bowen (Eds.), *Competitive social media marketing strategies* (pp. 1–19). Hershey, PA: IGI Global. doi:10.4018/978-1-4666-9776-8.ch001

Melander-Wikman, A., Jansson, M., & Gard, G. (2005). The mobihealth usability evaluation questionnaire. *E-health. International Journal (Toronto, Ont.)*, *2*(1), 9–14.

Mell, P., & Grance, T. (2011). *The NIST definition of Cloud Computing*. National Institute of Standards and Technology, NIST Special Publication 800-145. Retrieved from: http://nvlpubs.nist.gov/nistpubs/Legacy/SP/nistspecialpublication800-145.pdf

Mell, P., & Grance, T. (2011). *The NIST Definition of Cloud Computing*. National Institute of Standards and Technology. doi:10.6028/NIST.SP.800-145

Mels, G., Christo, B., & Deon, N. (1997). The dimensions of service quality: The original European perspective revisited. *Service Industries Journal*, *17*(1), 173–189. doi:10.1080/02642069700000009

Mendez, D., Papapanagiotou, I., & Yang, B. (2015). *Internet of Things: Survey on Security and Privacy*. Available at: https://arxiv.org/pdf/1707.01879.pdf

Mercer, C. (2016). Li-Fi explained | What is it, how does it work and why Wi-Fi might be a thing of the past. *Techworld*. Available at: http://www.techworld.com/big-data/what-is-li-fi-everything-you-need-know-3632764

Merisavo, M. (2009). The effects of digital marketing on customer relationships. In I. Lee (Ed.), *Electronic business: Concepts, methodologies, tools, and applications* (pp. 1430–1441). Hershey, PA: IGI Global. doi:10.4018/978-1-60566-056-1.ch088

Merlo, C., Vicien, G., & Ducq, Y. (2014). Interoperability modelling methodology for product design organisations. *International Journal of Production Research*, *52*(15), 4379–4395. doi:10.1080/00207543.2013.774484

Meuter, M. L., Ostrom, A. L., Roundtree, R. I., & Bitner, M. J. (2000). Self-service technologies: Understanding customer satisfaction with technology-based service encounters. *Journal of Marketing*, *64*(3), 50–64. doi:10.1509/jmkg.64.3.50.18024

Micheal, M. (2009). *Cloud computing: Web- based applications that change the way you work and collaborate online*. Que Publishing.

Miller, K. (1999). *Simple steps - Developmental Activites for infants, toddlers, and two-year-olds*. Beltsville, MD: Gryphon House, Inc.

Millman, C., & El-Gohary, H. (2013). New digital media marketing and micro business: A UK perspective. In H. El-Gohary (Ed.), *Transdisciplinary marketing concepts and emergent methods for virtual environments* (pp. 59–81). Hershey, PA: IGI Global. doi:10.4018/978-1-4666-1861-9.ch005

Milošević, M., Lukić, D., Borojević, S., Šimunović, G., & Antić, A. (2017). A model of collaborative process planning system (e-CAPP). *Tehnicki Vjesnik (Strojarski Fakultet)*, *24*(1), 97–103. Retrieved from https://hrcak.srce.hr/file/257838

Milošević, M., Todić, V., & Lukić, D. (2011). Web-based collaborative environment for process planning. *34th International Conference On Production Engineering, 28.*

Minelli, M. Chambers, M., & Dhiraj, A. (2013). Big Data, Big Analytics: Emerging Business Intelligence and Analytic Trends for Today's Businesses. Wiley & Sons.

Mingay, S. (2002). *Outlining the Gartner BCP maturity model.* Retrieved 26 March 2017, from https://www.gartner.com/doc/367647/outlining-gartner-bcp-maturity-model

Mitchell, J. (2000). Increasing the cost-effectiveness of telemedicine by embracing e-health. *Journal of Telemedicine and Telecare, 6*, S16-S19.

Mobasher, B., Dai, H., Luo, T., Sun, Y., & Zhu, J. (2000). *Integrating web usage and content mining for more effective personalization.* Paper presented at the International Conference on Electronic Commerce and Web Technologies. doi:10.1007/3-540-44463-7_15

Monk, E. F., & Wagner, B. J. (2006). *Concepts in Enterprise Resource Planning* (2nd ed.). Thomson.

Monkova, K., & Monka, P. (2014). Newly developed software application for multiple access process planning. Advances in Mechanical Engineering.

Mont, D. (2007). *Measuring disability prevalence. SP Discussion Paper No. 0706.* Retrieved 22 July 2017: http://siteresources.worldbank.org/DISABILITY/Resources/Data/MontPrevalence.pdf

Montecchi, M., & Nobbs, K. (2017). Let it go: Consumer empowerment and user-generated content – An exploratory study of contemporary fashion marketing practices in the digital age. In A. Vecchi (Ed.), *Advanced fashion technology and operations management* (pp. 294–317). Hershey, PA: IGI Global. doi:10.4018/978-1-5225-1865-5.ch013

Moon, J., & Kim, Y. (2001). Extending the tam for a word wide web context. *Information & Management, 38*(4), 217–230. doi:10.1016/S0378-7206(00)00061-6

Mourad, M., Nassehi, A., & Schaefer, D. (2016). Interoperability as a key enabler for manufacturing in the cloud. *Procedia CIRP, 52,* 30-34. Retrieved from: http://www.sciencedirect.com/science/article/pii/S2212827116308010

Mourad, M., Nassehi, A., & Schaefer, D. (2016). Interoperability as a key enabler for manufacturing in the cloud. *Procedia CIRP, 52*, 30-34. Retrieved from: https://www.sciencedirect.com/science/article/pii/S2212827116308010

Mourtzis, D., Vlachoua, E., Xanthopoulos, N., Givehchi, M., & Wang, L. (2016). Cloud-based adaptive process planning considering availability and capabilities of machine tools. *Journal of Manufacturing Systems, 39*, 1–8. doi:10.1016/j.jmsy.2016.01.003

MPULSE. (2015). *The future of maintenance management.* Retrieved from: https://www.mpulsesoftware.com/sites/default/files/MPulse_Future_of_Maintenance_Management_eBook.pdf

Mu, E., Kirsch, L. J., & Butler, B. S. (2015). The assimilation of enterprise information system: An interpretation systems perspective. *Information & Management, 52*(3), 359–370. doi:10.1016/j.im.2015.01.004

Mukherjee, A., & Nath, P. (2003). A model of trust in online relationship banking. *International Journal of Bank Marketing, 21*(1), 5–15. doi:10.1108/02652320310457767

Mukherjee, A., & Nath, P. (2007). Role of electronic trust in online retailing: A re-examination of the commitment-trust theory. *European Journal of Marketing, 41*(9/10), 1173–1202. doi:10.1108/03090560710773390

Mulvenna, M. D., Anand, S. S., & Büchner, A. G. (2000). Personalization on the Net using Web mining: Introduction. *Communications of the ACM, 43*(8), 122–125. doi:10.1145/345124.345165

Muppidi, H. (2016). *Email archive retention using enterprise content management* (Master's thesis). St Cloud State University, St. Cloud, MN. Retrieved from http://repository.stcloudstate.edu/cgi/viewcontent.cgi?article=1052&context=mme_etds

Murray, A., & Ward, M. (2007). *Improving project performance using the PRINCE2 Maturity Model (P2MM)*. Norwich, UK: The Stationary Office.

Mutesi, J., Mutingi, M., & Charkraborty, A. (2017). Service quality assessment for internet banking: Empirical evidences from Namibia. *e-Service Journal*, *10*(1), 42–65. doi:10.2979/eservicej.10.1.03

Mutingi, M. (2015). Enhancing Decision Support in Health-Care Systems through mHealth. In mHealth Multidisciplinary Verticals (pp. 511 – 526). Taylor & Francis Group, LLC.

Mutingi, M., & Mbohwa, C. 2016. Sustainable Process Innovation for Customer Experience: A Conceptual Framework. IEEE Industrial Engineering and Engineering Management, 2182-2187.

Myers, M. D. (1997). Qualitative research in information systems. *Management Information Systems Quarterly*, *21*(2), 241–242. doi:10.2307/249422

Nah, F., Lau, J., & Kuang, J. (2001). Critical factors for successful implementation of enterprise systems. *Business Process Management Journal*, *7*(3), 285–297. doi:10.1108/14637150110392782

Namiot, D. (2016). On Internet of Things and Smart Cities educational courses. *International Journal of Open Information Technologies*, *4*(5), 26–38.

Narayanasamy, K., Rasiah, D., & Tan, T. M. (2011). The adoption and concerns of e-finance in Malaysia. *Electronic Commerce Research*, *11*(4), 383–400. doi:10.1007/s10660-011-9081-5

Nasri, W. (2011). Factors influencing the adoption of internet banking in Tunisia. *International Journal of Business and Management*, *6*(8). doi:10.5539/ijbm.v6n8p143

Nath, A.K., & Singh, R. (2010). Evaluating the Performance and Quality of Web Services in Electronic Marketplaces. *e-Service Journal*, *7*(1), 43-59.

Ngu, A. H., Carlson, M. P., Sheng, Q. Z., & Paik, H.-y. (2010). Semantic-based mashup of composite applications. *IEEE Transactions on Services Computing*, *3*(1), 2–15. doi:10.1109/TSC.2010.8

Nikolaidou, M., & Alexopoulou, N. (2009). Frameworks for model-based design of enterprise information systems. In M. Syed & S. Syed (Eds.), *Handbook of research on modern systems analysis and design technologies and applications* (pp. 326–356). Hershey, PA: IGI Global. doi:10.4018/978-1-59904-887-1.ch021

Nishant Srivastava, S., & Hashank, S. (2016). E- Business Scope and Challenges in India. *International Journal of Business and Management Invention.*

Niu, D. D., Liu, L., Zhang, X., Lü, S., & Li, Z. (2016). Security analysis model, system architecture and relational model of enterprise cloud services. *International Journal of Automation and Computing*, *13*(6), 574–584. doi:10.1007/s11633-016-1014-2

Norazah, M. S., Ramayah, T., & Norbayah, M. S. (2008). Internet shopping acceptance: Examining the influence of intrinsic versus extrinsic motivations. *Direct Marketing: An International Journal*, *2*(2), 97–110. doi:10.1108/17505930810881752

Nordheim, S., & Päivärinta, T. (2006). Implementing enterprise content management: From evolution through strategy to contradictions out-of-the-box. *European Journal of Information Systems*, *15*(6), 648–662. doi:10.1057/palgrave.ejis.3000647

Nor, K. M., & Pearson, J. M. (2008). An exploratory study into the adoption of internet banking in a developing country: Malaysia. *Journal of Internet Commerce*, *7*(1), 29–73. doi:10.1080/15332860802004162

North, C., & Shneiderman, B. (2000). Snap-together visualization: a user interface for coordinating visualizations via relational schemata. *Proceedings of the working conference on Advanced visual interfaces*. doi:10.1145/345513.345282

Npd-Solutions. (2011). Retrieved from http://www.npd-solutions.com/pdm.html

Nunnally, J. C., & Bernstein, I. H. (1994). *Psychometric Theory* (3rd ed.). New York: McGraw-Hill.

Nysveen, H., Pedersen, P. E., & Thorbjørnsen, H. (2005). Intentions to use mobile services: Antecedents and cross-service comparisons. *Journal of the Academy of Marketing Science*, *33*(3), 330–346. doi:10.1177/0092070305276149

O'cass, A., & Frenech, T. (2003). Web retailing adoption: Exploring the nature of Internet users web retailing behavior. *Journal of Retailing and Consumer Services*, *10*(2).

Obregon, L. (2015). Secure architecture for industrial control systems. *Technet Magazine*.

Oh, H., Rizo, C., Enkin, M., & Jadad, A. (2005). What is e-health? A systematic review of published definitions. *Journal of Medical Internet Research*, *24*.

Oliveira, T., Faria, M., Thomas, M. A., & Popovič, A. (2014). Extending the understanding of mobile banking adoption: When UTAUT meets TTF and ITM. *International Journal of Information Management*, *34*(5), 689–703. doi:10.1016/j.ijinfomgt.2014.06.004

Oluwaseyi, A. (2008). *Service oriented architecture and web services: guidelines for migrating from legacy systems and financial consideration* (Master's thesis). Blekinge Institute of Technology. Retrieved from: https://www.diva-portal.org/smash/get/diva2:828526/FULLTEXT01.pdf

Openclinical. (2005). Retrieved from: http://www.openclinical.org/e-Health.html

OpenIoT Consortium. (2013). *OPENIoT project description*. Available at: http://www.openiot.eu/

Otieno, J. (2008). ERP implementation challenges: A Kenyan case study. *Lecture Notes in Business Information Systems*, 399-409.

Otieno, J., & Abeysinghe, G. (2006). A framework for ERP implementation in Kenya: An empirical study. *Proceedings of Information Resource Management Association (IRMA) International Conference*.

Ozdemir, S., Trott, P., & Hoecht, A. (2008). Segmenting internet banking adopter and non-adopters in the Turkish retail banking sector. *International Journal of Bank Marketing*, *26*(4), 212–236. doi:10.1108/02652320810884777

Päivärinta, T., & Munkvold, B. E. (2005). Enterprise content management: An integrated perspective on information management. *Proceedings of the 38th Annual Hawaii International Conference on System Sciences (HICSS)*. doi:10.1109/HICSS.2005.244

Parasuraman, A., Zeithaml, V. A., & Berry, L. L. (1985). A conceptual model of service quality and its implications for future research. *The Journal of Marketing*, 41-50.

Parasuraman, A. (1998). Customer service in business-to-business markets: An agenda for research. *Journal of Business and Industrial Marketing*, *13*(4/5), 309–321. doi:10.1108/08858629810226636

Parasuraman, A. (2004). Assessing and improving service performancefor maximum impact: Insights from a two-decade-long research journey. *Performance Measurement and Metrics*, *5*(2), 45–52. doi:10.1108/14678040410546064

Parasuraman, A., Berry, L. L., & Zeithaml, V. A. (1991). Refinement and reassessment of the SERVQUAL scale. *Journal of Retailing*, *67*(4), 420.

Parasuraman, A., Berry, L. L., & Zeithaml, V. A. (1993). More on improving service quality measurement. *Journal of Retailing*, *69*(1), 140–147. doi:10.1016/S0022-4359(05)80007-7

Parasuraman, A., Zeithaml, V. A., & Berry, L. L. (1988). Servqual. *Journal of Retailing*, *64*(1), 12–40.

Parasuraman, A., Zeithaml, V. A., & Berry, L. L. (1990). *Delivering quality service*. Balancing Customer.

Parasuraman, A., Zeithaml, V. A., & Berry, L. L. (1994). Alternative scales for measuring service quality: A comparative assessment based on psychometric and diagnostic criteria. *Journal of Retailing*, *70*(3), 201–230. doi:10.1016/0022-4359(94)90033-7

Parasuraman, A., Zeithaml, V. A., & Malhotra, A. (2005). E-S-QUAL – A multiple-item scale for assessing electronic service quality. *Journal of Service Research*, *7*, 213–233. doi:10.1177/1094670504271156

Park, C. S. (2003). Knowledge capturing methodology in process planning. *Computer Aided Design*, *35*(12), 1109–1117. doi:10.1016/S0010-4485(02)00182-3

Parthasarathy, S. (2010). *Enterprise information systems and implementing IT infrastructures: Challenges and issues*. Hershey, PA: IGI Global. doi:10.4018/978-1-61520-625-4

Paviot, T., Cheutet, V., & Lamouri, S. (2011). A PLCS framework for PDM / ERP interoperability. *International Journal of Product Lifecycle Management, 5*(2/3/4), 295-313. Retrieved from: https://hal.archives-ouvertes.fr/hal-00799829/document

Pavlou, P. A. (2003). Consumer acceptance of electronic commerce: Integrating trust and risk with technology acceptance model. *International Journal of Electronic Commerce*, *7*(3).

PDF3D. (2017). Retrieved from: https://www.pdf3d.com/u3d/

Pelz-Sharpe, A., Durga, A., Smigiel, D., Hartman, E., Byrne, T., & Gingras, J. (2010). *ECM 3 - ECM Maturity Model*. Retrieved 26 March 2017, from http://ecmmaturity.files.wordpress.com/2009/02/ecm3-v2_0.pdf

Penciuc, D., Durupt, A., Belkadi, F., Eynard, B., & Rowson, H. (2014). Towards a PLM interoperability for a collaborative design support system. *Procedia CIRP*, *25*, 369-376. Retrieved from: http://www.sciencedirect.com/science/article/pii/S2212827114010828

Perera, C., & Chakrabarti, R. (2015). mHealth and Medical Imaging. In mHealth Multidisciplinary Verticals (pp. 651 – 668). Taylor & Francis Group, LLC.

Pfeifer, T. (Ed.). (2002). *Total Quality Management: Strategies, methods, techniques*. Munich, Germany: Cal Hanser Verlag. doi:10.3139/9783446224025

Pikkarainen, T., Pikkarainen, K., Karjaluoto, H., & Pahnila, S. (2004). Consumer acceptance of online banking: An extension of the technology acceptance model. *Internet Research*, *14*(3), 224–235. doi:10.1108/10662240410542652

Pillat, R. M., & Freitas, C. M. (2006). Coordinating views in the InfoVis toolkit. *Proceedings of the working conference on Advanced visual interfaces*. doi:10.1145/1133265.1133368

Polatoglu, V. N., & Ekin, S. (2001). An empirical investigation of the Turkish consumers' acceptance of internet banking services. *International Journal of Bank Marketing*, *19*(4), 156–165. doi:10.1108/02652320110392527

Poli, C. (2001). *Design For Manufacturing: A Structured Approach*. London: Elsevier Science and Technology Books.

Poonam, M., & Priyanka, D. (2016). Challenges and Future Prospects for E-Governance in India. International Journal of Science, Engineering and Technology Research, 3(7).

Popoola, N. F., & Ibn Arshad, M. D. R. (2015). Strategic approach to build consumers trust in adoption of internet banking in Nigeria. *Journal of Internet Banking and Commerce*, *20*(1).

Powell, R. R., & Connaway, L. S. (2004). *Basic research methods for librarians*. Westport, CT: Libraries Unlimited.

Pullant, R. (2010). *Manufacturing feature recognition by 3D solid model slicing and contour based geometric reasoning* (Master of Science in Mechanical Engineering Thesis). University of Cincinnati, Department of Mechanical Engineering, College of Engineering and Applied Science.

Raines, G. (2009). *Cloud Computing and Service Oriented Architecture (SOA)*. The MITRE Corporation. Retrieved from: https://www.mitre.org/sites/default/files/pdf/09_0743.pdf

Rajaobelina, L., Ricard, L., Bergeron, J., & Toufaily, É. (2014). An integrative model of installed online trust in the financial services industry. *Journal of Financial Services Marketing*, *19*(3), 186–197. doi:10.1057/fsm.2014.18

Raman, A., & Goyal, D. P. (2017). Extending IMPLEMENT framework for enterprise information systems implementation to information system innovation. In M. Tavana (Ed.), *Enterprise information systems and the digitalization of business functions* (pp. 137–177). Hershey, PA: IGI Global. doi:10.4018/978-1-5225-2382-6.ch007

Ramarao, A. (2016). Analysis of e-manufacturing capabilities and characteristics. *International Research Journal of Engineering and Technology*, *3*(9).

Rameshbabu, V., & Shunmugam, M. S. (2009). Hybrid feature recognition method for setup planning from STEP AP-203. *Robotics and Computer-integrated Manufacturing*, *25*(2), 393–408. doi:10.1016/j.rcim.2007.09.014

Ramseook-Munhurrun, P., & Naidoo, P. (2011). Customers' perspectives of service quality in internet banking. *Services Marketing Quarterly*, *32*(4), 247–264. doi:10.1080/15332969.2011.606753

Ranaweera, C., & Prabhu, J. (2003). The influence of satisfaction, trust and switching barriers on customer retention in a continuous purchasing setting. *International Journal of Service Industry Management*, *14*(4), 374–395. doi:10.1108/09564230310489231

Ranjan, K. R., Sugathan, P., Rossmann, A., & Martin, C. (2015). A narrative review and metaanalysis of service interaction quality: New research directions and implications. *Journal of Services Marketing*, *29*(1), 3–14. doi:10.1108/JSM-01-2014-0029

Rathore, S., & Panwar, A. (2016). *Capturing, analyzing, and managing word-of-mouth in the digital marketplace*. Hershey, PA: IGI Global. doi:10.4018/978-1-4666-9449-1

Real Story Group, & DAM Foundation. (2014). *DAM maturity model*. Retrieved 3 March 2016, from http://dammaturitymodel.org/

Real Story Group. (2016). *Enterprise content management and cloud file sharing: Comprehensive product evaluations*. Olney, MD: Real Story Group.

Reichheld, F. F., & Schefter, P. (2000, July). E-loyalty: Your secret weapon on the Web. *Harvard Business Review*, 105–113.

Reid, M., & Levy, Y. (2008). Integrating trust and computer self-efficacy with TAM: An empirical assessment of customers' acceptance of Banking Information Systems (BIS) in Jamaica. *Journal of Internet Banking and Commerce*, *13*(3), 1–19.

Remi, A., & Tony, P. (2007). *Developing web application with COLLADA and X3D*. Retrieved from: https://www.khronos.org/collada/presentations/Developing_Web_Applications_with_COLLADA_and_X3D.pdf

Rezaei, S., Amin, M., & Wan Ismail, W. K. (2014). Online repatronage intention: An empirical study among Malaysian experienced online shoppers. *International Journal of Retail & Distribution Management*, *42*(5), 390–421. doi:10.1108/IJRDM-03-2012-0026

Rezvani, A., Dong, L., & Khosravi, P. (2017). Promoting the continuing usage of strategic information systems: The role of supervisory leadership in the successful implementation of enterprise systems. *International Journal of Information Management*, *37*(5), 417–430. doi:10.1016/j.ijinfomgt.2017.04.008

Ribbink, D., Van Riel, A. C., Liljander, V., & Streukens, S. (2004). Comfort your online customer: Quality, trust and loyalty on the internet. *Managing Service Quality*, *14*(6), 446–456. doi:10.1108/09604520410569784

Rina, D. (n.d.). Challenges and Future Scope of E-commerce in India. *International Journal of Emerging Trends & Technology in Computer Science*.

Ringle, C. M., Wende, S., & Will, A. (2005). *SmartPLS 2.0 (M3) Beta*. Retrieved from http://www.smartpls.de

Ringle, C. M., Sarstedt, M., & Straub, D. M. (2012). Editor's Comments: A Critical Look at the Use of PLS-SEM in MIS Quarterly. *Management Information Systems Quarterly*, *36*(1), iii–xiv.

River Publishers Series in Communication. (2013). *Internet of Things: Converging Technologies for Smart Environments and Integrated Ecosystems*. Available at: http://www.internet-of-things-research.eu/pdf/Converging_Technologies_for_Smart_Environments_and_Integrated_Ecosystems_IERC_Book_Open_Access_2013.pdf

Rod, M., & Ashill, N. J. (2010). Management commitment to service quality and service recovery performance: A study of frontline employees in public and private hospitals. *International Journal of Pharmaceutical and Healthcare Marketing*, *4*(1), 84–103. doi:10.1108/17506121011036042

Rod, M., Ashill, N. J., Shao, J., & Carruthers, J. (2009). An examination of the relationship between service quality dimensions, overall internet banking service quality and customer satisfaction: A New Zealand study. *Marketing Intelligence & Planning*, *27*(1), 103–126. doi:10.1108/02634500910928344

Rodriguez, T. J. (2009). *Understanding human behavior - A psychology worktex*. Manila, Philippines: Rex Publishing.

Rogers, C., Malmas, S., & Enns, L. (2011). *CS14 - City of Surrey: (Surrey) policies, guidelines and procedures for a drive migration project as part of an enterprise content management program*. Retrieved 26 March 2017, from http://interpares.org/ip3/ip3_cs14_report.cfm

Rose, K., Eldridge, S., & Chapin, L. (2015). *The internet of things: An overview. The Internet Society*. ISOC.

Rousseau, D., Sitkin, S., Burt, R., & Camerer, C. (1998). Not so different after all: A cross-discipline view of trust. *Academy of Management Review*, *23*(3), 393–404. doi:10.5465/AMR.1998.926617

Rowe, W.-A., & Miller, C. (2011). *My skills, my money, my brighter future in Rwanda - An assessment of economic strengthening interventions for adolescent girls*. Baltimore, MD: Catholic Relief Services.

Rowley, J. (2006). An analysis of the e-service literature: Towards a research agenda. *Internet Research, 16*(3), 339–359. doi:10.1108/10662240610673736

Rowley, J., & Edmundson-Bird, D. (2013). Brand presence in digital space. *Journal of Electronic Commerce in Organizations, 11*(1), 63–78. doi:10.4018/jeco.2013010104

Rozenes, S., & Cohen, Y. (2017). *Handbook of research on strategic alliances and value co-creation in the service industry*. Hershey, PA: IGI Global. doi:10.4018/978-1-5225-2084-9

Rullis, H., & Sloka, B. (2010). Marketing within Internet banking in Latvia: Possibilities and challenges in commercial banks. *Economics and Management, 15*(1), 705–710.

Rust, R. T., & Kannan, P. K. (2003). E-service: A new paradigm for business in the electronic environment. *Communications of the ACM, 46*(6), 36–42. doi:10.1145/777313.777336

Saavedra, F. U., Criado, J. R., & Andreu, J. L. (2013). Is social media marketing really working?: Its impact on the relationships among market orientation, entrepreneurial orientation, and business performance. In H. El-Gohary & R. Eid (Eds.), *E-marketing in developed and developing countries: Emerging practices* (pp. 174–193). Hershey, PA: IGI Global. doi:10.4018/978-1-4666-3954-6.ch011

Saenz, B., Artiba, A., & Pellerin, R. (2009). Manufacturing Execution System- A Literature Review. *Production Planning and Control, 20*(6), 525–539. doi:10.1080/09537280902938613

Saha, R., & Grover, S. (2011). Identifying enablers of e-manufacturing. *ISRN Mechanical Engineering, 2011*, 1–6. doi:10.5402/2011/193124

Said, O., & Masud, M. (2013). Towards Internet of Things: Survey and future vision. *International Journal of Computer Networks, 5*(1), 1–17.

Salamntu, L. T. P., & Seymour, L. F. (2014). *A review of organisational benefits through the use of enterprise content management (ECM) System in public sector organisations*. Paper presented at the Third International Conference on Informatics Engineering and Information Science (ICIEIS), Lodz, Poland.

Saleh, Y., & Alshawi, M. (2005). An alternative model for measuring the succes of IS projects: The GPIS model. *Journal of Enterprise Information Management, 18*(1), 47–63. doi:10.1108/17410390510571484

Salinas & Ben Saied. (2013). Dissemination Level. Internet of Things Architecture Concepts and Solutions for Privacy and Security in the Resolution Infrastructure. (257521).

Sangeetha, J., & Mahalingam, S. (2011). Service quality models in banking: A review. *International Journal of Islamic and Middle Eastern Finance and Management, 4*(1), 83–103. doi:10.1108/17538391111122221

Santos, Â. S. S., & Pinto, M. M. (2015). *O repositório digital seguro: Diagnóstico e especificação*. Paper presented at the 12 Congreso Nacional BAD, Évora, Portugal. Retrieved from http://bad.pt/publicacoes/index.php/congressosbad/article/viewFile/1329/pdf_100

Santos, J. (2003). E-service quality: A model of virtual service quality dimensions. *Managing Service Quality*, *3*(3), 233–246. doi:10.1108/09604520310476490

Santouridis, I., Trivellas, P., & Tsimonis, G. (2012). Using E-S-QUAL to measure internet service quality of e-commerce web sites in Greece. *International Journal of Quality and Service Sciences*, *4*(1), 86–98. doi:10.1108/17566691211219751

Sathye, M. (1999). Adoption of internet banking by Australian consumers: An empirical investigation. *International Journal of Bank Marketing*, *17*(7), 324–334. doi:10.1108/02652329910305689

Sayar, C., & Wolfe, S. (2007). Internet banking market performance: Turkey versus the UK. *International Journal of Bank Marketing*, *25*(3), 122–141. doi:10.1108/02652320710739841

SearchTelecom. (2009). *What is 3G (third generation of mobile telephony)?* Available at: http://searchtelecom.techtarget.com/definition/3G

Şengül, Ü., & Eren, M. (2017). Selection of digital marketing tools using fuzzy AHP-fuzzy TOPSIS. In I. Management Association (Ed.), Advertising and branding: Concepts, methodologies, tools, and applications (pp. 587–616). Hershey, PA: IGI Global. doi:10.4018/978-1-5225-1793-1.ch027

Serope, K., & Schmid, S. R. (2009). *Manufacturing Engineering and Technology* (6th ed.). Singapore: Prentice Hall.

Shegda, K. M., Chin, K., Logan, D., & Lundy, J. (2004). *Building the magic quadrant for ECM 2004*. Retrieved 26 March 2017, from https://www.gartner.com/doc/461108/building-magic-quadrant-ecm-

Shih, H. (2004). Extended technology acceptance model of Internet utilization behavior. *Information & Management*, *41*(6), 719–729. doi:10.1016/j.im.2003.08.009

Shih, K.-H., Hung, H.-F., & Lin, B. (2010). Assessing user experiences and usage intentions of m-banking service. *International Journal of Mobile Communications*, *8*(3), 257–277. doi:10.1504/IJMC.2010.032974

Shivanand, H. K., Nanjundaradhya, N. V., Kammar, P., Divya, S. S., & Keshavamurthy, Y. C. (2008). E-Manufacturing: A technology review. *Proceedings of the World Congress on Engineering 2008*, *2*.

Siller, H. R., Estruch, A., Vila, C., Abellan, J. V., & Romero, F. (2008). Modeling workflow activities for collaborative process planning with product lifecycle management tools. *Journal of Intelligent Manufacturing*, *19*(6), 689–700. doi:10.1007/s10845-008-0120-6

Simón, V., Galar, D., & Baglee, D. (2015). eMaintenance platform for performing data Fusion mutation on machine tools. *Proceeding of Big Data Mining in real world applications,* 24-30. Retrieved from: http://www.worldcomp-proceedings.com/proc/proc2015/DMIN15_Final_Edition/DMIN15_Papers.pdf

Simon, A., Schoeman, P., & Sohal, A. S. (2010). Prioritised best practices in a ratified consulting services maturity model for ERP consulting. *Journal of Enterprise Information Management, 23*(1), 100–124. doi:10.1108/17410391011008923

Smaglik, P., Hawkins, R., Pingree, S., & Gustafson, D. (1998). The quality of interactive computer use among HIV infected individuals. *Journal of Health Communication, 3*(1), 53–68. doi:10.1080/108107398127508 PMID:10947374

Smits, M., & O'Callaghan, R. (2014). Strategy development for enterprise content management. In J. vom Brocke & A. Simons (Eds.), *Enterprise content management in information systems research* (pp. 91–107). Berlin: Springer. doi:10.1007/978-3-642-39715-8_6

Sohail, M. S., & Shaikh, N. M. (2008). Internet banking and quality of service: Perspectives from a developing nation in the Middle East. *Online Information Review, 32*(1), 58–72. doi:10.1108/14684520810865985

Srivastava, J., Cooley, R., Deshpande, M., & Tan, P.-N. (2000). Web usage mining: Discovery and applications of usage patterns from web data. *Acm Sigkdd Explorations Newsletter, 1*(2), 12–23. doi:10.1145/846183.846188

Statistical Year Book India. (2015). Ministry of Statistics and Programme Implementation, Government of India. Retrieved from www.mospi.nic.in/statistical-year-book-india/2015

Stolte, C., Tang, D., & Hanrahan, P. (2002). Polaris: A system for query, analysis, and visualization of multidimensional relational databases. *IEEE Transactions on Visualization and Computer Graphics, 8*(1), 52–65. doi:10.1109/2945.981851

Stone, M. (1974). Cross-Validatory Choice and Assessment of Statistical Predictions. *Journal of the Royal Statistical Society. Series A (General), 36*(2), 111–147.

Stroetmann, K., Jones, T., & Dobrev, A. (2006). Evaluating the economic impact of e-health applications approach and method. *E-health Conference 2006.*

Suh, B., & Han, I. (2002). Effect of trust on customer acceptance of internet banking. *Electronic Commerce Research and Applications, 1*(1).

Sukhoo, A., Barnard, A., Eloff, M. M., & Van der Poll, J. A. (2007). An evolutionary software project management maturity model for Mauritius. *Interdisciplinary Journal of Information, Knowledge, and Management, 2*, 99–118. doi:10.28945/103

Sun, Z., Firmin, S., & Yearwood, J. (2012). Integrating online social networking with e-commerce based on CBR. *International Proceedings of the 23rd ACIS 2012,* 3–5.

Sun, Z., Strang, K., & Yearwood, J. (2014). Analytics service oriented architecture for enterprise information systems. International Proceedings of iiWAS2014, 506-518. doi:10.1145/2684200.2684358

Sundmaeker, H., Guillemin, P., Friess, P., & Woelfflé, S. (2010). *Vision and challenges for realizing the Internet of Things. In Cluster of European Research Projects on the Internet of Things*. European Commission.

Sun, Y., & Mouakket, S. (2015). Assessing the impact of enterprise systems technological characteristics on user continuance behavior: An empirical study in China. *Computers in Industry*, *70*, 153–167. doi:10.1016/j.compind.2015.01.003

Suresh, K. (2015). *Past, Present & Future of IoT. Internet of Things*. Retrieved December 3, 2015, from http://www.tips2secure.com

Surjadjaja, H., Ghosh, S., & Antony, J. (2003). Determining and assessing the determinants of e-service operations. *Managing Service Quality: An International Journal*, *13*(1), 39–53. doi:10.1108/09604520310456708

Swearingen, K., Majkowski, W., Bruggeman, B., & Gilbertson, D. (2007). *An open system architecture for condition based maintenance overview* (pp. 1–8). IEEE Aerospace Conference. doi:10.1109/AERO.2007.352921

Swink, D. (2014). Smart Manufacturing Leadership Coalition (SMLC). Available at http://cdn.osisoft.com/corp/en/media/presentations/2014/UsersConference2014/PDF/UC2014_SMLC_Swink_AVisionforInfusingIntelligenceintoManufacturingEnterprises.pdf

Szymanski, D. M., & Hise, R. T. (2000). E-satisfaction: An initial examination. *Journal of Retailing*, *76*(3), 309–322. doi:10.1016/S0022-4359(00)00035-X

Tan, O., Ng, J., Wong, A., & Koh, W. K. (2012). Bridging gaps between three generation family's Needs and Attitudes towards e-health technologies. *IEEE 14th International Conference on e-Health Networking, Applications and Services*.

Tan, M., & Teo, T. S. H. (2000). Factors influencing the adoption of internet banking. *Journal of the Association for Information Systems*, *1*(5), 1–38.

Tao, F., Zhang, L., Venkatesh, V. C., Luo, Y., & Cheng, Y. (2011). Cloud manufacturing: A computing and service-oriented manufacturing model. *Proceedings of the Institution of Mechanical Engineers. Part B, Journal of Engineering Manufacture*, *225*(10), 1969–1976. doi:10.1177/0954405411405575

Tao, F., Zuo, Y., Xu, L. D., & Zhang, L. (2014). IoT-based intelligent perception and access of manufacturing resource toward cloud manufacturing. *IEEE Transactions on Industrial Informatics*, *10*(2), 1547–1557. doi:10.1109/TII.2014.2306397

Tarnanidis, T., Vlachopoulou, M., & Papathanasiou, J. (2017). *Driving agribusiness with technology innovations*. Hershey, PA: IGI Global. doi:10.4018/978-1-5225-2107-5

Tavana, M. (2012). *Enterprise information systems and advancing business solutions: Emerging models*. Hershey, PA: IGI Global. doi:10.4018/978-1-4666-1761-2

Tavana, M. (2017). *Enterprise information systems and the digitalization of business functions*. Hershey, PA: IGI Global. doi:10.4018/978-1-5225-2382-6

Taylor, S., & Todd, A. (1995a). Assessing information technology usage: The role of prior experience. *Management Information Systems Quarterly*, *16*(4).

Taylor, S., & Todd, P. A. (1995b). Understanding information technology usage: A test of competing models. *Information Systems Research*, *6*(2), 144–176. doi:10.1287/isre.6.2.144

Teller, C., & Gittenberger, E. (2011). Patronage Behaviour of Elderly Supermarket Shoppers -Antecedents and Unobserved Heterogeneity. *International Review of Retail, Distribution and Consumer Research*, *21*(5), 483–499. doi:10.1080/09593969.2011.619701

Tenenhaus, M., Esposito Vinzi, V., Chatelin, Y.-M., & Lauro, C. (2005). PLS Path Modeling. *Computational Statistics & Data Analysis*, *48*(1), 159–205. doi:10.1016/j.csda.2004.03.005

Teo, T. S. H., Lim, V. K. G., & Lai, R. Y. C. (1999). Intrinsic and extrinsic motivation in internet usage. *International Journal of Management Sciences*, *27*(1), 25–37.

Terry Ma, H., & Zaphiris, P. (2003). *The Usability and Content Accessibility of the E-government in the UK, Centre for Human-Computer Interaction Design*. City University, London. Retrieved 20 July 2017 from www.soi.city.ac.uk/zaphiri/Papers/HCII2003/HCII2003-Accessibility.pdf

Thaichon, P., Lobo, A., & Mitsis, A. (2014). An empirical model of home internet services quality in Thailand. *Asia Pacific Journal of Marketing and Logistics*, *26*(2), 190–210. doi:10.1108/APJML-05-2013-0059

Thakur, A. D., & Pande, S. S. (2006). Web-based feature modeling and process planning system for sheet metal components. *First International and 22nd All India Manufacturing Technology Design and Research Conference (22nd AIMTDR)*. IIT Roorkee.

Theron, W., & Hendry, G. (2015). *The value of systems-driven product development: Case studies from the automotive & aerospace industries. Proceedings of the Eleventh (11th) SA INCOSE Conference*. Retrieved from https://www.incose.org.za/pubs/2015/Papers/SA_INCOSE_2015_submission_5.pdf

ThingWorx. (2016). Retrieved from https://www.thingworx.com/

Tim, O. (2005). *What is web 2.0? design patterns and business models for the next generation of software*. Academic Press.

Triandis, H. C. (1979). Values, attitudes and interpersonal behavioural. *Nebraska Symposium on Motivation*, *27*(1), 195–259. PMID:7242748

Trigo, A., Varajão, J., Barroso, J., Soto-Acosta, P., Molina-Castillo, F. J., & Gonzalvez-Gallego, N. (2013). Enterprise information systems adoption in Iberian large companies: Motivations and trends. In I. Management Association (Ed.), Enterprise resource planning: Concepts, methodologies, tools, and applications (pp. 1035–1052). Hershey, PA: IGI Global. doi:10.4018/978-1-4666-4153-2.ch060

Tufte, E. R. (1997). *Visual explanations: images and quantities, evidence and narrative* (Vol. 36). CT: Graphics Press Cheshire.

Tung, H., Marshall-Bradley, T., Wood, J., Sofge, D. A., Grayson, J., Bergman, M., & Lawless, W. (2010). Enterprise information systems: Two case studies. In M. Cruz-Cunha (Ed.), *Social, managerial, and organizational dimensions of enterprise information systems* (pp. 434–458). Hershey, PA: IGI Global. doi:10.4018/978-1-60566-856-7.ch022

Uckelmann, D., Harrison, M., & Michahelles, F. (2011). An architectural approach towards the future Internet of Things. In D. Uckelmann, M. Harrison, & F. Michahelles (Eds.), *Architecting the Internet of Things* (pp. 1–24). Berlin: Springer-Verlag. doi:10.1007/978-3-642-19157-2_1

Unnamalai, V. E., & Thresphine, J. R. (2014). Service-Oriented Architecture for Cloud Computing. *International Journal of Computer Science and Information Technologies*, *5*(1), 251–255. Retrieved from http://ijcsit.com/docs/Volume%205/vol5issue01/ijcsit2014050153.pdf

Urban, W. (2009). Service quality gaps and their role in service enterprises development. *Technological and Economic Development of Economy*, *15*(4), 631–645. doi:10.3846/1392-8619.2009.15.631-645

Usman, M., Muzaffar, A. W., & Rauf, A. (2009). *Enterprise content management (ECM): Needs, challenges and recommendations*. Paper presented at the 2nd IEEE International Conference on Computer Science and Information Technology (ICCSIT), Beijing, China. doi:10.1109/ICCSIT.2009.5234541

Valdes, G., Solar, M., Astudillo, H., Iribarren, M., Concha, G., & Visconti, M. (2011). Conception, development and implementation of an e-government maturity model in public agencies. *Government Information Quarterly*, *28*(2), 176–187. doi:10.1016/j.giq.2010.04.007

Van der Meulen, R., & Rivera, J. (2014). *Gartner Says Worldwide Business Intelligence and Analytics Software Market Grew 8*. Academic Press.

Van der Wal, R. W. E., Pampalis, A., & Bond, C. (2002). Service quality in a cellular telecommunications company: A South African experience. *Managing Service Quality*, *12*(5), 323–335. doi:10.1108/09604520210442119

Van Mien, A. D. (2001). *The Gartner security process maturity model*. Retrieved 26 March 2017, from https://www.gartner.com/doc/326006/gartner-security-process-maturity-model

Van Ossel, G. (1998). Service guarantees and service-level agreements. In *Services Management. An Integrated Approach*. Financial Times Pitman Publishing.

Vargiu, E., & Urru, M. (2012). Exploiting web scraping in a collaborative filtering-based approach to web advertising. *Artificial Intelligence Review*, *2*(1), 44.

Vatanasombut, B., Igbaria, M., Stylianou, A. C., & Rodgers, W. (2008). Information systems continuance intention of web-based applications customers: The case of online banking. *Information & Management*, *45*(7), 419–428. doi:10.1016/j.im.2008.03.005

Venkatesan, D., & Sridhar, S. (2017). A novel programming framework for architecting next generation enterprise scale information systems. *Information Systems and e-Business Management*, *15*(2), 489–534. doi:10.1007/s10257-016-0330-y

Venkatesh, V. (2000). Determinants of perceived ease of use: Integrating control, intrinsic motivation and emotion into the technology acceptance model. *Information Systems Research*, *11*(4), 342–365. doi:10.1287/isre.11.4.342.11872

Venkatesh, V., & Davis, F. D. (1996). A model of the antecedents of perceived ease of use: Development and test. *Decision Sciences*, *27*(3), 451–481. doi:10.1111/j.1540-5915.1996.tb01822.x

Venkatesh, V., & Davis, F. D. (2000). A theoretical extension of the technology acceptance model: Four longitudinal field studies. *Management Science*, *46*(2), 186–204. doi:10.1287/mnsc.46.2.186.11926

Venkatesh, V., Morris, M. G., Davis, G. B., & Davis, F. D. (2003). User acceptance of information technology: Toward a unified view. *Management Information Systems Quarterly*, *27*(3), 425–478.

Venua, B., Komma, V. R., & Srivastava, D. (2015). A methodology to recognize manufacturing features of prismatic parts from STEP AP 203 file. *Materials Today: Proceedings*, *2*(4-5), 3170–3177. doi:10.1016/j.matpr.2015.07.111

Verdouw, C. N., Beulens, A. J. M., Reijers, H. A., & van der Vorst, J. G. A. J. (2015). A control model for object virtualization in supply chain management. *Computers in Industry*, *68*, 116–131. doi:10.1016/j.compind.2014.12.011

Vermesan, O., Friess, P., Guillemin, P., Gusmeroli, S., Sundmaeker, H., Bassi, A., . . . Doody, P. (2011). *Internet of Things – Global technological and societal trends*. River Publishers. Retrieved from: https://www.riverpublishers.com/pdf/ebook/chapter/RP_9788793609105C3.pdf

Vesset, D., McDonough, B., Schubmehl, D., & Wardley, M. (2013). Worldwide Business Analytics Software. 2013–2017 Forecast and 2012 Vendor Shares.

Vital Wave Consulting. (2009). *Health information systems in developing countries: A landscape analysis*. Retrieved 26 March 2017, from http://vitalwave.com/wp-content/uploads/2015/09/Gates-Foundation-HIS-Analysis-2009.pdf

Vom Brocke, J., Herbst, A., & Urbach, N. (2014). *How to measure ECM success*. Retrieved 26 March 2017, from http://www.bptrends.com/bpt/wp-content/uploads/06-03-2014-COL-ClassNotes-How-to-Measure-ECM-Success-vom-Broke-et-al-final-1.pdf

Vom Brocke, J., Simons, A., & Cleven, A. (2011). Towards a business process-oriented approach to enterprise content management: The ECM-blueprinting framework. *Information Systems and e-Business Management*, *9*(4), 475–496. doi:10.1007/s10257-009-0124-6

Wagner, G., Schramm-Klein, H., & Steinmann, S. (2013). Effects of Cross-Channel Synergies and Complementarity in a Multichannel E-commerce System — An Investigation of the Interrelation of E-Commerce, M-commerce and IETV-Commerce. *International Review of Retail, Distribution and Consumer Research*, *25*(5), 571–581. doi:10.1080/09593969.2013.832697

Walker, A., McBride, T., Basson, G., & Oakley, R. (2012). ISO/IEC 15504 measurement applied to COBIT process maturity. *Benchmarking: An International Journal*, *19*(2), 159–176. doi:10.1108/14635771211224518

Walker, R. H. (1996). Towards identifying how visitors to Tasmania define and assess service quality in the hospitality industry. *Australian Journal of Hospitality Management*, *3*(2), 27–39.

Wang, J., Zhang, L., Duan, L., & Robert, X. G. (2017). A new paradigm of cloud-based predictive maintenance for intelligent manufacturing. *Journal of Intelligent Manufacturing*, *28*(5), 1125–1137. doi:10.1007/s10845-015-1066-0

Wang, L., Bretschneider, S., & Gant, J. (2005). Web-based e-government services with a citizen-centric approach. *Proceedings of 38th Annual Hawaii International Conference on Systems Sciences*. doi:10.1109/HICSS.2005.252

Wang, X. V., Givehchi, M., & Wang, L. (2017). Manufacturing system on the cloud: A case study on cloud-based process planning. *Procedia CIRP*, *63*, 39–45. doi:10.1016/j.procir.2017.03.103

Wang, Y. S., Wang, Y. M., Lin, H. H., & Tang, T. I. (2003). Determinants of user acceptance of internet banking: An empirical study. *International Journal of Service Industry Management*, *14*(5), 501–519. doi:10.1108/09564230310500192

Weichhart, G., Molina, A., Chen, D., Whitman, L. E., & Vernadat, F. (2016). Challenges and current developments for sensing, smart and sustainable enterprise systems. *Computers in Industry*, *79*, 34–46. doi:10.1016/j.compind.2015.07.002

Weightless.org. (2016). *What is Weightless - Weightless*. Available at: http://www.weightless.org/about/what-is-weightless

Weir, C. S., Anderson, J. N., & Jack, M. A. (2006). On the role of metaphor and language in design of third party payments in e-banking: Usability and quality. *International Journal of Human-Computer Studies*, *64*(8), 770–784. doi:10.1016/j.ijhcs.2006.03.003

Weiser, M. (1991). The computer for the 21st century. *Scientific American*, *265*(3), 94–104. doi:10.1038/scientificamerican0991-94 PMID:1675486

Welbourne, E., Battle, L., Cole, G., Gould, K., Rector, K., Raymer, S., & Borriello, G. et al. (2009). Building the Internet of Things using RFID: The RFID ecosystem experience. *IEEE Internet Computing*, *13*(3), 1089–7801. doi:10.1109/MIC.2009.52

Wendler, R. (2012). The maturity of maturity model research: A systematic mapping study. *Information and Software Technology*, *54*(12), 1317–1339. doi:10.1016/j.infsof.2012.07.007

Wetzels, M., Odekerken-Schroder, G., & Oppen, C. V. (2009). Using PLS path modeling for assessing hierarchical construct models: Guidelines and empirical illustration. *Management Information Systems Quarterly*, *33*(1), 177–195.

Wicaksono, S., Shihab, M. R., & Sandhyaduhita, P. (2015). *Formulating implementation strategy for enterprise content management system using soft system methodology: A case of a marine logistics company in Indonesia.* Paper presented at the 2015 International Conference on Advanced Computer Science and Information Systems (ICACSIS), Depok, Indonesia. doi:10.1109/ICACSIS.2015.7415178

Wickramasinghe, N., & Misra, S. (2004). A wireless trust model for healthcare. *International Journal e-Health*, *1*(1), 60-77.

Wickramasinghe, N., & Goldberg, S. (2004). How M 1/4 EC2 in healthcare. *International Journal of Mobile Commerce*, *2*(1), 22–47.

Wikipedia. (2016). *GSM*. Available at: https://en.wikipedia.org/wiki/GSM#GSM_carrier_frequencies

Wikipedia. (2017). *Cloud computing*. Retrieved from: https://en.wikipedia.org/wiki/Cloud_computing

Williams, M. D., Dwivedi, Y. K., Lal, B., & Schwarz, A. (2009). Contemporary trends and issues in IT adoption and diffusion research. *Journal of Information Technology*, *24*(1), 1–10. doi:10.1057/jit.2008.30

Williamson, O. E. (1979). Transaction-cost economics: The governance of contractual relations. *The Journal of Law & Economics*, *22*(2), 233–261. doi:10.1086/466942

Wold, H. O. (1974). Causal Flows with Latent Variables: Partings of the Ways in the Light of NIPALS Modeling. *European Economic Review*, *5*(1), 67–86. doi:10.1016/0014-2921(74)90008-7

Wolfinbarger, M., & Gilly, M. C. (2003). eTailQ: Dimensionalizing, measuring and predicting etail quality. *Journal of Retailing*, *79*(3), 183–198. doi:10.1016/S0022-4359(03)00034-4

World Health Organization (WHO). (2011). *Introduction to medical equipment inventory management.* Retrieved from http://apps.who.int/iris/bitstream/10665/44561/1/9789241501392_eng.pdf

Wu, D., Rosen, D. W., Wang, L., & Schaefer, D. (2015). Cloud-based design and manufacturing: A new paradigm in digital manufacturing and design innovation. *Computer Aided Design*, *59*, 1–14. doi:10.1016/j.cad.2014.07.006

Xie, S. Q., & Xu, X. (2006). A STEP-compliant process planning system for sheet metal parts. *International Journal of Computer Integrated Manufacturing*, *19*(6), 627–638. doi:10.1080/09511920600623708

Xue, C., & Xu, Y. (2017). Influence factor analysis of enterprise IT innovation capacity based on system dynamics. *Procedia Engineering*, *174*, 232–239. doi:10.1016/j.proeng.2017.01.125

Xue, J. (2014). Research on ERP enterprise management information system in the cloud computing model. *BioTechnology: An Indian Journal*, *10*(20), 12556–12562.

Xu, X. (2012). From cloud computing to cloud manufacturing. *Robotics and Computer-integrated Manufacturing*, *28*(1), 75–86. doi:10.1016/j.rcim.2011.07.002

Yahia, E., Aubry, A., & Panetto, H. (2012). Formal measures for semantic interoperability assessment in cooperative enterprise information systems. *Computers in Industry*, *63*(5), 443–457. doi:10.1016/j.compind.2012.01.010

Yahia, E., Lezoche, M., Aubry, A., & Panetto, H. (2012). Semantics enactment for interoperability assessment in enterprise information systems. *Annual Reviews in Control*, *36*(1), 101–117. doi:10.1016/j.arcontrol.2012.03.008

Yang, H., & Yoo, Y. (2004). It's all about attitude: Revisiting the technology acceptance model. *Decision Support Systems*, *38*(1), 19–31. doi:10.1016/S0167-9236(03)00062-9

Yang, Z., & Jun, M. (2002). Consumer perceptions of e-service quality: From internet purchaser and non-purchaser perspectives. *The Journal of Business Strategy*, *1*, 19–41.

Yang, Z., Peterson, T., & Huang, L. (2001, Summer). Taking the pulse of internet pharmacies. *Marketing Health Services*, 5–10. PMID:11406905

Yap, K. B., Wong, D. H., Loh, C., & Bak, R. (2010). Offline and online banking-where to draw the line when building trust in e-banking? *International Journal of Bank Marketing*, *28*(1), 27–46. doi:10.1108/02652321011013571

Yoo, B., & Donthu, N. (2001). Developing a scale to measure the perceived quality of an internet shopping site (Sitequal). *Quarterly Journal of Electronic Commerce*, *2*, 31–46.

Yoon, H. S., & Steege, L. M. B. (2013). Development of a quantitative model of the impact of customers' personality and perceptions on internet banking use. *Computers in Human Behavior*, *29*(3), 1133–1141. doi:10.1016/j.chb.2012.10.005

Yoshikawa, H., Tsubokura, T., Toida, S., Kawai, Y., & Hatori, F. (2015). Cloud-based equipment maintenance and facility management service platform. *Hitachi Review*, *64*(4). Retrieved from: http://www.hitachi.com/rev/pdf/2015/r2015_04_108.pdf

Yousafzai, S. Y., Pallister, J. G., & Foxall, G. R. (2009). Multi-dimensional role of trust in internet banking adoption. *Service Industries Journal*, *29*(5), 591–605. doi:10.1080/02642060902719958

Yu, C. S. (2012). Factors affecting individuals to adopt mobile banking: Empirical evidence from the UTAUT model. *Journal of Electronic Commerce Research*, *13*(2), 104–121.

Yu, J., Benatallah, B., Casati, F., & Daniel, F. (2008). Understanding mashup development. *IEEE Internet Computing*, *12*(5), 44–52. doi:10.1109/MIC.2008.114

Yu, P. L., Balaji, M., & Khong, K. W. (2015). Building trust in internet banking: A trustworthiness perspective. *Industrial Management & Data Systems, 115*(2), 235–252. doi:10.1108/IMDS-09-2014-0262

Yusof, Y., Zakiah, N., Tan, Z., & Kasim, N. (2009). Exploring the ISO14649 (STEP-NC) for intelligent manufacturing system. *European Journal of Scientific Research, 36*(3), 445–457.

Zacharewicz, G., Diallo, S., Ducq, Y., Agostinho, C., Jardim-Goncalves, R., Bazoun, H., & Doumeingts, G. et al. (2017). Model-based approaches for interoperability of next generation enterprise information systems: State of the art and future challenges. *Information Systems and e-Business Management, 15*(2), 229–256. doi:10.1007/s10257-016-0317-8

Zdravković, M., & Panetto, H. (2017). The challenges of model-based systems engineering for the next generation enterprise information systems. *Information Systems and e-Business Management, 15*(2), 225–227. doi:10.1007/s10257-017-0353-z

Zeballos, L., & Quiroga, O. (2017). From computer integrated manufacturing to cloud manufacturing. *SII, Simposio Argentino de Informatica Industrial*. Retrieved from: http://www.clei2017-46jaiio.sadio.org.ar/sites/default/files/Mem/SII/sii-16.pdf

Zeithaml, V. A., Berry, L. L., & Parasuraman, A. (1996). The behavioral consequences of service quality. *The Journal of Marketing*, 31-46.

Zeithaml, V. A., Parasuraman, A., & Malhotra, A. (2000). *A Conceptual Framework for Understanding e-Service Quality: Implications for Future Research and Managerial Practice*. Marketing Science Institute, Cambridge, MA, Report No. 00-115.

Zeithaml, V. A. (2000). Service quality, profitability and the economic worth of customers: What we know and what we need to learn. *Journal of the Academy of Marketing Science, 28*(1), 67–85. doi:10.1177/0092070300281007

Zeithaml, V. A., Berry, L. L., & Parasuraman, A. (1996). The behavioral consequences of service quality. *Journal of Marketing, 60*(2), 31–46. doi:10.2307/1251929

Zeithaml, V. A., Parasuraman, A., & Malhotra, A. (2002). Service quality delivery through web sites: A critical review of extant knowledge. *Academy of Marketing Science Journal, 30*(4), 362–375. doi:10.1177/009207002236911

Zemblyte, J. (2015). The Instrument for Evaluating E-Service Quality. *Procedia: Social and Behavioral Sciences, 213*, 801–806. doi:10.1016/j.sbspro.2015.11.478

Zeng, X., & Parmanto, B. (2004). Web content accessibility of consumer health information web sites for people with disabilities: A cross sectional evaluation. *Journal of Medical Internet Research, 6*(3), e19. doi:10.2196/jmir.6.2.e19 PMID:15249268

Zhang, L. (2015). *Big Data Analytics for e-Maintenance: Modeling of high-dimensional data streams* (Licentiate Thesis). Operation and Maintenance Engineering Luleå University of Technology. Retrieved from: https://www.diva-portal.org/smash/get/diva2:990005/FULLTEXT01.pdf

Zhang, C., Zhou, X., & Li, C. (2009). Automatic recognition of intersecting features of freeform sheet metal parts. *Journal of Zhejiang University. Science A*, *10*(10), 1439–1449. doi:10.1631/jzus.A0820705

Zhang, D., Liu, Y., Sheng, X., Li, J., Wu, T., Xue, C. J., & Yang, H. (2015). Deadline-aware task scheduling for solar-powered nonvolatile sensor nodes with global energy migration. In *Proceedings of the 52nd Annual Design Automation Conference*. ACM. Retrieved from: doi:10.1145/2744769.2744815

Zhao, Y. Z., Zhang, J. B., Zhuang, L., & Zhang, D. H. (2005). *Service-Oriented Architecture and technologies for automating integration of manufacturing systems and services.* 10th International Conference on Emerging Technologies and Factory Automation (ETFA), Catania, Italy. Retrieved from: http://fileadmin.cs.lth.se/ai/Proceedings/etfa2005/html/articles/CF-000361.pdf

Zhao, A. L., Koenig-Lewis, N., Hanmer-Lloyd, S., & Ward, P. (2010). Adoption of internet banking services in China: Is it all about trust? *International Journal of Bank Marketing*, *28*(1), 7–26. doi:10.1108/02652321011013562

Zhou, L., Zhang, Y., & Xu, J. (2002). *A critical assessment of SERVQUAL's applicability in the banking context of China*. Valdosta, GA: Asia Pacific Advances in Consumer Research.

Zhou, T. (2011). An empirical examination of initial trust in mobile banking. *Internet Research*, *21*(5), 527–540. doi:10.1108/10662241111176353

Zhou, T., Lu, Y., & Wang, B. (2010). Integrating TTF and UTAUT to explain mobile banking user adoption. *Computers in Human Behavior*, *26*(4), 760–767. doi:10.1016/j.chb.2010.01.013

Zhou, X., Qui, Y., Hua, G., Wang, H., & Ruan, X. (2007). A feasible approach to the integration of CAD and CAPP. *Computer Aided Design*, *39*(4), 324–338. doi:10.1016/j.cad.2007.01.005

Zussman, E., & Horsch, T. (1994). A planning approach for robot-assisted multiple-bent profile handling. *Robotics and Computer-integrated Manufacturing*, *11*(1), 35–40. doi:10.1016/0736-5845(94)90005-1

About the Contributors

Norman Gwangwava is a professional Engineer with experience from industry and academia. He is currently a lecturer at the Botswana International University of Science and Technology (BIUST), department of Mechanical, Energy and Industrial Engineering. He has presented and published many research papers at conferences and refereed journals. Research interests are in; Reconfigurable Manufacturing Systems (RMS), Cyber-Physical Production Systems (CPS), Collaborative Product Design and Closed Loop Life-Cycle Systems, Manufacturing Information and Database Systems.

Michael Mutingi is a Senior Lecturer in Industrial Engineering at the Namibia University of Science and Technology, Namibia. He is also a Senior Visiting Research Associate at the University of Johannesburg, South Africa. He obtained his PhD in Engineering Management from the University of Johannesburg, South Africa. He also holds a MEng and a BEng in Industrial Engineering from the National University of Science and Technology, Zimbabwe, where he served as a Research Fellow and a Lecturer in Industrial Engineering. Michael Mutingi also served as a Research Associate at the National University of Singapore, Singapore, and a Lecturer at the University of Botswana, Botswana. His research interests include operations management, supply chain management, quality management, multi-criteria decision making, and operational excellence in healthcare. He has published two books and more than 90 articles in international journals and conference proceedings.

* * *

Yakup Akgül was born on March 22, 1977. He studied Department of Information Management at the university of Hacettepe, Ankara (Turkey), from which he graduated in 2001. He received Master (2010) and Ph.D. (2015) in Business Administration at Süleyman Demirel University, Isparta, Turkey. He works as an Ass. Prof. at the Alanya Alaaddin Keykubat University, Alanya/ANTALYA, Turkey.

Ayon Chakraborty is currently working as an Associate Professor in the Operations Management and Quantitative Techniques area at IIM Tiruchirappalli. He obtained his doctorate from Industrial and Systems Engineering Department, National University of Singapore after completing his masters in Manufacturing Systems from BITS, Pilani with distinction. He is also a qualified Six Sigma Green Belt. Previously he was Post-Doctoral Research Fellow, Services Science Discipline, Information Systems School at Queensland University of Technology, Brisbane, Australia. Prof. Chakraborty has more than eleven years of experience in teaching business students and executives at various institutes such as Queensland University of Technology, James Cook University Australia (Singapore Campus). He has taught MBA courses on service operations and innovation management, project management, quality planning and management, and quantitative techniques. As a researcher, Prof. Chakraborty has worked closely with public sector organizations in Singapore, and with Suncorp, Infosys Australia, and Queensland government.

Kijpokin Kasemsap received his BEng degree in Mechanical Engineering from King Mongkut's University of Technology, Thonburi, his MBA degree from Ramkhamhaeng University, and his DBA degree in Human Resource Management from Suan Sunandha Rajabhat University. Dr. Kasemsap is a Special Lecturer in the Faculty of Management Sciences, Suan Sunandha Rajabhat University, based in Bangkok, Thailand. Dr. Kasemsap is a Member of the International Economics Development and Research Center (IEDRC), the International Foundation for Research and Development (IFRD), and the International Innovative Scientific and Research Organization (IISRO). Dr. Kasemsap also serves on the International Advisory Committee (IAC) for the International Association of Academicians and Researchers (INAAR). Dr. Kasemsap is the sole author of over 250 peer-reviewed international publications and book chapters on business, education, and information technology. Dr. Kasemsap is included in the TOP 100 Professionals–2016 and in the 10th edition of 2000 Outstanding Intellectuals of the 21st Century by the International Biographical Centre, Cambridge, England.

Shadrack Katuu completed his Undergraduate Degree in Kenya, Masters Studies in Canada and Doctoral Studies in South Africa. Since 2013 he has been Head of the Records Unit at the International Atomic Energy Agency in Vienna (Austria) and between 2010 and 2012 he was an archives/records officer at the International Monetary Fund in Washington DC. (US). He has previously held several positions including manager of information systems in Mr. Nelson Mandela's post-presidential office at the Nelson Mandela Foundation. He has worked as information analyst at the South African History Archive in Johannesburg (South Africa). He has also been full time lecturer at the University of Botswana as well as guest-lectured in several universities in Barbados, Canada, South Africa, and the USA. He is a Research Fellow at the University of South Africa and has authored or co-authored several publications that have been cited more than 270 times (see http://goo.gl/sJ3qBG).

Iqbal Uddin Khan is Assistant Professor at Hamdard Institute of Engineering and Technology, Faculty of Engineering Sciences and Technology, Hamdard University, Karachi, Pakistan. He received MS – Telecommunication degree from Graduate School of Engineering Sciences and Technology (GSESIT) Hamdard University Karachi in 2013, B-Tech Electronics from Indus Institute of Higher Education Karachi Pakistan in 2011. Currently, he is pursuing Ph.D. in Computer Science from Graduate School of Engineering Sciences and Technology (GSESIT) Hamdard University. His research Interests are IoT, Computer and Communication Networks, Computer Network Security and Cloud Computing.

Siti Aishah Mohd Selamat is a Ph.D. candidate with particular interests in data science, strategic studies, and business management. Prior to enrolling at Bournemouth University, she has had 10 years of working experience in both general and IT project management in Singapore, Indonesia, and Malaysia. She holds an M.B.A in IT Management and B.A. in Business Management from the University of Bedfordshire and also, an Advanced Diploma and Diploma in IT from Informatics Institute Singapore. Her Ph.D. delves in developing an automatised analytic solution for Small-Medium Enterprises (SMEs) in the context of the transportation sector.

Ahmar Murtaza is Lecturer at Hamdard Institute of Engineering and Technology, Faculty of Engineering Sciences and Technology, Hamdard University, Karachi, Pakistan. He received MS – Telecommunication degree from Graduate School of Engineering Sciences and Technology (GSESIT) Hamdard University Karachi in 2014. Currently, he is pursuing Ph.D. in Computer Science from Graduate School of Engineering Sciences and Technology (GSESIT) Hamdard University. His research interests are Computer Networks, IoT, Sensor Networks, Mobile Computing, Electronic Engineering, etc.

Uzma Naz is Lab Engineer at Hamdard Institute of Engineering and Technology, Faculty of Engineering Sciences and Technology, Hamdard University, Karachi, Pakistan. She received BE – Computer Systems degree from HIET, FEST, Hamdard University Karachi in 2016. Currently, she is pursuing ME in Electrical Engineering from Graduate School of Engineering Sciences and Technology (GSESIT) Hamdard University. Her research interests are Computer Networks, Embedded Systems, Computer Security, IoT, Computer Engineering, etc.

Manoharan Ramachandran joined Bournemouth University in September 2015 as a full time post graduate researcher. His research is funded by Bournemouth University and Country Coaches. He received his M.Eng in computer science and B.Tech in information technology from Anna University – India. His interest in transportation systems and his decision to continue research in this field is the outcome of his profound interest in intelligent transportation systems. He has been involved in a number of research projects in the field of intelligent transportation systems to create innovative applications for the safety of drivers and passengers. Manoharan has published his research progress as part of various national and international conferences. For his research work he have been awarded best paper award, best project award, young investigator award and International Society for Technical Education's best student award of the year 2012.

Aqeel ur Rehman is a Professor, Deputy Director (Admin)-HIET and Chairman of Department of Computing at Hamdard Institute of Engineering and Technology, Faculty of Engineering Sciences and Technology, Hamdard University, Karachi, Pakistan. He is associated with Hamdard University since last 18 years. Dr. Aqeel-ur-Rehman has over 19 years of teaching, research, and academic administration experience. He obtained the Ph.D. degree in Computer Science with specialization in Ubiquitous Computing from National University of Computer and Emerging Sciences, Karachi, Pakistan in 2012. He completed MS in Information Technology from Hamdard University, Karachi, Pakistan and BS in Electronic Engineering from Sir Syed University of Engineering and Technology, Karachi, Pakistan in 2001 and 1998 respectively. His current research interests include IoT, Sensor Networks, Ubiquitous Computing, Computer Networks and Smart Agriculture.

Karthiga Shankar currently serves as an Assistant Professor at the Department of Information Technology at Thiagarajar College of Engineering, Madurai. Her areas of interest include Big data Analytics, Internet of Things, Analyzing Telecom data, Wireless Sensor Networks and Automation. Ms.S. Karthiga completed her B. E. (Computer Science and Engineering) in Karunya University, Coimbatore 2009 and she did her M.E (Computer Science and Engineering) in SNS College of Engineering and Technology, 2011. She is pursuing her PhD in Information and Communication Engineering in the area of Big data and Internet of Things. She is working with Thiagarajar College of Engineering, Madurai since 2012. Prior to that she was working as an Assistant Professor at Bannari Amman Institute of Technology, Sathyamangalam. She has done various Consultancy Projects at Thiagarajar college of Engineering such as Design and Development of RFID based application for Human Machine Interfaces (HMI), Automation Systems for Travels website and Departmental Stores. She has undergone Various Faculty Development programme and Short Term Training Programmes at reputed Institutions such as Big data Analytics at IIT Roorkee, Parallel Programming at CDAC Bangalore, Advanced Data Analytics at IIT Delhi, etc. She has conducted and delivered Lectures in many Faculty development training programmes such as TEQIP II Sponsored Hands on Training – Hadoop, DST SERB Sponsored Short Term Training Programme on Big data & Internet of Things: The Strategies for Smart Environments. She received Young Woman in Engineering Award in Information Technology from Venus International Foundations, Chennai She is currently doing her research in IoT and Big Data in the domain of Telecommunication and Healthcare Monitoring systems. She is a Professional member of IEEE, ACM and ISACA and serving as a IEEE Faculty Advisor and Academic Advocate for ISACA Professional Organization.

R. Suganya currently serves as an Assistant Professor at the Department of IT at Thiagarajar College of Engineering, Madurai. Her areas of interest include Medical imaging, Soft computing, Big data Analytics, Internet of Things and Wireless Sensor Networks. She is a reviewer for many Journals. She is working with Thiagarajar College of Engineering, Madurai since 2006. She has undergone ten days Big Data Analytics Training programme in IIT Roorkee and CDAC Chennai. She has conducted Faculty development training programme in Big Data and Hadoop for the faculty members from various colleges around Tamilnadu. She delivered her lecture and did hands on in Big data and Hadoop in DST Sponsored Programme for Big Data and Analytics organized by Anna University, Trichy. She is currently doing research work in IoT and Big Data in the Health care monitoring system and intelligent transportation system. She is a Professional member of Association for Computing Machinery and conducted several technology based and societal projects.

Abhishek Vashishth has done his Bachelors in Production and Industrial Engineering from Rajasthan Technical University, Kota (India). He later joined the Indian Institute of Management Tiruchirappalli as Fellow Program scholar. His area of research is in operations management with a focus on service quality.

Index

H

I

L

M

O

P

Q

R

S

T

U

W

X

www.ingramcontent.com/pod-product-compliance
Lightning Source LLC
LaVergne TN
LVHW080340011026
849161LV00009B/2136